Wide-Area Data Network Performance Engineering

For a listing of recent titles in the *Artech House Telecommunications Library*, turn to the back of this book.

Wide-Area Data Network Performance Engineering

Robert G. Cole
Ravi Ramaswamy

Artech House
Boston • London

Library of Congress Cataloging-in-Publication Data
Cole, Robert G., 1955–
    Wide-area data network performance engineering / Robert G. Cole, Ravi Ramaswamy.
      p.   cm. — (Artech House telecommunications library)
    Includes bibliographical references and index.
    ISBN 0-89006-569-1 (alk. paper)
      1. Wide-area networks (Computer networks)   2. Computer network protocols.
    I. Ravi Ramaswamy.   II. Title.   III. Series.
    TK5105.87.C65   2000
    004.67—dc21   99-052404
                                                              CIP

British Library Cataloguing in Publication Data
Cole, Robert G.
    Wide-area data network performance engineering
    1. Wide-area networks (Computer networks)   2. Data
    transmission systems   3. Systems engineering
    I. Title   II. Ramaswamy, Ravi
    004.6'7

    ISBN 0-89006-569-1

Cover design by Andrew Ross

Copyright © 2000 AT&T. All rights reserved.

Printed and bound in the United States of America. No part of this book may be reproduced or utilized in any form or by any means, electronic or mechanical, including photocopying, recording, or by any information storage and retrieval system, without permission in writing from the publisher.

All terms mentioned in this book that are known to be trademarks or service marks have been appropriately capitalized. Artech House cannot attest to the accuracy of this information. Use of a term in this book should not be regarded as affecting the validity of any trademark or service mark.

International Standard Book Number: 0-89006-569-1
Library of Congress Catalog Card Number: 99-052404

10 9 8 7 6 5 4 3 2 1

To
Alexandria and Melanie
and
Anjali, Ashwin, and Ratna

Contents

	Preface	***xv***
	Acknowledgments	*xvii*
1	**Introduction**	**1**
1.1	Enterprise Networking	1
1.2	Our Approach to Performance Engineering	4
1.3	Layout of the Book	5
	Part I	
	General Principles	**9**
2	**Wide-Area Networking Technologies**	**11**
2.1	Introduction	11
2.2	Time-Division Multiplexing Networks	13
2.3	X.25 Networks	15
2.4	Frame Relay	19
2.4.1	Frame Relay Network Architecture	19
2.4.2	Frame Relay and Enterprise Networks	29
2.5	Asynchronous Transport Mode	30
2.6	Internetworking Protocol	37
2.6.1	IP Architecture	38

2.6.2   IP on Broadcast Networks   42

2.6.3   IP on ATM (and Other NBMA) Networks   43

2.7   Multiprotocol Encapsulation: Agreements for PPP,
        X.25, Frame Relay, ATM, and IP   47

2.8   Link Level Interworking Agreements   49

2.9   Summary   52

3   Performance Analysis: Some Basic Tools   55

3.1   Introduction   55

3.2   Network Delays   56

3.2.1   Delay and Latency   57

3.2.2   Propagation   59

3.2.3   Transmission, or Insertion Delays, on Serial Lines   61

3.2.4   Processing   64

3.2.5   Queuing   66

3.2.6   Delay Synopsis   69

3.3   Timing Diagrams: A Data Communications
        Score Card   69

3.4   Pipelining   74

3.5   Throughput   79

3.5.1   Windowing Systems   82

3.5.2   Lossless Throughput Systems Summary   93

3.5.3   Optimal Window Sizes and Bulk Data Transfer
        Times   94

3.5.4   Throughput Summary   98

3.6   Summary   99

4   Techniques for Performance Engineering   101

4.1   Introduction   101

4.2   Load Engineering   102

4.3   Latency-Sensitive and Bandwidth-Sensitive
        Applications   106

4.4	Methods to Discriminate Traffic in a Multiprotocol Network	108
4.4.1	Window Size Tuning	109
4.4.2	Type of Service Routing	113
4.4.3	Priority Queuing	115
4.4.4	Processor Sharing	121
4.4.5	Adaptive Controls	124
4.4.6	Selective Discards	126
4.5	Data Collection	127
4.5.1	LAN/WAN Analyzers	127
4.5.2	Network Management Systems and RMON Probes	128
4.5.3	A Taxonomy of Commercially Available Tools for Performance Engineering Data Networks	130
4.6	An Example: Deploying New Applications	132
4.7	Summary	135
5	**Frame Relay Performance Issues**	**137**
5.1	Introduction	137
5.2	Private Line Versus Frame Relay	138
5.2.1	Dual Insertion Delay	140
5.2.2	Delay Variation	147
5.2.3	Application Impact of Migration to Frame Relay	148
5.3	Global Frame Relay Connections	150
5.4	Bandwidth sizing	152
5.4	Bandwidth sizing	153
5.4.1	Sizing Ports and PVCs	151
5.5	Traffic Discrimination	165
5.5.1	Congestion Shift From Routers Into the Network	165
5.5.2	Response to the Congestion shift	166
5.6	Global Versus Local DLCI	174

| 5.7 | Virtual Circuit Scaling Issues | 176 |
| 5.8 | Summary | 177 |

6	**Using Pings for Performance Analysis**	**181**
6.1	Introduction	181
6.2	Pings	182
6.3	Calculating Ping Delays	184
6.3.1	Leased Line Connection	184
6.3.2	Frame Relay Connection	186
6.3.3	Observations	187
6.4	Using Pings to Verify Network Latency	189
6.5	General Comments Regarding the Use of Pings to Estimate Network Latency	191
6.6	Calculating Delays for Large Pings	192
6.6.1	Example 1: Leased Lines	193
6.6.2	Example 2: Calculating Large Ping Delays Over Frame Relay	194
6.6.3	Some Comments Regarding the Use of Large Pings to Calculate Throughput	195
6.7	Summary	196

| **Part II** | | |
| **Specific Application/Protocol Suites** | | **197** |

7	**WAN Performance Analysis of TCP/IP Applications: FTP, HTTP, and Telnet**	**199**
7.1	Introduction	199
7.2	Some Essential Aspects of TCP Operation	201
7.3	Calculating TCP Bulk Data Transfer Times and Throughput	204
7.3.1	Variables Affecting TCP File Transfer Performance	205
7.3.2	Computing File Transfer Times for a Simple Point-to-Point Connection	205

7.3.3	Private Line Analysis	208
7.3.4	Frame Relay Analysis	215
7.3.5	General Formulas for Calculating TCP Throughput for Private Lines and Frame Relay	222
7.4	Calculating TCP Throughput for Loaded WAN Links	225
7.4.1	Private Line Case	226
7.4.2	Frame Relay Case	227
7.5	WAN Performance Issues for HTTP	230
7.5.1	Summary of HTTP WAN Performance Issues	232
7.5.2	A Sample Trace of an HTTP Transaction	232
7.5.3	Estimating HTTP Performance Over a WAN Connection	234
7.6	TCP Telnet Performance Issues	242
7.7	Review of Methods to Provide Traffic Discrimination for TCP/IP Applications	247
7.7.1	Prioritize Telnet Over Bulk Data Transfers at the Router	247
7.7.2	Separate Telnet on Its Own PVC	249
7.7.3	Traffic Shaping at the Router	249
7.7.4	More General Bandwidth Management Techniques	250
7.8	Summary	250
8	**WAN Performance Considerations for Novell NetWare Networks**	**253**
8.1	Introduction	253
8.2	Overview	254
8.3	Overhead and Bandwidth Considerations	256
8.4	Novell Windowing Schemes	260
8.4.1	NetWare Pre-Release 3.11	261
8.4.2	NetWare Release 3.11	262
8.4.3	NetWare Releases 3.12 and 4.0	264

8.5	Private Line and Frame Relay Formulas	265
8.5.1	Private Line Formulas	267
8.5.2	Frame Relay Formulas	271
8.5.3	Cross-Application Effects: Mixing Novell and TCP/IP	276
8.6	Summary	277
9	**WAN Performance Issues for Client/Server Applications**	**279**
9.1	Introduction	279
9.2	Client/Server Overview	281
9.3	Client/Server Application WAN Traffic Characterization	283
9.3.1	Examples of Two-Tier Application Traffic Patterns	284
9.3.2	Example of a Three-Tier Transaction	288
9.4	Data Collection	290
9.5	Bandwidth Estimation Guidelines	292
9.5.1	Applications With a Ping-Pong Traffic Characteristic	292
9.5.2	What Happens When Think Times Are Not Available?	293
9.5.3	Bandwidth Estimation for Bulk Data Transfers	294
9.5.4	Bandwidth Estimation for Hybrid Transactions	295
9.5.5	Example of an SAP R3 Application	296
9.5.6	An Approach to Computing Response Times	298
9.5.7	Response Times for Bulk Data Transfer Transactions	300
9.5.8	Response Times for Hybrid Transactions	300
9.6	The Thin Client Solution	300
9.6.1	The Remote Presentation Approach	300
9.6.2	The Network Computing Approach	307
9.7	Summary	308

10	**WAN Design and Performance Considerations for SNA Networks**	**311**
10.1	Introduction	311
10.2	SNA Transport Methods: A Review	312
10.2.1	TCP/IP Encapsulation: Data Link Switching	312
10.2.2	Emulation Using SNA Gateways	314
10.2.3	Direct Encapsulation Over Frame Relay: RFC1490	315
10.2.4	SNA Translation: TN3270	316
10.2.5	Web Access to Mainframe Applications	316
10.3	Data Center Architecture Issues for Large SNA Networks	319
10.4	Quality of Service Issues for SNA	322
10.4.1	Delay Trade-Offs in SNA Migration to IP	324
10.4.2	FEP-to-FEP Issues	326
10.4.3	Traffic Discrmination	327
10.5	Summary	328

	Part III Case Studies	**331**

11	**Case Studies**	**333**
11.1	Introduction	333
11.2	TCP/IP Case Studies	334
11.2.1	Validating Network Latency and Throughput	334
11.2.2	TCP Bulk Data Transfer	337
11.2.3	Sizing Bandwidth for a TCP/IP Application	339
11.3	Client/Server Application Case Studies	342
11.3.1	Troubleshooting Response Times for a Sales-Aid Application Over a Global Frame Relay Network	343
11.3.2	Troubleshooting Response Times for Custom Client/Server Application Over a Frame Relay Network	347

11.3.3   Troubleshooting Performance Problems for an Oracle Financials Application Using Citrix Winframe Over a Global Frame Relay Network   349

11.4   Novell Networking Case Studies   353

11.4.1   How Novell SAPs Can Impact Performance   353

11.4.2   Comparing Leased Line and Frame Relay Performance for Novell File Transfers   357

11.4.3   A Paradox: Increasing Bandwidth Results in Worse Performance   359

11.5   SNA-Related Case Studies   360

11.5.1   Migration From SNA Multidrop to DLSw   361

11.5.2   Insuring SNA Performance in the Presence of TCP/IP Traffic   365

11.6   Quantifying the WAN Bandwidth Impact of SNMP Polling   369

Appendix A: Queuing: A Mathematical Digression   375

Appendix B: Throughput in Lossy Environments   381

B.1   Introduction   381

B.2   Transmission Errors   382

B.3   Buffer Overflows   385

B.4   Transmitter Time-Outs   387

B.5   Out-of-Sequence Receptions   388

B.6   Impact of Packet Losses on Throughputs   388

Appendix C: Definitions   393

List of Acronyms   395

About the Authors   401

Index   403

Preface

This book has its origins in our data network performance engineering work at the erstwhile AT&T Bell Laboratories in the late 1980's and continuing into the better part of this decade in AT&T Solutions. Two distinct, but related, aspects of that work can be singled out as the major contributors to this book: the AT&T Datakit fast packet switch, and the rapid deployment of public frame relay services. Few in the networking industry have heard of Datakit, but it was one of AT&T's early attempts at providing a vehicle to use packet technologies to integrate separate wide-area networks. A handful of AT&T's large customers built private frame relay-like networks based on the Datakit family of switches and end devices. Subsequently, public frame relay services and multi-protocol routers gained popularity in the early 1990s. Wide-area network integration has since matured tremendously. The next phase of the integration, that is, combining voice, data, video, and multi-media applications over IP, is currently under way.

In the days of Datakit and early frame relay services, we spent a great deal of time and effort trying to understand and quantify the performance impact of integrating multiple protocols and applications on a packet network. In those days, the important protocols were IBM's SNA/SDLC and various bisynchronous and asynchronous protocols. Corporate networks also carried IP, Novell IPX, AppleTalk and others, initially over local area networks, but soon over wide-area networks. We soon realized that much of the battle in modeling application performance was in developing a detailed understanding of the packet processing within the protocol stacks and the packet flows across LANs and WANs. A thorough understanding of issues such as protocol overhead, windowing, time-out, retransmission issues in TCP/IP, Novell IPX, and SNA was required.

Equally important is the development of methods to quantify performance. Here we recognized two approaches—build atomistic and deterministic models of packet processing and packet flows and use simple queuing theory formulas to capture the effects of congestion, or build sophisticated queuing models with complex mathematical assumptions about traffic patterns, packets size variability, and so on. It did not take us very long to realize that the former approach yielded favorable results when compared to test results and measurements, as long as the networks were operating within their engineered limits. It also did not take us long to realize that the formal queuing theory approach was too impractical to be of value to network managers trying to get quick answers to pressing performance issues.

During the years since Datakit and the early days of frame relay, we and others at AT&T have continued this work. We have given training sessions to numerous internal and external clients in the area of wide-area network performance engineering from an end-user application perspective. We have continued to find success in our approach towards modeling, engineering, and architecting numerous corporate enterprise data networks.

This book is an effort to bring together all the methods and ideas we have developed and the experiences we have gained over the years. It combines the three aspects of performance engineering we discussed—protocol and application details, deterministic models of packet flows, and a very small dose of queuing theory to capture the effects of congestion. Our target audience are those individuals responsible for designing, maintaining, and trouble shooting performance problems in data networks, as well as those responsible for capacity planning, performance assurance, application testing, and other responsibilities related to data networking. The treatment is technical in nature ("bit and bytes") and every effort has been taken to explain technical terms. It relies on back-of-the-envelope analysis, and does not employ complex techniques in mathematics, although where appropriate, there are some digressions into more complex analysis. These are mainly in the form of appendices and references. While there are no absolute prerequisites, a background in data communications will certainly help.

We realize that, despite our efforts to make the book as up to date as possible, some of the protocol implementations and capabilities will change. However, we believe that the methodology described in this book will survive the years, because, to a large extent, they are independent of specific implementations and protocols. It is our sincere hope that in writing this book we have been able to convey to the reader the art and science of wide-area network performance engineering.

Acknowledgments

Many have contributed to this book and it is with great pleasure that we acknowledge their contributions. First, we would like to thank Diane Sheng, our manager during the late 80s and early 90s, for introducing us to data networking and performance modeling, and for her support during this period. This book would not have been possible but for the support and encouragement of our management at AT&T, in particular, Bill Strain and Bill Gewirtz. We owe much thanks to Bill Strain and Lawrence Brody for their thorough and frequent reviews and comments of the entire manuscript, including multiple rewrites. Thanks are also due to Rick Oberman and Lalitha Parameswaran for their review of specific chapters and for their insightful comments, discussions and suggestions. And to Larry Booth and Bill Zimmerman for their early reviews of our initial chapters. The staff at Artech House has been a great help during the somewhat trying times of manuscript development. In particular, we wish to thank Barbara Lovenvirth for her help and her patience throughout. We know we put her in numerous difficult situations. The technical referees provided many suggestions to improve the book in both its technical accuracy and its organization. We thank them for their efforts. We also wish to thank Mark Walsh at Artech House for his support and encouragement early in the development of the manuscript.

Finally, to our families. This has been a hectic time in our lives and theirs and the writing of this book has only exacerbated the situation.

I, Bob Cole, wish to acknowledge the loving support of my wife Melanie and daughter Alexandria. I dedicate the book to them. Without their encouragement, understanding and support, I never would have been able to complete the writing of this book.

I, Ravi Ramaswamy, dedicate this book to my wonderful wife Ratna, and my two beautiful children, Anjali and Ashwin, without whose love and support, this book could not have been written. Thanks again for enduring all those innumerable nights and weekends away from home.

Bob Cole
Havre de Grace, Maryland

Ravi Ramaswamy
Belle Mead, New Jersey

1

Introduction

1.1 Enterprise Networking

It seems like a long time ago, but in the early to mid-1980s, there were few wide-area network (WAN) data network architectures from which to choose other than leased lines and X.25. The dominant protocol suite was IBM's System Network Architecture® (SNA). Most corporate data resided in mainframes and AS/400s. E-mail was virtually unknown. The Internet was primarily a tool for academia. Novell was the leader in local-area network (LAN) operating systems. LANs were LANs and WANs were WANs, and the twain never met. Routers were a quaint curiosity.

How times have changed! Today, just over a decade later, the buzz words in the networking industry are no longer *X.25*, *Novell*, or *SNA*, but *frame relay*, *ATM*, *VPNs*, *intranets*, *Internet*, and so on. The Internet Protocol (IP) has emerged as the de facto networking standard. Routers and switches are ubiquitous, client/server applications are everywhere, corporations live and die by e-mail, and the Internet is being seriously considered as a tool for corporate communications.

Everything over IP is the vision of the future.

Many trends have contributed to this facelift, but some of the technology trends of the 1990s seem to have had more impact than others—public frame relay services, multiprotocol router technology, and an emerging set of applications built on the Transmission Control Protocol/Internetwork Protocol (TCP/IP). The subsequent introduction of Web technologies has also had a tremendous impact on data networks as well.

Performance engineering of wide-area networks is the topic of this book. We examine this issue against the backdrop of migration of legacy networks to high-speed *multiprotocol* and *multiapplication* networks. In the old days of SNA networking, performance engineering of low-speed multidrop SDLC and high-speed front-end processor (FEP)-to-FEP was a complex task, involving intricate details of the SNA protocol suite. However, the principles and methods were well understood, and sufficient field experience and expertise existed. With the introduction of multiprotocol routers, legacy networks can be integrated and traditional LAN-based applications and protocols such as Novell, AppleTalk®, and NetBIOS® can begin to compete for WAN resources. Packet switching technologies, like frame relay and asynchronous transfer mode (ATM), introduce another significant element in the performance engineering equation. With the proliferation of client/server applications, intranet Web-based applications, and increasing demand for Internet access, the task of performance engineering becomes a daunting one. What was an almost exact science in SNA networking, performance engineering in today's environments requires a skillful combination of art and science.

This concept can be better understood in the context of one of the prominent trends in data networking, that is, the convergence of packet layer transport and the integration of high-level protocol suites onto a common transport layer. The common packet transport technologies we refer to include frame relay, ATM, and the IP technologies. Riding on these common packet transport methods are the high-level application/protocol suites such as TCP/IP, Novell NetWare®, numerous client/server applications, SNA, and others commonly found in corporate enterprise networks. Figure 1.1 is a schematic showing the mapping of these applications and protocols onto this common network infrastructure. The figure shows the high-level protocols being mapped onto the IP layer or directly onto the link layer technologies, for example, frame relay and ATM. Highlighted between these two layers is a *multiprotocol integration layer*. This is not meant to represent a strict protocol layer; rather, it represents the collection of tools, techniques, and protocols necessary to ensure a successful mapping for the multiple high-level protocols onto the underlying packet transport.

Traditional 3270 applications are inquiry–response in nature, following a "small in, large out" transaction pattern, that is, a 100- to 200-byte inquiry into the host and a 700- to 1000-byte response from the host. The network typically has low traffic volumes. Contrast that with a frame relay network, as shown in Figure 1.1, carrying a multitude of applications such as legacy traffic,

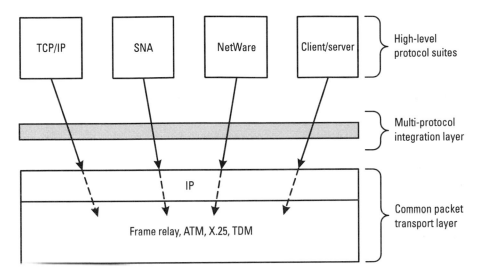

Figure 1.1 High-level protocol integration onto a common packet transport infrastructure.

client/server applications, Internet access, and e-mail. Legacy traffic has low traffic volumes, but needs consistent response times. Client/server applications are frequently chatty, but can also send large bulk data transfers. Internet/intranet traffic is bursty in nature. Considering all of these issues, it is clear why performance engineering is not only important but also at once more complicated. In this book, we examine in great detail the tools, techniques, and methodologies available today for performance engineering complex multiprotocol and multiapplication wide-area networks. The objective is to provide the necessary analytical skills to the network engineer, architect, and administrator so they persons can accomplish protocol integration in as smooth and seamless a fashion as possible.

To illustrate the various tools and techniques used in performance engineering, we have chosen to rely throughout the book on frame relay and private lines as the underlying transport technology. Frame relay is representative of modern packet technology services, another being ATM, and is an increasingly popular transport service being deployed within corporate enterprise networks. The growth of frame relay services during the 1990s has been tremendous. As Figure 1.1 shows, this multiprotocol integration often occurs through an intermediate IP layer, but sometimes direct mapping onto frame relay is performed. However, the techniques and issues discussed throughout this book apply to the other transport services as well, for example, X.25, ATM, and Virtual Private Networks (VPNs). In addition, even though we focus on the four high-level protocol suites shown in Figure 1.1 in this book, the performance engineering methodology we discuss can be applied to other protocol suites and applications.

1.2   Our Approach to Performance Engineering

In the process of performance engineering wide-area networks, the network engineer, architect, or administrator needs to understand in some detail how applications behave over various underlying data network architectures. We approach this issue by building an atomistic-level understanding of the packet flows across data networks. We then build a deterministic model of the packet flows. Finally we augment this model with some simple statistical reasoning, as appropriate. From our experience, the process of building an atomistic model of packet flows forces several important thought processes on the network engineer. First, one is forced to extract the most important aspects of the high-level protocols involved in the analysis. This means understanding packet size and protocol overhead characteristics, windowing and flow control mechanisms, and packet fragmentation. Next, the engineer must understand the details of mapping of the high-level protocols onto the underlying transport mechanisms. Once this knowledge is built, we provide tools for representing this information in the form of timing diagrams. Timing diagrams explicitly show the timing relationships between the numerous activities that must occur in order to transport data over the network. Once this atomistic-level timing diagram is developed, it is relatively straightforward to augment this base model to capture the effects of statistical traffic multiplexing or cross applications effects, as discussed throughout the book.

Although the emphasis in this book is on building an analytical understanding of network dynamics, we strongly believe that analysis is not the end but the beginning point in performance engineering. One needs to be able to recognize first-order performance effects, make reasonable assumptions regarding the network and applications, perform "back-of-the-envelope" or "quick-and-dirty" calculations, interpret the results, and draw intelligent conclusions about the network. The mere use of off-the-shelf analytical or simulation modeling tools is not sufficient. These tools also require information about the network and applications that is often hard to obtain. For instance, think of a new client/server application that needs to be deployed enterprise wide. Information regarding transaction characteristics and usage patterns is difficult to obtain. However, the network manager needs to plan for bandwidth upgrades ahead of time.

One way to confirm assumptions used to build models is to perform test measurements. By collecting data from live test results, the engineer can confirm the assumptions used to build the model of the expected performance of the applications being deployed or integrated onto the underlying transport network. Once the application is deployed, it is necessary to constantly monitor key aspects of the network in order to ensure that satisfactory performance is maintained. Here again the models are useful to help in the determination of what to monitor and how to determine if modifications to the network architecture or design are required.

Finally, although this book focuses on an analytical understanding of WANs and of applications deployed on them, we make no pretense that these techniques will resolve all problems in a data network. Needless to say, improperly configured routers, hardware and software errors in end systems, and faulty WAN links contribute significantly to poor performance in a network. Nevertheless, an understanding of the set of basic principles that governs the performance wide-area networks and applications is important when troubleshooting problems in a network.

Clearly, there are numerous aspects to successfully deploying or integrating new applications onto a corporate enterprise-wide network. These include integration testing, deployment and migration planning, network monitoring and maintenance, and life cycle management. Although we briefly discuss these activities, the main goal of this book is to provide the network engineer, architect, and administrator with the skills and tools needed to develop simple performance models to aid in network design, engineering, and troubleshooting these seemingly complex multiprotocol, enterprise-wide data networks.

1.3 Layout of the Book

We have laid out the book in the following parts:

Part One: General Principles

The first part of the book lays the foundation for performance engineering multiprotocol enterprise networks. The discussions in the first part of the book remain at a protocol/application independent level. The techniques and principles presented in this part of the book are generic, and are applicable to performance engineering network applications in general. Although the following part of the book concentrates on specific high-level protocol suites, the material in this part can be applied to any application protocol being mapped onto a packet transport infrastructure. The following chapters comprise this first part of the book:

- *Chapter 2: Wide-Area Networking Technologies.* This chapter provides an overview of the underlying packet transport technologies found in corporate data networks.

- *Chapter 3: Performance Analysis: Some Basic Tools.* This chapter covers the basic aspects of performance modeling and performance issues necessary to build simple models of data application behavior.

- *Chapter 4: Techniques for Performance Engineering.* This chapter discusses tools and techniques to consider when building and supporting multiprotocol data networks.

- *Chapter 5: Frame Relay Performance Issues.* This chapter expands on the previous chapter by focusing on frame relay-specific issues.

- *Chapter 6: Using Pings for Performance Analysis.* This brief chapter analyzes the results from this simple tool and demonstrates its utility in testing and troubleshooting various networking situations.

Part Two: Specific Application/Protocol Suites

The second part of the book concentrates on specific high-level application/protocol suites. We discuss these because of the dominant role they play in enterprise data networking. This material is subject to the development schedules of the corporations responsible for the development of specific networking protocols. These chapters discuss issues specific to the various protocols when they are mapped onto underlying packet transport technologies and when they are multiplexed with other protocol suites on a common transport medium. The following chapters comprise this second part of the book:

- *Chapter 7: WAN Performance Analysis of TCP/IP Applications: FTP, HTTP, and Telnet.* This first chapter on a specific high-level protocol suite identifies the salient features of TCP/IP necessary to build simple models of TCP/IP performance.

- *Chapter 8: WAN Performance Considerations for Novell NetWare Networks.* This chapter presents aspects of the Novell NetWare networking suite.

- *Chapter 9: WAN Performance Issues for Client/Server Applications.* Although not a specific protocol suite, this chapter discusses a set of issues related to the development, evolution, and performance of client/server applications.

- *Chapter 10: WAN Design and Performance Considerations for SNA Networks.* This chapter covers issues with the IBM SNA applications and integration onto IP and frame relay networks.

Part Three: Case Studies

This last part of the book contains various case studies that the authors have been involved in over the years. They were chosen because we felt they were representative of common situations occurring in the evolution of enterprise networks. They also help to emphasize some of the prominent points discussed earlier in the book. A single chapter, Chapter 11, comprises this part of the book.

Part I
General Principles

2

Wide-Area Networking Technologies

2.1 Introduction

In this chapter we present an overview of the prominent high-speed, wide-area networking (WAN) technologies which have been deployed, or are being deployed, in today's multiprotocol, enterprise network environments. These technologies include time-division multiplexing (TDM) or private line, X.25, frame relay (FR), asynchronous transfer mode (ATM), and the internet protocol (IP). We consider other networking technologies, for example, IBM's systems network architecture (SNA), Novell's NetWare, and Apple's AppleTalk, as single-protocol networks or end-system protocols and as such these are discussed elsewhere in the book. Here we focus on those technologies which are often used to integrate (or multiplex) multiple end-system protocols.

The development and deployment of high-speed network technologies has accelerated during the last decade, and we expect this trend to continue at an even higher pace into the turn of the century. Prior to the mid-1980s most multiprotocol network environments physically separated the individual protocols onto separate, "private-line based" networks. These networks were rather inefficient in their bandwidth utilization and were rigid and inflexible in their ability to be modified based on changing end-system requirements. Further, interoperability between applications developed to run over different networks required cumbersome gateways.

In response, packet networking emerged in the 1980s, primarily in the form of X.25 networks, but also in ARPA's support for the development of IP networking. X.25 standards are based on virtual circuit technology, where a logical connection is established through the network prior to sending any data.

11

X.25 networking placed a high protocol processing demand on the network switches to handle circuit establishment during the call setup phase and to handle error recovery and network congestion issues in this statistically multiplexed network during the data transport phase. In contrast, IP technology placed a premium on end-system protocol processing, relying on end systems to handle session establishment, to respond to network congestion, and to accomplish packet error recovery. While the ARPA IP network remained primarily a network of interest to the research community at this time, X.25 networking was widely promoted by public network service providers and (in a more limited extent) began to support multiprotocol environments. This multiprotocol integration relied on packet assemblers/dissemblers (PADs) with built-in terminal adaptation[1] (TA) functionality to wrap the native protocols into the X.25 protocol stack and provide the WAN interface to the X.25 WAN networks. Due to the relatively low speed (limited to roughly 56 Kbps) of these X.25 networks, the utility of integrating multiple protocols onto this single technology never fully developed.

In the late 1980s fast packet technologies were developed that increased the speed of the WANs. Fast packet technologies are virtual circuit based, like X.25, but they streamlined the protocol processing required in the network during the data transport phase. This in turn allowed the switches to operate at higher speeds (limited initially to roughly 45-Mbps trunking[2]). Frame relay networks were one realization of these emerging fast packet technologies. Due to the relatively high speed of frame relay networks and their attractive economics in public network deployments, there was a huge push in the data communications industry to integrate multiple networks running different protocol stacks onto this common, layer 2 technology.

So what is the impetus for the emergence of ATM switching? Frame relay is capable of supporting the integration of multiple data networking applications and protocols on a common WAN, but does not naturally[3] support integration of non-real-time and real-time applications[4] onto a single platform. Just as the industry expects cost savings by integrating multiple data applications onto a single WAN, it also expects cost savings by integrating real-time and non-real-time applications onto a single WAN. ATM was developed to

---

1. Terminal adaptation is the operation of converting, or wrapping, the native protocol, for example, IBM's SNA, into a protocol suitable for transport over a WAN.

2. It is always dangerous to state (and forever rely on) rate limitations that are technology based, that is, not based on some physical law. Technology evolves at an ever increasing rate and this soon invalidates earlier assertions.

3. This does not imply that it cannot be done. In fact, frame relay networks are being used to carry real-time applications.

naturally support both real-time and non-real-time applications. ATM is based on virtual circuit technology but is unique in that it bases data transport on fixed size cells. This fixed sized cell format allows for simple algorithms to handle constant bit rate applications, such as PCM encoded digital voice. This fixed sized cell format also has generated much criticism of ATM in inefficiently carrying variable sized data packets.

More recently, IP has emerged as a network technology for multiprotocol network integration. Not only is IP being used in this fashion, it is also directly replacing the underlying network layers of legacy protocols like Novell's NetWare 5.0. Also, like ATM, there is a strong push to have IP support the integration of real-time and non-real-time applications.

In summary, the WAN technologies used for multiprotocol integration have evolved from TDM networking into datagram transport such as IP, and virtual circuit transport such as X.25, frame relay, and ATM. We provide a more detailed discussion of each of the dominant WAN technologies in the following sections of this chapter. Where possible, we try to identify the salient features of the technologies and networking environments where they might be applied. Further we identify reference materials for further reading in each of the sections.

2.2 Time-Division Multiplexing Networks

TDM networks are still a prominent WAN technology even today. Their primary advantage is that they provide guaranteed delay and throughput performance to data applications. These networks are often used to keep separate the data applications running over different protocol stacks, for example, SNA, IP, and AppleTalk. The job of network engineer is simplified in this networking scenario. Engineers can analyze the applications requirements in buckets separated by the protocol families and engineer the TDM network accordingly. There are no issues to consider regarding the cross applications/protocol effects when merging different protocol stacks onto a lower layer network layer. There are, however, significant network topology design issues to consider in order to

---

4. By real-time transport we mean the ability to deliver information at a constant delay, with negligible delay variation, to a user of the network. A real-time application is one that requires real-time transport. A non-real-time application does not require real-time transport. An example of a real-time application is PCM voice telephony or H.260 video distribution, where the information must be presented to the user at precise time points in order to maintain a smooth picture or a steady audio stream. Non-real-time applications are data applications such as e-mail transport.

minimize the network costs due to the distance-dependent pricing structure for these private line services.

TDM networks consist of private lines interconnecting TDM multiplexers (see Figure 2.1). These multiplexers are switches that transfer time slots on incoming lines to other time slots on outgoing lines. The time slots are fixed duration on any given line and are typically a bit or a byte in length. The relationship between an incoming time slot and an outgoing time slot is manually provisioned into the TDM switches. The frequency of the time slot determines the bandwidth dedicated to the individual application. The bandwidth is allocated, typically, in multiples of 64 Kbps or in submultiples of 64 Kbps, for example, 2.4, 4.8, and 9.6 Kbps. Because TDM switches rely on time relationships to determine the time slot positions, they do not depend on detecting information within the time slots. In fact, the switches have no mechanism to determine if the time slots are idle or if they contain application information. Because of this, the TDM switches cannot take full advantage of the bandwidth available by allowing different applications to share facility bandwidth in real time. An advantage, however, is that no special equipment is necessary to support the different data protocols; protocol multiplexing occurs at the physical layer of the OSI protocol stack [1].

Once engineered, these TDM networks are relatively static in nature. Modern TDM multiplexers support dynamic reconfigurations; however, the reconfigurations are typically initiated in response to facility failures. Time-of-day reconfigurations are also feasible to take into account the time-of-day variations in traffic, although this is rarely implemented in practice.

One dominant advantage of TDM networks today is their ability to support data, video, and voice applications. In fact, most large TDM networks carry all three types of traffic. These networks provide some bandwidth sharing

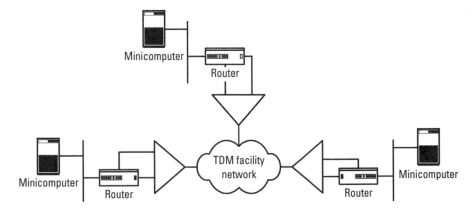

Figure 2.1 TDM networks do not support sharing of bandwidth into the wide-area network.

between voice and data traffic for disaster recovery scenarios. Critical data applications can preempt voice bandwidth during disaster recovery scenarios, and the voice traffic can overflow to public voice networks until the failure is repaired. TDM networking supports disparate networks and ensures that interference between applications on these networks will not occur. The disadvantage of TDM networking is that it maintains disparate networks and hence bandwidth sharing is not possible.

Our primary interest in TDM or private line networks is to function as a baseline network configuration. Due to price pressures, that is, due to public packet services being priced lower than comparable TDM networks, many engineers are considering replacing their TDM networks with newer packet networks. In the process, it is necessary to develop a model of the behavior of the data applications on the new packet network and compare this to the current behavior on the TDM networks. If the engineer is not careful, performance can be degraded when moving to the packet network. Therefore, network engineers must be cautious when considering a migration to packet network technologies to integrate data applications.

2.3 X.25 Networks

X.25 is an International Telecommunications Union (ITU) standard for a virtual circuit packet network technology developed in the mid-1970s. This standard was developed primarily by the telecommunications industry and public services providers. This technology is widely deployed around the world today. X.25 switches employ statistical multiplexing of data packets and operate at relatively low speeds, for example, 2.4 Kbps up to 1.5 Mbps. Virtual circuit technology relies on the preestablishment of a circuit across the network. This circuit can be manually provisioned, as in permanent virtual circuits (PVCs), or they can be signaled dynamically by the premises equipment, called switched virtual circuits (SVCs). A circuit is established between two data terminal equipment (DTE) interfaces. Once a virtual circuit (VC) is established, packets travel across the network, over the VCs, and maintain their relative ordering over the circuit. Further, X.25 provides a reliable, VC service, in the sense that packet delivery is guaranteed at the data link level and packet streams follow the same predefined path through the network.

X.25 technology was developed at a time when the quality of transmission systems was suspect. Digital transmission systems were not widely deployed and in many parts of the world the quality of the analog transmission systems (measured in terms of bit or packet error rates) was poor. As a result, packet retransmissions were common in these environments and the X.25

technology was designed to maximize data throughput over these relatively noisy transmissions facilities by implementing data link level retransmissions. Further, due to the relatively low speed of the transmission facilities and that the technology was a statistical multiplexing format, significant network congestion control mechanisms were required. To maximize data throughput, a very conservative approach to congestion control was adopted, that of employing both link-by-link flow control and end-to-end flow control (as shown in Figure 2.2). Quality of service (QoS) parameters are supported in X.25 networks but are limited to the notion of a throughput class. The throughput class is the expected throughput, measured in kilobits per second (Kbps), over a given X.25 virtual circuit. The mechanism typically employed to control this QoS parameter is the size of the layer 3 flow control window. This parameter is negotiated either out of band for PVC provisioning of circuits or through the X.25 signaling protocol for SVC establishment. However, this QoS parameter is not an accurate estimate of circuit throughput (as discussed in Chapter 3 on estimating window limited throughputs).

In contrast to TDM switches, packet switches, such as X.25 switches, detect the existence of information on incoming links. In packet technologies, data are carried in variable size packets. The existence of these packets, for the most part, is determined by the existence of special flag bytes. These flags, which are guaranteed to be unique by a technique called *byte stuffing*, delimit the beginning and ending of a data packet. Therefore, an X.25 packet switch will monitor the incoming facility searching for a beginning of packet flag. Once found it will search for the end-of-packet flag and will place the data between these flags in a packet buffer. Packet processing (e.g., forwarding, retransmissions, flow control, etc.) is then based on protocol information that

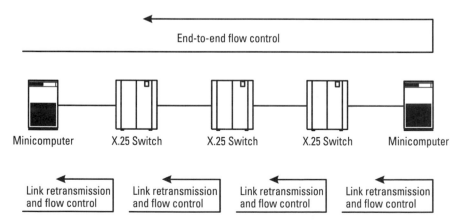

Figure 2.2 An X.25 connection showing both link retransmission and flow control and end-to-end flow control.

is either prepended (in a packet header) or appended (in a packet trailer) to the packet. Packet forwarding is based on the virtual circuit identifier (VCI) that is carried in the packet header. Errored packets are determined based on error checking information carried in the packet trailer. If the packet is determined to be errored, then the packet is discarded and a retransmission is requested across the data link. The format of the X.25 data packet is shown in Figure 2.3.

The flags help to delimit the X.25 packet on the physical transmission facility. The level 2 header control contains the level 2 window and flow control information. The VC identifier is contained in the level 3 control header, as is the level 3, end-to-end, flow control information. Finally, the trailer contains a cyclic redundancy check (CRC) for error detection. (For an excellent discussion of error detection and correction technologies, refer to [2].)

The flow control will limit the amount of data stored in the switch for any given VCI, hence reducing the probability of buffer overflows. Statistical multiplexing on the outgoing transmission facility is made possible because of the switches' ability to distinguish packet information from transmission idle. This allows DTE to share access bandwidth between multiple destinations, as identified in Figure 2.4. This is referred to as a *multiplexed network interface*. The X.25 network relies on the VCI to determine the network egress point.

To support end user applications riding on a variety of end-system data protocols, it is necessary to deploy a terminal adaptation device. The TA performs a transformation between the end-system application protocols and the X.25 networking protocols. This conversion comes in two varieties:

1. *Encapsulation,* where the end systems' protocols are wrapped in the X.25 protocol and are carried across the X.25 network intact. In this case the X.25 network can be considered to be like a pipe, albeit a pipe with some unusual characteristics, which transports the native protocol. (See the later section on multiprotocol encapsulation standards.)

2. *Conversion,* where the end systems' protocols are terminated and replaced by the X.25 protocol. In this case, each TA on the edge of the X.25 network appears to the respective end system as its corresponding end device.

Flag	Level 2 header	Level 3 control	Data field	CRC 16	Flag

Figure 2.3 An X.25 packet format showing flags, level 2 header and trailers, level 3 control, and the data field.

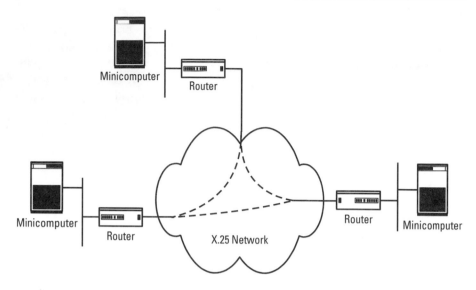

Figure 2.4 A typical X.25 configuration demonstrating the multiplexed interface supported by X.25.

Numerous X.25 TA devices are available in the market. There are encapsulating TAs for SDLC®, bisync protocols, polled asynchronous protocols, and Borroughs polled protocols, DEC's DDCMP® protocol, to name a few. Also, encapsulations are defined for the various internetworking protocols, for example, IP, AppleTalk's IP, Novell's IPX®, XNS's® IP, SNA's® FID 2 and FID 4 networking layer, onto X.25. Conversion TAs have been developed that convert, for example, SDLC and bisync polled protocols and character asynchronous protocols to X.25. In fact, the X.3 standard defines the mapping of character asynchronous protocols to X.25 and these TAs are referred to typically as packet assemblers/disassemblers. It is common to refer to TAs (both encapsulating and converting) that support the polled protocols as PADs as well.

An extensive list of X.25 network service providers exists. These providers offer public X.25 transport services to end users. These services are offered by private telecommunications companies and by public and government-owned PTTs. To interconnect these numerous, autonomous X.25 networks and to provide worldwide X.25 services, the ITU developed an interworking standard, called *X.75*. The X.75 standard defines the network interface between autonomous X.25 networks. Consistent and unique addressing is ensured by the ITU X.121 standard. With the development of these standards, the various X.25 service providers were able to work out pair-wise interconnection agreements and to build a worldwide X.25 network infrastructure. A high-level view of this infrastructure is shown in Figure 2.5. In this example, two DTEs are interconnected via an SVC or PVC that spans several autonomous X.25 service providers' networks. Here link-by-link flow control and error recovery occurs on each of the internal X.25 links and the X.75 gateway links. The level 3 controls still extend end to end between the two respective DTEs.

In summary, there was some interest in using X.25 networks to integrate multiple data applications onto this technology in the 1980s but this activity has greatly diminished with the development and wide deployment of frame relay networks. Because this trend never fully materialized, there was little work in developing capabilities into X.25 to improve enterprise network deployment, such as improved QoS capabilities through fair sharing or priority queuing algorithms.

2.4 Frame Relay

2.4.1 Frame Relay Network Architecture

As the deployment of digital transmission systems progressed in the early 1980s, the quality, in terms of bit and burst error statistics, improved to the point that the need for packet switches to perform extensive error recovery mechanisms became unnecessary (or rather was rarely invoked). In fact, the software to perform error recovery mechanism and extensive hop-by-hop flow control limited the throughput of the individual link processors. Performance studies at the time showed that optimum end-to-end performance was achieved by eliminating this unnecessary protocol processing, which allowed link processor data throughput to increase to 1.5-Mbps rates and higher. Error recovery and flow control were relegated to end-to-end transport protocols, for example, TCP. In this case, the network simply discarded errored frames and relied on the transport protocols to detect frame losses and to handle frame

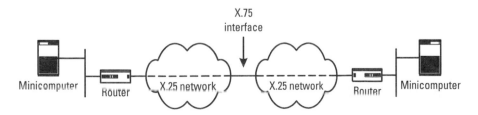

Figure 2.5 An example of multiple X.25 networks interconnected through X.75 interface gateways.

retransmissions. Instead of relying on windowing to manage congestion, access control mechanisms were employed in the network. This is illustrated in Figure 2.6.

Like X.25 technology, frame relay is based on virtual circuit, connection-oriented forwarding. Further it relies on statistical multiplexing and supports both PVCs and SVCs. Figure 2.7 shows the packet level header and trailer for a frame relay packet. The frame header consists of two bytes for virtual circuit addressing, or data link connectionl identifiers (DLCIs). The frame trailer consists of two bytes of a frame check sequence, that is, a 16-bit CRC, for error detection. The frame is delimited by a beginning and ending flag sequence, consisting of one byte each.

Unlike X.25, congestion control in frame relay networks had to be handled in a fundamentally different method due to the lack of layer 2 and 3 flow control feedback. At the frame relay layer, the sending device could not rely on a windowing scheme with real-time acknowledgments to derive the status of

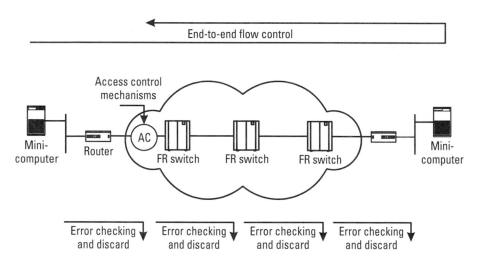

Figure 2.6 A frame relay connection showing error checking and discards on a link-by-link basis and end-to-end flow control.

Flag	DLCI	Data	CRC-16	Flag

Figure 2.7 Packet format for a frame relay packet.

the individual virtual circuits. Also, the frame relay network could not rely on a layer 2 window to throttle network traffic in periods of network congestion. Without a throttling capability, a throughput collapse could occur [3].

Frame relay implemented several new networking mechanisms to address these issues, as discussed in the following paragraphs.

A *link management protocol* was implemented to communicate the status of the access link and the individual virtual circuits. DTE devices rely on this protocol and the information provided by it to determine when it considers the virtual circuits and access links as functional and when to initiate failure recovery mechanisms. Without this protocol, the DTE would have no mechanism to monitor the frame relay network, the network access and egress facilities, or the individual VCs and would continue transmitting frames into the network even in conditions of network failure.

In *credit management schemes,* an access control mechanism was developed to control the rate at which the DTE's frames can be transmitted into the network on an individual virtual circuit basis. Again, without layer 2 windowing, the network would not be able to flow control the rate of incoming frames without a credit management scheme. A QoS parameter, termed the committed information rate (CIR), is defined for each individual virtual circuit. The definition of the CIR is rather ambiguous, but in some instances is interpreted as the minimum throughput on the virtual circuit supported by the frame relay network. The frame relay network is thus engineered to support this VC-based throughput during periods of network congestion. During slack periods on the network, the individual VC throughputs can exceed the CIR. If the network is congested, and the DTE is attempting to send in excess of the available network bandwidth, then the network may discard these frames at the congestion point in order to protect other users' traffic in the network.

Various congestion control feedback mechanisms were developed. These include (1) backward explicit congestion notifications (BECNs) and forward explicit congestion notifications (FECNs), (2) a discard eligible (DE) bit in the frame relay header, and (3) an R-bit setting in the link management protocol. The FECNs (BECNs) are sent downstream (upstream) on individual VCs in response to the DTE sending in excess of their CIR. The frames that the network determines to exceed the CIR are tagged as such through the DE bit and are subject to network congestion. The R bit is sent over the link management interface in the upstream direction and is to be interpreted by the DTE as an indication to slow down transmission across all VCs on the interface. In reality, there is little that the DTE can do to throttle the data sources. This is indicated in Figure 2.8. The tagging of FECN, BECN, and DE bits is counted by management systems and used as a capacity management tool to help determine

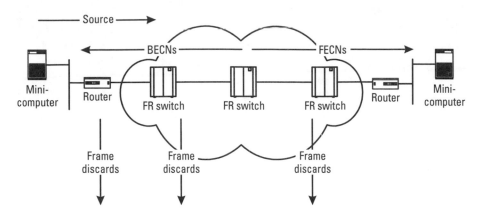

Figure 2.8 A frame relay reference connection demonstrating the lack of end-to-end flow control mechanisms in many important networking scenarios.

when to upgrade CIR and/or port capacity or take other appropriate actions. These are discussed further in the following subsections.

2.4.1.1 Link Management Interface

Frame relay standards define a link management interface (LMI), which provides the attached CPE capabilities to monitor and manage the frame relay links, ports, and VCs. The definition of a separate LMI for frame relay was required due to the fact that frame relay provides the user with a best effort service. Unlike X.25, where the user is provided with acknowledgments at both the link level and the network level that their packets were successfully received, frame relay service provides no such indication. Instead the user relies on information obtained through the LMI to determine the status of the access and egress links and the provisioned VCs.

Various versions of the link management capabilities are defined in the various standards and de facto standards. In general, LMI defines message formats and exchanges between the network interface and the CPE. The capabilities defined through the link management protocols include:

- PVC status, for example, what VCs are provisioned and the status of the provisioned VCs, that is, are they active or inactive;
- Notification of adds, changes, or deletes of VCs;
- Local link status polling and notification of far end link status; and
- CPE access to configuration parameters, for example, maximum frame size and FECN/BECN and DE bit setting thresholds.

The CPE can pass this information to higher layers and initiate various actions based on the link and VC status. For example, in the event that a link or VC fails, this information is received from LMI and the CPE can initiate a network layer rerouting. Also, some CPE use the add, change, delete information in an autoconfiguration process. For example, on notification of a new VC, the attached router can initiate the inv-ARP protocol [4] to configure IP connectivity to the far end router on the new VC.

2.4.1.2 Credit Management Schemes

Frame relay networks define a bandwidth management policy for the VC established on the network interface from the customer access device to the frame relay network. The network interface is multiplexed, meaning that multiple VCs are simultaneously established on the access facilities. For each of the VCs, whether they are permanent or switched VCs, a bandwidth policy is defined. Policies generally define the long-term bandwidth associated with the VC, referred to as the CIR. The policy also defines the nature of the bursting—an information rate higher than the subscribed CIR—associated with the VC. Frame relay service providers generally allow traffic to burst into their networks at speeds higher than the subscribed CIR of the VC for a limited period of time. The effectiveness of this bursting depends on overall network congestion control implementations and network engineering policies. By definition, "bursting" is not guaranteed at all times. This introduces an element of variability relative to available bandwidth.

It is convenient to define the frame relay bandwidth management policy in terms of a "leaky bucket" algorithm [5]. Figure 2.9 shows a schematic of a frame relay access control filter, based on a leaky bucket scheme. This access control filter (ACF) maintains a token buffer, which determines the treatment received by incoming data frames. When a data frame is received by the ACF, it compares the size of the data packet to the equivalent token pool in reserve. Each token has an equivalent packet size associated with it; for example, each token is worth 24 bytes. If a sufficient number of tokens is stored in the buffer to cover the size of the data packet, then the entire packet is allowed entry into the frame relay network, the equivalent number of tokens is removed from the token buffer, and no further processing of the data packet is required. If there are not enough tokens in the buffer to fully cover the size of the incoming data packet, then the ACF generally implements one of two different strategies:

1. The packet is tagged as exceeding the token rate through the DE bit on the frame header and is sent into the frame relay network. (We refer to this scheme as the "tag-and-send" ACF.)

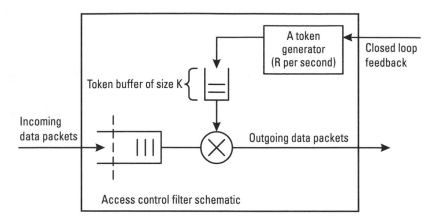

Figure 2.9 A schematic of an implementation of a frame relay access control filter.

2. It is buffered in the access module and delayed until the necessary number of tokens is built up in the token buffer. (We refer to this as the "delaying" ACF.)

Each VC on the network interface has a separate ACF and its associated token buffer. These separate ACFs on the network interface run independently of each other.

The bandwidth policy for each VC is defined by the parameters specifying the leaky bucket. The parameters defining the leaky bucket are the CIR (the rate at which the token buffer is replenished), K (the maximum number of tokens allowed in the token buffer), and the size of the data buffer (for the delaying ACF).

The ACF results in the following output characteristics. The long-term, untagged output from the ACF has a maximum data rate equal to the CIR, the rate at which the token buffer is replenished. This has a smoothing effect on the delaying ACF output stream, that is, the data traffic is effectively metered out at the rate of the CIR. If the input stream to the ACF has been idle for a long enough period of time, the output will consist of an initial "burst" of data equal to or less than the total size of the token buffer, that is, the K parameter. Therefore, except for the initial burst of data, the untagged output from the delaying ACF is smoothed in comparison to the input stream. This is illustrated in Figure 2.10. In this figure, the token generation process is shown at the top. The token rate is constant and determines the CIR. The second line shows a given arrival process.

It is assumed in Figure 2.10 that one token is equivalent to a single data packet. (Often several tokens are necessary to allow a given data packet to pass

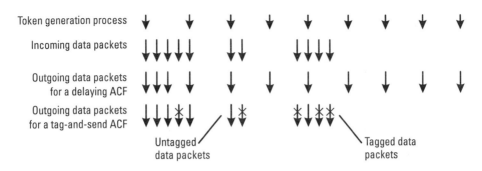

Figure 2.10 Input and output process from several types of ACFs.

through the filter when the frame relay network performs internal pipe-lining.) The incoming stream shows several bunches of packets arriving. The third line shows the expected output stream from a delaying ACF. It is assumed that the incoming stream had been idle for a sufficient amount of time to allow the token buffer to completely fill. Further, it is assumed that the size of the token buffer is three, that is, $K = 3$. Therefore, the first three packets are transmitted into the frame relay network. From then on, the remaining packets from the arrival process are buffered and metered into the network at a constant rate as determined by the CIR. This is in contrast with the fourth line, which shows the output process from a tag-and-send ACF. Here, the untagged packets (those indicated by the arrows) show similar output characteristics as the output from the delaying ACF from the line above. However, the tag-and-send ACF does not buffer packets that arrive in excess of the ACF bandwidth policy; it instead tags those packets and sends them without delay into the frame relay network. These are shown as arrows with X's through them.

Most public frame relay service providers allow traffic to burst into their networks at speeds higher than the subscribed CIR. This introduces an element of variability relative to the available bandwidth. Again, the effectiveness of this bursting depends on overall network congestion and, by definition, is not guaranteed at all times. The network congestion is affected by the specific congestion control schemes implemented in the frame relay network and the nature of the network engineering. In the following subsections, we discuss two different frame relay network congestion control strategies: closed-loop and open-loop feedback schemes.

2.4.1.3 Closed-Loop versus Open-Loop Feedback Schemes

At first blush, it appears that the tag-and-send ACF results in improved performance over that of the delayed ACF. Certainly in periods of no traffic congestion in the frame relay network, the tag-and-send ACF provides better delay performance than the delaying ACF simply due to the fact that the tag-and-send ACF will never delay packet access to the frame relay cloud, whereas the delaying ACF will.

To improve the low load performance of the delaying ACF scheme, these frame relay networks also implement a type of closed-loop feedback algorithm in order to dynamically adjust the token arrival rate. The closed-loop schemes typically have the network switches indicate in the network packet headers the status of the buffers along the VC path. The egress switch, that is, the last frame relay switch on the VC path, uses this information to send back to the ACF the status of the VC. For example, these could indicate that the components are lightly loaded, moderately loaded, or heavily loaded. In periods of low load along the VC path, the delaying ACF can increase the token arrival rate and hence increase the flow of data packets into the network over the VC. This reduces the buffer delay experienced by the data packets in the access module. In the event that the load on the VC path increases, the delaying ACF can reduce the rate of the token generation process in order to lessen the load on the VC and, in turn, manage the VC congestion. Therefore, the token generation rate for the closed-loop feedback scheme will vary from a low equal to the CIR to a maximum information rate (MIR). The MIR will be bounded by the minimum of the access and egress facility speed associated with the VC, but may be further bounded at a lower speed depending on the specific implementation. If the VC path becomes congested, the ACF is notified through the closed-loop feedback mechanism. The ACF then reduces the effective rate by diminishing the token generation rate, potentially down to the minimum CIR value. This reduces the load on the congested network resource and places a greater burden on the ACF buffer by throttling the rate of admission of the data packets into the frame relay network. Because the delaying ACF in this method is capable of buffering the incoming data and then can meter the data into the network at the current token rate, the triggers within the network that sense network congestion can be extremely sensitive, for example, 10 data packets in queue. By crossing a threshold, the closed-loop systems do not discard the data as the first reaction. They instead indicate to the ACF that it should reduce the current token generation rate.

In contrast, the tag-and-send ACF is associated with an open-loop scheme. Here, the data packets are sent without delay into the frame relay network. Those packets exceeding the CIR are tagged as such. The DE bit is used to indicate that this packet exceeded the CIR according to the ACF policy. There is no feedback derived from the network to influence the behavior of the ACF. If the VC path becomes congested, the congested resource reacts by discarding packets with the DE bit set first. By discarding data packets from the "offending" VCs, as indicated by the DE bit, the end systems' transport protocol will hopefully throttle back the rate at which they send data into the network. This reduces the congestion on the overloaded frame relay network resource.[5] In the open-loop congestion control systems, the triggers within the network, for example, buffer occupancy thresholds, have to be set at fairly high values, say, 75% occupancy, because they immediately initiate a data discard strategy. This is a fairly harsh response and should be avoided in all but the most extreme cases. Therefore, the buffer threshold is set fairly high.

The actions of both the tag-and-send and the delaying ACF are prompted by congestion at overloaded resources within the network. As mentioned earlier, there are numerous resources along a given VC path, any of which can become overloaded. Each of these resources is front ended with a data buffer (in a statistical multiplexing environment). For example, Figure 2.11 shows a high-level schematic of a frame relay switch. The example highlights the input and output buffers on the various interface modules. Each interface module contains an input buffer, which stores incoming data off the transmission facility until it can be transmitted over the switch data bus. Each interface module also contains an output buffer, which stores data pulled from the switches' data buses until they can be transmitted onto the outgoing transmission facility. Typically, each of these buffers will maintain thresholds that trigger specific actions, if exceeded. In the event that the buffer occupancy exceeds this threshold, specific actions will be triggered. A common action taken by a resource in a frame relay network relying on an open-loop congestion control architecture is to discard packets with the DE bit set.

2.4.1.4 CPE Participation: FECNs, BECNs, and Other Mechanisms

Let us now examine situations of extreme congestion that cannot be fully handled within the wide-area network. We will find that it is necessary to invoke the support of the various other data communications equipment within the data path. For extreme congestion situations, various mechanisms have been built into the frame relay network standards. By indicating to the other data communication equipment that the frame relay network is experiencing network congestion, it is hoped that this other equipment can slow down the flow

---

5. There is no guarantee that the load on the network will be reduced. By dropping packets the network may just be forcing end systems to retransmit the dropped packets (and often additional packets within the transport window). However, the hope is that the end system is running a transport protocol, which will dynamically adjust its window in response to packet discards, such as TCP. This would then reduce the congestion on the network generated by this data source.

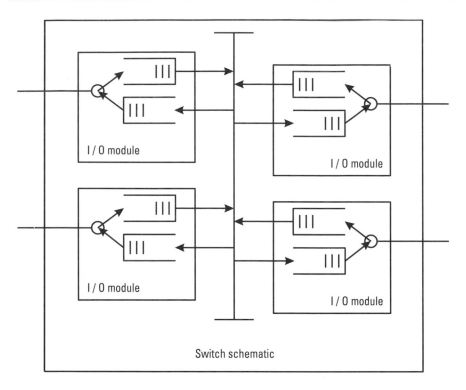

Figure 2.11 High-level schematic of a frame relay switch.

of data into the already congested network. Unfortunately, this is not always the case.

The various explicit congestion notifications are communicated either through several bits defined in the frame relay packet headers or through a separate link management protocol run over the frame relay interface. FECNs and BECNs are communicated through the FECN and BECN bits in the frame headers, respectively. FECN/BECN bits are set on a per-VC basis by the frame relay carrier when the VC experiences congestion. The R bit is set in response to congestion in the interface module within the ingress frame relay switch and is communicated to the CPE through the link management protocol. As such, the R bit throttles traffic on the entire frame relay interface.

Figure 2.12 shows the ECNs found within frame relay network implementations. In the figure, it is assumed that the direction of the traffic flow is from the left to the right. In this case, the forward (downstream) direction is from left to right and the backward (upstream) direction is from right to left. Therefore, the FECN indicates to the receiver that there is congestion on the forward direction VC, while the BECN indicates to the transmitter that there is

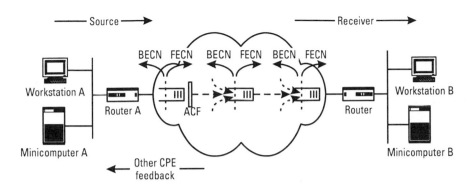

Figure 2.12 Explicit congestion notifications within the frame relay standards.

congestion on the VC. The R bit tells the transmitter to halt transmission—a basic form of flow control.

Essential to the value of this type of communication is the ability of the CPE attached to the frame relay network to be able to respond in a useful way. Evident in Figure 2.12 is the fact that the attached CPE, in this example a router, is not the source of the traffic, but rather an intermediate switch in the path. Therefore, to be an effective means of alleviating congestion, either the router requires a significant buffer or it needs some means of pushing back on the ultimate traffic sources. Unfortunately, this is not always the case. Routers typically support various different networking protocols, not all of which allow for the router to "push back" on the traffic sources to flow control their output.

2.4.2 Frame Relay and Enterprise Networks

Frame relay is the first packet-based technology which has received large support in enterprise networks for integration of multiple data applications onto a single backbone. This is due primarily to two effects: (1) By minimizing the protocol processing, the throughput of the link processors increased dramatically over previous packet switching technologies; and (2) efficient link utilization of a statistical multiplexing technology provides economic incentives for corporations to migrate their data networks to this new WAN technology. Link throughputs of 45 Mbps are now generally available on frame relay switching products.

Numerous networking devices, or TAs, are available in the market that will adapt the native networking protocols, for example, SDLC, bisync, and IP, onto a frame relay interface. These devices can be categorized as encapsulators and protocol converters as for the X.25 PADs discussed in the previous section. In fact, some vendors have simply upgraded their X.25 PADs to similar frame relay devices, known as FR access devices (FRADs). Router vendors helped to stimulate the demand for frame relay networking by developing FR interfaces on their products. Initially, routers simply performed IP, and other layer 3 protocol, encapsulation onto their FR interfaces. However, routers have also incorporated FRAD functionality into their products in order to support legacy protocols such as SDLC and bisync. Standards are developed to ensure interworking of these encapsulations and conversion methods. These are discussed in Section 2.7.

2.5 Asynchronous Transport Mode

The deployment of frame relay networks is providing an economic and performance incentive to integrate disparate data networks onto a common WAN environment. However, for numerous years, the vision has been of a common WAN environment to integrate voice, video, and data applications. The ATM technology was developed by the telecommunications industry as the tech- nology to achieve this integration. The International Telecommunications Union–Telecommunications Sector (ITU-TS), formally CCITT, laid the groundwork for the ATM standards in the late 1980s. During these discussions, various architectures were analyzed, with the ATM method of carrying traffic from voice, video, and data applications being chosen.

In our earlier discussion of TDM networking, we saw that the fundamental transport entity is the time slot. Incoming time slots are mapped to outgoing time slots on TDM switches resulting in time slot mapping across the TDM WAN. These mappings are deterministic and so is the performance characteristics of the connection. Thus a deterministic QoS channel is established and is capable of transporting real-time traffic (that is, traffic that has a fixed delay requirement in its delivery to the end system) with little effort. These fixed time slots carry a fixed length information entity, typically a bit or a byte. However, TDM transport is relatively inefficient in its ability to carry statistical data traffic.

In contrast, packet switches operate by mapping variable length packets of data from incoming ports to outgoing ports. The switch cannot rely on temporal methods to determine the boundaries of the information. Instead, the information is delimited by data flags (a redefined bit pattern that is unique to this function on the link). Once an information packet is identified, the switch determines its forwarding based on a circuit identifier, for example, the VCI in X.25 or DLCI in frame relay, or a datagram address in IP, which is carried in the packet header. Because the information is explicitly delimited with flags and carries its own forwarding information, it is possible for the packet switches to perform statistical multiplexing on the outgoing links.

The ITU-TS decided that, in order to develop a technology designed to carry both real-time and non-real-time traffic, it must combine the above aspects of TDM and packet switching technologies. Toward this end, the decision was made that (like TDM) ATM would base its transport on a fixed length data unit, termed an *ATM cell*. Further, to support statistical multiplexing, this ATM cell would contain a header field, which among other functions, carried a virtual circuit identifier for cell forwarding. The choice of the cell length was 53 bytes (48 bytes of payload and 5 bytes of cell header), which represented a compromise between a desire for small cells optimized to carry PCM encoded voice and large cells optimized to carry data.[6] This cell format is shown in Figure 2.13.

The ATM switches organize these cells into physical level frame structures, for example, SONET frames and DS-3 physical layer convergence protocol (PLCP) frames, as shown in Figure 2.14.

Early on, there was much discussion of whether the respective cell slots in the physical layer frames should be preallocated (i.e., synchronous transport mode), not preallocated (i.e., asynchronous transport mode), or a hybrid of both where the first N cell slots are preallocated and the remaining M cell slots are not, which is termed hybrid transfer mode (HTM). Obviously, the outcome of these discussions was that ATM was the preferred choice.

With the framing structure identified earlier, it is easy to develop a simple conceptual model describing how ATM switches could carry both real-time and non-real-time traffic over a common trunking interface. In reality, other algorithms are used (based primarily on priority queuing and bandwidth accounting methods at circuit setup), but our model is a useful one for discussion purposes. The PLCP frame in Figure 2.14 has a duration of 125 sec.[7] A cell payload is 48 bytes, and 12 cell slots exist per PLCP. Therefore, a single cell slot per PLCP frame amounts to 3.0 Mbps of traffic. To carry a number of, say,

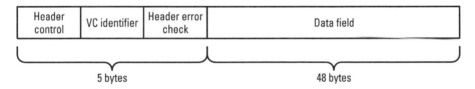

Figure 2.13 The ATM cell format.

---

6. We are using the term "optimal" very loosely in this discussion.

7. The 125-sec period was chosen to match the 8-kHz timing frequency of the current digital hierarchy. In fact, the period of the DS-3 PLCP is traceable to a Stratum 1 clock and can be used to derive timing for networking equipment.

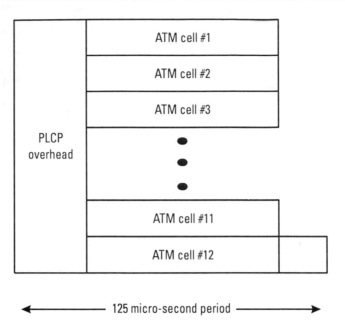

Figure 2.14 The physical layer structure of the DS-3 PLCP framing for ATM trunking.

1.5-Mbps constant bit rate video streams or 1.5-Mbps circuit emulation traffic streams, then the ATM switch could reserve a single cell slot per every other PLCP frame.[8] For the variable bit rate data traffic, their cells can be statistically multiplexed into the remaining, unreserved cell slots in the PLCP frames. This is a conceptually simple algorithm, yet it is capable of multiplexing onto a single facility both real-time and non-real-time traffic and in the process achieve high facility utilization.

Like frame relay and X.25, ATM is a connection-oriented technology relying on the establishment of virtual connections. ATM standards are defined for both permanent virtual connections and switched virtual connections. ATM interfaces are defined to operate from as low as 1.5 Mbps to 2.4 Gbps and higher rates. ATM is a fast packet-based technology, like frame relay, in the sense that error recovery and flow control mechanisms are relegated to the end systems. The network-based congestion control mechanisms are rate-based, access controls similar to those discussed for frame relay networks.

---

8. This assumes that real-time traffic from a video source is encoded into a constant bit rate cell stream, that is, that the 1.5-Mbps video encoder will generate a full ATM cell every 250 sec. However, video encoders do not have to generate a constant bit rate stream.

Because ATM is intended to support all application types, the definitions and types of QoS classes for ATM networks are numerous. Fundamentally, ATM supports constant bit rate (CBR) and variable bit rate (VBR) connections. CBR connections are characterized by delivering cells across the network with extremely low cell losses and extremely low cell delay variation (or cell jitter). VBR connections are characterized with a higher cell loss probability and with little concern regarding the cell delay variation. Both CBR and VBR connections are further characterized by a minimum sustainable cell rate (SCR), which is interpreted in a fashion similar to the CIR for a frame relay virtual circuit. ATM networks control the access rate by implementing a credit management scheme.

Figure 2.15 shows various layers in the ATM suite of standards. The *ATM layer* defines the precise structure of the fixed size ATM cell. This layer defines the cell-based transport and multiplexing capability. Below the ATM layer resides the *physical layer*, composed of the physical medium standard and a PLCP. This PLCP layer defines how the fixed size ATM cells are mapped onto

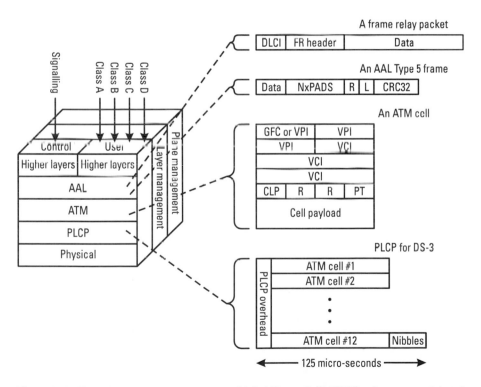

Figure 2.15 Broadband Integrated Services Digital Network (B-ISDN) reference model and individual layer protocols with an example of frame relay over ATM (see discussion of FR-ATM interworking in Section 2.8).

the various physical transmission standards available to the ATM layer. These range from DS-1 (1.544-Mbps) up to STS-12c (622-Mbps) signals. For example, Figure 2.14 shows a PLCP for mapping 12 ATM cells onto a 125-sec framing structure defined on a DS-3 signal format.

The layers above the ATM layer, that is, the ATM adaptation layers (AALs), are responsible for mapping basic service classes onto a common ATM layer. The basic services defined are:

- *Class A:* Connection-oriented CBR transport for applications requiring bit synchronization.

- *Class B:* Connection-oriented VBR transport for applications requiring bit synchroppnization.

- *Class C:* Connection-oriented VBR transport for applications not requiring bit synchronization.

- *Class D:* Connectionless VBR transport for applications not requiring bit synchronization.

- *Class X:* An unspecified class of service.

Typical applications using Class A services include circuit emulation of private lines and CBR video, and applications using Class B services include packet voice and video. Applications using Class C services include frame relay and X.25 packet applications, and applications using Class D services include SMDS. Different AALs are defined for the various service classes, for example, AAL Type 5 is defined for Class C services. Functions performed by one or more of the AALs include converting higher level, variable size protocol data units into 48-byte ATM cell payloads (a function termed *segmentation and reassembly,* or SAR), error detection and/or correction, cell sequencing, buffering, and isochronous playout of data. Figure 2.15 shows an example of a frame relay application being carried over an ATM AAL Type 5 Class C service. Here the frame relay frame header and data are carried in an AAL Type 5 data unit that is padded out to an integral multiple of 48 bytes (including the AAL Type 5 trailer). This is then segmented into an ATM cell stream, which is then mapped onto a DS-3 signal for transmission.

The protocol layers just discussed, for example, the physical layers up to the AALs, define the data transport (or user plane as identified in Figure 2.15) aspects of the ATM standards. In addition to these, the standards define control, layer, and plane management protocols. The control plane contains the signaling specification. This defines the protocol exchanges between, for example, a user and the network, required to invoke the network to establish a

SVC for the end user. In the event that the CPE and network have not implemented the ATM signaling protocol, then PVCs can be manually administered for the end user. The layer management protocols are responsible for management of the specific protocol layers, such as operations, administration, and maintenance (OAM) cell flows to manage ATM virtual connections. Plane management extends across individual layer boundaries, such as ILMI's capability to manage ATM interfaces. For an excellent overview of the ATM standards, refer to [6].

Initial deployment of ATM switching was in local-area networking environments, where ATM offers a higher speed alternative to switched Ethernet and FDDI LANs. To support the relatively dynamic and high point-to-point throughput requirements of the LAN environment, extensive ATM SVC capabilities were developed. This is in distinct contrast with the development of frame relay switches in the WAN environment. The ATM Forum, an industry consortium developed to define interoperability standards for ATM networks, has defined numerous networking capabilities for ATM switches in LAN and WAN environments. In the LAN environment, the ATM Forum has defined the LAN emulation capability. This capability gives an ATM switch or network the appearance of a MAC layer LAN, for example, Ethernet or token ring. The LAN emulation capability relies on ATM SVCs and the definition of registration and multicast servers to emulate the multiaccess, broadcast capabilities of an Ethernet, token ring, and FDDI LANs. An advantage of MAC layer emulation is the ability to support existing, nonrouted applications, for example, DEC's LAT® or Microsoft's NetBEUI®, on ATM networks.[9]

Because of the perception by many that ATM network deployment will be ubiquitous in the future, there is a concerted effort by the ATM Forum to extend the flexibility of ATM LAN capabilities into a vast WAN infrastructure. This has led to some rather interesting addressing and routing capabilities defined for ATM networks, borrowed from concepts developed by ISO and the IETF. First, a flexible addressing scheme was adopted by the ATM Forum, based on the network services access point (NSAP) addressing plan developed by ISO. The advantages of this addressing plan lie in (1) the large size of the address space, that is, 20 bytes; (2) the capability of embedding legacy addressing plans, for example, X.121, E.164, and IP, within the NSAP format; (3) the capability of supporting geographically based and organizationally based addressing plans; and (4) the existence of an organizational infrastructure defined for the administration of this addressing plan.

---

9. An alternative implementation of LAN capabilities on ATM is defined in the Internet Engineering Task Force's (IETF) Classical IP on ATM standard, discussed in the following section on the internetworking protocol.

Coupled with the adoption of a flexible addressing scheme, the ATM Forum has defined a flexible and dynamic routing capability for large ATM infrastructures. This capability is termed the private network-to-network interface (P-NNI) standard. The P-NNI routing capability borrows from the dynamic routing capabilities defined by the IETF and ISO during the 1990s. The P-NNI architecture includes the definitions of autonomous systems (ASs), which are segments of the ATM infrastructure under the administration of a separate organization, routing metrics based on various QoS parameters, dynamic exchanges of routing updates, and flexible initialization and configuration discovery capabilities.

The joint definitions of flexible addressing and dynamic routing capabilities has led to the term *ATM internets* being applied to this overall architecture. (Refer to the following section on the discussion of the IP architecture to see the similarities.) This is illustrated in Figure 2.16.

In addition to the flexibility being defined into ATM in the areas of routing, addressing, and QoS support, TA functions are being standardized for

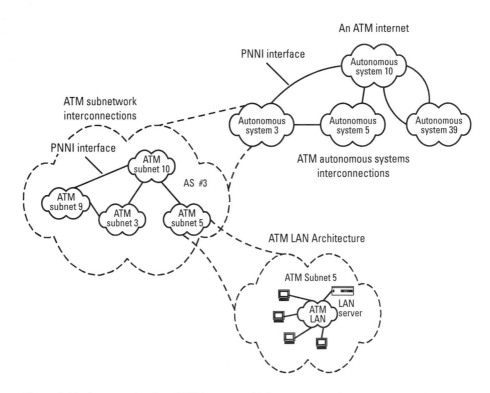

Figure 2.16 An example of an ATM internet with interconnected autonomous systems consisting of interconnected ATM LANs.

numerous applications. These standard adaptations include encapsulation TAs to map the following:

- A digital bit stream, for example, a 1.5-Mbps DS-1 signal onto an ATM CBR virtual circuit, termed *circuit emulation;*
- A frame relay, or an X.25, connection onto a VBR ATM connection;
- $n \times$ 64-Kbps video standards onto a CBR ATM circuit;
- PCM encoded voice onto a CBR ATM circuit;
- Routed protocols, for example, IP and IPX, onto a VBR ATM circuit;
- Nonrouted protocols, for example, DEC LAT and NetBEUI, by mapping MAC layer protocols (Ethernet, token ring, FDDI) onto VBR ATM circuits, that is, the ATM Forum's LAN emulation service; and
- MPEG 2 encoded video onto a VBR ATM circuit.

They also include conversion TAs to map:

- Frame relay circuits into VBR ATM circuits.

This list is not meant to be all-inclusive and, due to the rate at which new ATM-related standards are being defined, could not be inclusive.

Clearly, an advantage of ATM is the range of terminal adaptations being developed that will allow for a broad range of integration capabilities. However, this comes with a price, namely, the relatively high overhead associated with small cell sizes. Another downside to ATM networking is the relatively high complexity of the technology, primarily in the signaling capabilities required to support all of the various services being demanded of the technology.

2.6 Internet Protocol

The internet protocol (IP) suite was developed in the 1970s and early 1980s in response to a networking project funded by the Defense Advanced Research Projects Agency. The purpose of the project was to develop a computer networking technology, which was robust under a wide range of network failure scenarios. There are many excellent books on IP, including [7, 8]. Reference [9] presents an excellent overview of the issues when setting up an Internet connection from a corporate network.

2.6.1 IP Architecture

The IP architecture that evolved from this project consists of hosts connected to subnetworks and subnetworks interconnected through gateways, also called *routers*. These routers communicate through routing protocols in order to learn dynamically the topology of the subnetwork interconnectivity. This collection of subnetworks, hosts, and routers is termed an *internet*. The total collection of all interconnected internets is the Internet. The Internet is further divided into numerous ASs, which are internets under the administrative domain of a single organization. This hierarchical topology is shown in Figure 2.17.

The basic transport entity defined is the IP packet (see Figure 2.18).[10] Packets are transported as individual messages, or datagrams. Instead of predefining a connection, real or virtual, as in connection-oriented networks like frame relay or ATM, the IP packets contain in their data header all the

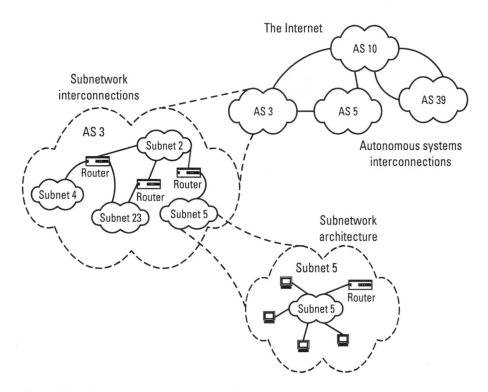

Figure 2.17 The high-level overview of the IP Internet architectural model.

---

10. Throughout this book we assume IP version 4 to be the protocol version. For an excellent overview of IP version 6, see [10].

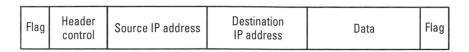

Flag	Header control	Source IP address	Destination IP address	Data	Flag

Figure 2.18 Format of an IP packet.

information, that is, the destination interface's IP address, required for the routers in the path to determine the correct forwarding in order to correctly deliver the packet to its destination. The IP address is divided into a network portion and a host interface portion. The network portion identifies the subnetwork to which the host is attached and the host portion identifies the interface over which the host is attached to the subnetwork. The IP header processing is performed by the hosts and gateways along the data path.

The routers are interconnected by links (also subnets). The links can be categorized into broadcast subnetworks or nonbroadcast multiple access (NBMA) subnetworks. Broadcast subnetworks support an efficient mechanism to deliver a packet to all hosts/gateways attached to the same network. NBMA subnetworks do not support an efficient method to deliver a packet to all attached hosts/gateways. Examples of broadcast subnetworks are Ethernet, token ring, FDDI, and point-to-point links. (Note that a point-to-point link, for example, a T1 or 56-Kbps private line facility, is a degenerate case of a broadcast network in that there exists only one other host/gateway in addition to the sending host/gateway.) Examples of NBMA subnetworks are X.25 networks, frame relay networks, ATM, and SMDS. We elaborate on these IP concepts in the following subsections.

2.6.1.1 Internets

As mentioned earlier, internets are a connected collection of subnetworks and routers. An internet is, for example, a corporation's private-based router data network. Recently the term *intranet* has come into vogue to describe a corporation's private internet. By subnetworks, we mean a collection of hosts directly connected with one another and sharing a common IP network address. By routers, we mean gateways connected to two or more subnetworks. Hosts are identified by their interfaces (a host can have more than one interface) and the addresses assigned to their interfaces. Internet addresses are hierarchical and are composed of a network portion, say, A, and a host (or host interface) portion, say, 1. We will refer to this address as A.1. All Internet addresses are of this form, and the basic complexity relates to how the 32 bits of an address are divvied up into network space and host space. This is also true when subnetting and supernetting is used, which are mechanisms that allow corporations more flexibility in determining the subnetwork/host address boundary.

So for the sake of further discussion, just remember the following: addresses consist of network and host portions, they will be identified as A.1 or C.5 or whatever, and all hosts with the same network address are assumed (by the routers) to be directly connected. (It is also true that all hosts that are directly connected, in the IP sense, have the same network address.) This last point is extremely important and has important implications when working through topology design considerations in NBMA networks.

2.6.1.2 IP Packet Forwarding

A router's primary function is packet forwarding. Routers receive packets from one subnetwork interface, search a forwarding table for the destination address (from the packet header), determine the next hop address (e.g., G.5) to forward the packet to, and determine the interface to send it out over. Therefore, routers have forwarding tables that consist of a destination address, next hop address, and a router interface. At a minimum, the forwarding tables consist of a destination address prefix, next hop address, and a router interface, where the destination address prefix is the first N bits of the address. The router then finds the longest prefix match to the destination address of the packet to be forwarded. This allows the routers to aggregate multiple destination address entries into a single address prefix in order to save memory in the router (for example, the Internet in 1996 had roughly 2 million subnet addresses defined that would make for an extremely large forwarding table in the absence of *route aggregation*). The router enters the table with the destination address and returns with the next host address, and interface pair. With this information the packet is handed to the interface, which determines the subnet-layer address (this may be an Ethernet address, an E.164 address, or a frame relay DLCI), and sends the packet onto the subnetwork with the appropriate subnet-layer address. This continues, hop by hop, until the packet reaches its destination host address.

2.6.1.3 IP Routing

Somehow, the packet forwarding tables have to get populated. This is accomplished in one of two ways, statically or dynamically. In static routing, an administrator generates the table entries and they forever remain the same (or until the administrator logs in and changes them again). Basically two types of routing protocols are used to accomplish dynamic routing: *distance vector* and *link state*. In addition, dynamic routing protocols are defined to support two levels of administration: *intradomain* and *interdomain* routing. Intradomain routing protocols address routing within a single AS. Interdomain routing protocols address routing between different ASs. Table 2.1 categorizes the various

Table 2.1

A Classification of Popular Routing Protocols in Internets Today

	Intradomain	Interdomain
Distance vector	RIP, RIP II, IGRP, EIGRP	BGP, IDRP
Link state	OSPF, IS-IS	

open and proprietary routing protocols typically found in today's intranets and Internet.

In dynamic distance vector routing (for example, IGRP, EGRP, RIP, or BGP), the routers pass to their neighbors information regarding networks to which they are connected and their distance from those networks. (If they are directly connected to a network, then they list a distance of one.) For interdomain, distance vector protocols, for example, BGP4, the routers pass AS names of the ASs they can reach, the AS path of the intermediate ASs required to reach the specified AS, and the list of the network addresses within the specified AS. Because the entire AS path is carried, this is referred to as *path vector routing.* The specific form these messages take, their frequency of exchange, and the response routers take on receipt of these messages define the particular routing protocol the routers implement (e.g., RIP, IGRP). In link state routing, routers flood the network with information about their local links. From this information, routers build a link state map of the entire network (or at least that part of their network sharing this level of detail). This link state map is then used by the router to build the forwarding table. See [11] for an excellent discussion of routing protocols in today's internets.

2.6.1.4 IP Quality of Service Issues

Increasingly, pressures have been exerted on the IP architects to embed within the IP network support for QoS capabilities. More and more applications, which rely on various performance guarantees in order to function properly, are being run over IP networks. These include applications such as transaction applications relying on small network delays in order to process large volumes of transactions; high-throughput applications requiring bounds on the time to transport files across the network; and video, audio, and multimedia applications requiring bounds on the delay variability across the network.

To address these new pressures for QoS support, the IETF has been developing several additional protocols and mechanisms. These include a signaling protocol called reservation service protocol (RSVP), a new transport protocol called real-time transport protocol (RTP), and differential services as a means to provide priority treatment to selected packets. These capabilities and others are being developed to provide improved QoS transport within internets.

2.6.1.5 IP Multicasting

Up to now, our discussions have concentrated on pair-wise communications between a single transmitter and receiver pair. But for some applications it is useful to define a capability where a single source simultaneously transmits to multiple receivers. This capability is referred to as *multicasting*. Example applications where multicasting is useful are video and audio distribution for distance learning, video teleconferences, and brokerage trading applications. Multicasting is distinct from broadcasting in that the multicast receivers form a subset of the totality of receivers. IP is not unique in supporting multicasting. In fact, multicasting capabilities are found in frame relay and ATM technologies. However, because IP is a connectionless network protocol, it seems better suited to support multicasting in a scalable and relatively efficient fashion.

2.6.2 IP on Broadcast Networks

We now discuss the interface between the IP layer and the subnet technology when the subnet is a broadcast subnet. The IP architectural model consists of routers interconnecting various layer 2 networks (or subnets). The routers communicate with one another to build routing tables and the routers utilize these tables to determine the appropriate packet forwarding tables based on the destination IP address. Once the router determines the appropriate interface and next IP hop to forward the packet to, several tasks are left to the router prior to transmitting the packet out onto the broadcast subnet:

- *MAC address discovery:* The router must determine the MAC layer address of the next IP hop on the subnetwork.
- *MAC format encapsulation:* The router must encapsulate the IP packet into the MAC layer packet format with the MAC level destination address.

The IETF has defined a MAC address discovery protocol, referred to as the address resolution protocol (ARP). When a router interface receives from the router forwarding process an IP packet to send out to the attached subnet, it must know the subnet address of the next IP hop. It finds this address in the interface's local ARP cache. If not, then the interface will broadcast an ARP request packet onto the subnet, requesting the MAC address associated with the IP address of the next hop. When the end station associated with the IP next hop receives the ARP request, it sends back a unicast message with its local MAC address. The original router interface can then cache this address in its local ARP cache, and it can now forward the original IP packet. To do so it must encapsulate the original IP packet into the appropriate subnet technology frame format with the correct MAC address, which is now resident in its local ARP cache. The mapping of the IP packet into the appropriate subnet frame format is defined in an *encapsulation standard* for IP over the specific subnet technology, for example, IP over Ethernet or IP over frame relay. This is discussed later in Section 2.7.

2.6.3 IP on ATM (and Other NBMA) Networks

Commercial interest in a number of NBMA-based WANs has grown considerably during the last 5 years. This interest is primarily economic and was initially fueled by the widespread deployment of frame relay network services by a number of the local and interexchange carriers. Due to this interest, a number of IETF working groups have defined various architectures for IP interworking on NBMA subnets. In this section we focus primarily on IP over ATM subnets, but much of the discussion holds for other NBMA subnets as well, for example, frame relay, X.25, and SMDS technologies.

ATM has many capabilities and hence many different implementations (as discussed in the earlier section on ATM technologies). Wide-area network implementations today tend to be based on PVC capabilities. Local-area network implementations tend to be based on switched virtual circuit implementations. For IP to map effectively on a subnet technology, it relies on certain subnet features, for example, broadcasting. Because of the diversity of ATM implementations, multiple IP on ATM models are defined or are in the process of definition [12]. Several of these are discussed in the following subsections.

2.6.3.1 IP on PVC-Based ATM WANs

The first model developed is that of IP mapped onto a PVC-based NBMA subnet. Here all hosts and routers are not necessarily directly connected (as evidenced by the majority of corporate frame relay network designs being implemented today). Some NBMA network designs are fully interconnected (hence the routers on the network are directly connected) and some designs are sparsely interconnected (hence all routers are not directly connected). When the underlying VC topology is being designed, it must be cognizant of the IP subnet topology as well. Reference [13] discusses the various ways to configure IP overlays on frame relay networks.

2.6.3.2   IP on SVC-Based ATM LANs (ATM LIS and LAN Emulation)

ATM LANs are proliferating due to the capability of ATM to support rates in excess of 100 Mbps and provide that bandwidth on a point-to-point basis. To effectively support the dynamic LAN environment, the ATM LAN must support switched virtual circuits. The question then becomes: What is the right model to support IP in a LAN environment over ATM?

Laubach [14] proposed the *classical IP over ATM* model to support LAN environments. Classical IP over ATM defines the notion of the logical IP subnet (LIS) as a way to preserve the traditional IP/subnet architecture when the subnet technology is an NBMA network. To provide the necessary broadcast capability, the classical IP model defines a server on the LAN subnet to which workstations register, and to which workstations send their ARP request messages. The server is responsible for responding to the ARP request which the ATM address of the workstation with the appropriate IP address. The server maintains this IP-to-ATM address mapping table, which it builds through the workstation registration process. This is illustrated in Figure 2.19.

An alternative model for supporting IP traffic over an ATM LAN is the LAN emulation (LANE) standard developed by the ATM Forum. Pictorially, the LANE model is similar to that shown for the classical IP model of Figure 2.19. The difference lies in the protocol interface the two models show to the higher layer protocols. As discussed earlier, the classical IP model has the

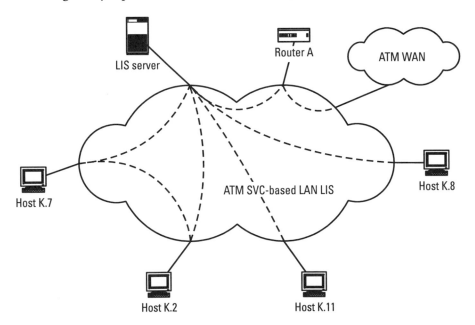

Figure 2.19   A Classical IP LAN showing a single server LIS.

IP layer attaching directly to the ATM stack, that is, the standards map IP, ARP, and so on directly to the ATM AAL Type 5 and the ATM signaling protocols. LANE, instead, presents an 802 MAC layer interface to the higher layer protocol, that is, the higher layer protocols are mapped onto the 802 MAC layer such as an Ethernet or token ring or FDDI subnet. This allows ATM to support various nonroutable protocols, for example, DEC's LAT or IBM's NetBEUI protocols, in contrast to the classical IP model, which supports only routable protocols, such as IP or IPX. LANE then provides a mapping of ATM capabilities onto the MAC layer capabilities. To accomplish this, LANE defines a multicast/broadcast server (as in classical IP) that provides for the LAN multicast/broadcast capabilities.

2.6.3.3 IP on SVC-Based ATM WANs (NHRP and Tag Switching)

Consider the typical IP over a NBMA subnet topological design shown in Figure 2.20. In this design, traffic between router B1 and router D1 must transit the NBMA network four times and through three separate tandem routers, even though routers B1 and D1 are both connected to the same NBMA network. It would be more efficient if, instead, the packets could flow directly between these two routers given the current topological design. Efficiency is gained through two aspects: (1) Minimizing the number of router hops should

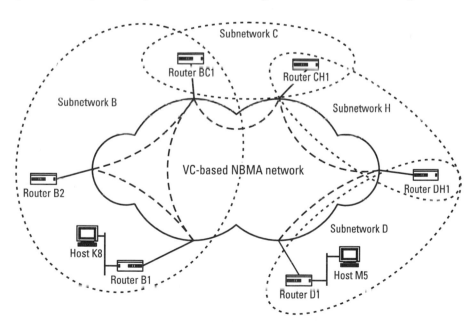

Figure 2.20 A typical topological design for an internet with four subnetwork addresses and interconnected over a NBMA subnetwork technology.

improve the end-to-end performance of the traffic flow both in the delay and the throughput sense and (2) reducing the number of IP packets flowing through the routers (in the "middle" of this path) should improve the overall combined IP and ATM network efficiency.

Several capabilities are under development to address these concerns, as discussed next.

The *next hop resolution protocol* (NHRP) is one approach being developed to address these issues. With this protocol implemented, router B1, when getting ready to forward the first packet of a flow between host K8 and host M5, will first issue an NHRP request to the next router. The purpose of the NHRP request is to identify the last, or exit, router from the NBMA network to host M5. This request packet gets forwarded eventually to router D1, which, knowing that it is the exit router, responds to the request packet with its subnetwork address, for example, its X.121 address for X.25 or its E.164 address for FR and SMDS, or its NSAP address for ATM, to the original issuing router, in this case router B1. On receipt of the response packet, router B1 signals the NBMA network to set up a virtual connection (in the case of SMDS no virtual connection request is necessary) to the interface address of router D1. Once this virtual connection is established between router B1 and D1, router B1 indicates in its forwarding table that this VC is for IP packets destined for host D5, and then it begins forwarding IP packets destined for host M5 onto this new VC. Once the packets stop flowing to host M5, the connection will eventually time out and it will be torn down.

This topological example is rather simple in that all stub networks, for example, subnetworks indicated as K and M, are directly connected to routers attached to the NBMA network. Running NHRP for other topologies is more difficult due to the possibility of routing loops. For a more complete discussion of NHRP see [15, 16].

Another capability under consideration is *tag switching.* There is a concern among some Internet engineers that the forwarding capacity of traditional routers will eventually be exhausted due to the growth in the size of the forwarding tables, the amount of Internet traffic, and the complexity of the search algorithms. NHRP can provide some relief to this problem by minimizing the number of routers each IP packet has to traverse. This reduces the load on intermediate routers. However, some Internet engineers are looking for more dramatic improvements. At the time the debate began, router technology was capable of forwarding roughly 200,000 IP packets per second. With the average IP packet size in the Internet being roughly 200 bytes, this amounts to forwarding roughly 320 Mbps per router. Assuming that we load trunk interfaces on a router at 50% utilization, this amounts to a single router supporting a maximum of four OC-3 facilities. Given the growth of typical Internet national backbone provider networks, this limit is rapidly being reached.

Tag switching was proposed as a new IP forwarding architecture to improve router forwarding throughputs. Tag switching defines a capability for routers to aggregate IP packets into flows through the router's switching fabric. The routers then append a tag onto the IP packet associated with the flow. Once a tag is appended, the routers can forward based on this simple tag, and thus smaller tag forwarding tables can be implemented in hardware. Thus, packet forwarding rates are improved over the packet forwarding methods that are based on software. Some tag switching proposals attempt to map these flows onto ATM VCI-like tags and then rely on an underlying ATM transport to handle most of the IP packet forwarding through the hybrid ATM/IP network.

As tag switching is being defined, improvements in forwarding rates of traditional routers have occurred as well. At this time, it is too early to tell whether tag switching will be widely deployed.

2.7 Multiprotocol Encapsulation: Agreements for PPP, X.25, Frame Relay, ATM, and IP

What are multiprotocol encapsulation standards all about? Consider the picture in Figure 2.21 of two devices, for example, a pair of routers or FRADs or terminal adapters, which multiplex and demultiplex multiple traffic streams over an underlying transport technology. The encapsulation device on the left

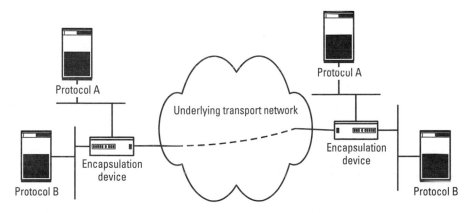

Figure 2.21 A typical encapsulation scenario.

takes in packets from its LAN interfaces (Figure 2.21 shows it to have two separate LAN interfaces supporting two different LAN protocols) and somehow wraps those LAN protocol packets into a packet format familiar to the underlying transport network. Here the transport network may be an X.25 network, a frame relay network, an ATM network, or even a simple private line network. The LAN protocols can be nonroutable protocols such as DEC's LAT or Microsoft's NetBEUI protocols or they may be routable protocols such as IP or IPX or AppleTalk.

Consider now the encapsulation device shown at the right in Figure 2.21, which must somehow demultiplex the incoming packet stream and determine what LAN interface to forward the packets onto. Once this encapsulation device opens up (or unwraps) the incoming packet from the underlying transport network, it needs to have some way to identify the protocol type of the packet inside. Otherwise the encapsulation device will have no way of knowing how to interpret the header information in the LAN protocol packets. It is the role of the encapsulation layer to identify to the demultiplexing device the type of protocol packet inside and hence what protocol process inside the demultiplexing device to hand the packet to for processing.

The encapsulation identifier is a protocol layer inserted between the encapsulated protocol and the protocol of the underlying transport protocol. This is shown generically in Figure 2.22. The encapsulation protocol field shown in this figure comes in various guises depending on the particular transport technology being discussed. If the transport technology is PPP, then this field is termed the *protocol field*. If the transport technology is Ethernet, then this field is referred to as the Ethernet *type field*.

Other technologies are identified in Table 2.2. Here the term NLPID refers to the network layer protocol ID and LLC/SNAP refers to the logical link control/subnetwork access point. Even though these various transport technologies have implemented different mechanisms, they all serve a similar function, that of identifying the protocol of the packet wrapped inside of the transport technology frame so that the receiving demultiplexer is capable of handing the inside packet to the appropriate protocol processor. It is absolutely critical that the multiplexing and the demultiplexing agents implement the identical encapsulation standard.

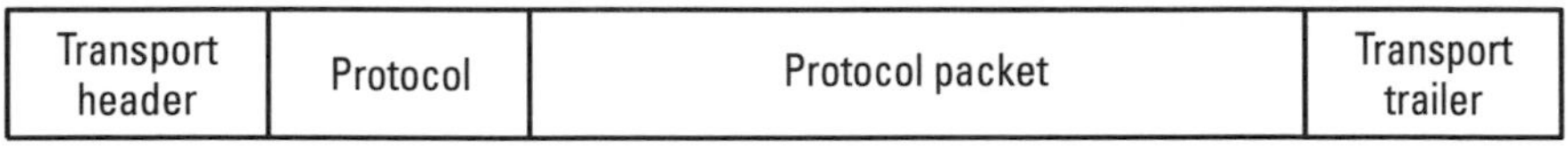

Transport header	Protocol	Protocol packet	Transport trailer

Figure 2.22 Packet encapsulation inside a protocol for transport.

Table 2.2

Various Examples of Encapsulation Standards and Their References

Transport Technology	Demultiplexing Field	Overhead (bytes)	Reference
Point-to-point protocol	Protocol field	8	RFC 1548
Ethernet	Ethernet type	18	RFC 894
Frame relay	NLPID	8*	RFC 1490
ATM	LLC/SNAP	16[†]	RFC 1483
802.3	LLC/SNAP	26[‡]	RFC 1042
X.25-SVC	Call setup	10[§]	RFC 877

*Overhead includes the contribution from the frame relay link level header and trailer as well as the NLPID field.

[†]Overhead includes the AAL5 trailer and LLC/SNAP encapsulation method. Does not include padding in order to fill out an integral multiple of 48 bytes, nor does this include the layer 2 cell header overhead of 5 bytes per each 53 byte cell.

[‡]Overhead includes the 802.3 MAC header and CRC and the LLC/SNAP fields.

[§]Overhead includes layer 2 and 3 headers, flags, and a 16-bit CRC. Assumes protocol type established in the call setup phase of the X.25 SVC establishment.

2.8 Link Level Interworking Agreements

We have discussed various, prominent WAN technologies in this chapter. Typical enterprise networks consist of several different WAN networking technologies. The reasons for this include are twofold.

First, the various WAN technologies have different price and performance points. X.25 networks generally support 9.6- to 56-Kbps access lines and have a rather ubiquitous presence worldwide. Frame relay networks generally support 56-Kbps to 1.5-Mbps access lines, but are less widely deployed worldwide. ATM supports access lines ranging from 1.5 to 155 Mbps and higher and is just starting to see more extensive deployment worldwide. Also because of the relative maturity of X.25 and frame relay services, they tend to be very competitively priced.

Second, enterprise networks are in a constant state of evolution and transition to newer technologies. Extensive router deployment has driven customers to deploy frame relay technologies. It is anticipated that new multimedia services will drive customers to deploy ATM networks.

Given the existence of these multiple WAN technologies deployed within single enterprise networks, it is natural to want to provide an interworking capability between these subnetworks. Figure 2.23 shows routers accessing the WAN via the FR protocol at speeds between 56 Kbps and 1.5 Mbps (DS-1),

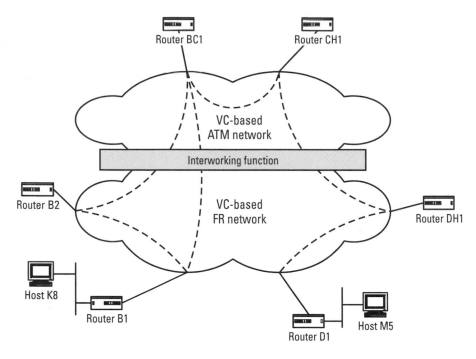

Figure 2.23 A typical enterprise network showing two subnetwork technologies, for example, FR and ATM, interconnected through an interworking function.

with interworking between the FR and ATM networks through an interworking function (IWF). The IWF is performing the link layer interworking between frame relay and ATM in this example. This allows customers whose data centers require network access bandwidths of DS-3 or greater to interwork with their regional and branch offices, which require network access bandwidths from 56 Kbps to T1. It further allows customers to migrate from FR-to-ATM on an as-needed basis, as their bandwidth requirements increase from sub-T1 to above T1 rates.

ITU-T standard I.555 defines two methods of interworking ATM and FR end points[11]: *network interworking* and *service interworking*. Network interworking defines the capability of carrying frame relay connections over an ATM connection. Service interworking defines a translation between ATM and FR connections.

---

11. ITU-T standard I.1555 also defines the capability to interwork X.25 end points with frame relay end points. However, because the dominant example involves FR-to-ATM interworking, we will not discuss X.25 link layer interworking.

For the network interworking option, a connection is established between the end user's FR UNI on the FR network to the end user's ATM UNI on the ATM service (see Figure 2.24). The IWF encapsulates the FR frame onto an ATM cell stream as identified in the I.555 standard. We discussed encapsulation protocols in an earlier section. However, in link layer interworking a different method is employed. Because a single link layer protocol is being encapsulated, one can simply preconfigure the demultiplexing agent to know the identity of the encapsulated protocol at the time at which the virtual connection is established. For PVCs this is implemented through manual provisioning. For SVCs, this must be indicated in the SVC signaling protocol.

Frame relay encapsulation on ATM is accomplished through the ATM AAL Type 5 Class C service. Here the frame relay frame header and data are carried in an AAL Type 5 data unit, which is padded out to an integral multiple of 48 bytes (including the AAL Type 5 trailer). This is then segmented into an ATM cell stream, which is then mapped onto a DS-3 signal for transmission. (This was detailed earlier in Figure 2.15.)

An advantage of network interworking is that the WAN network is transported to the higher level protocols, that is, the higher level protocols are multiplexed onto an end-to-end frame relay connection using the NLPID encapsulation define for frame relay networks. Over the ATM access facility, this frame relay connection is tunneled through an ATM AAL Type 5 connection.

Service interworking is illustrated in Figure 2.25. Although the ITU-T I.555 standard defines service interworking as a mapping from frame relay to ATM, this is not totally a link layer mapping. High-level protocol mappings are required for this to be of use to CPE. Referring to Table 2.2, we see that the higher layer encapsulation protocols used by routers connected to frame relay and ATM networks differ. (Routers utilize an NLPID encapsulation on frame

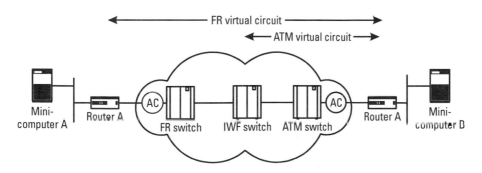

Figure 2.24 FR-to-ATM network interworking.

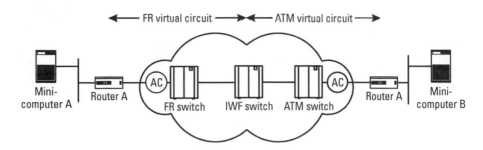

Figure 2.25 Frame relay-to-ATM service interworking.

relay interfaces and they utilize an LLC/SNAP encapsulation on native ATM interfaces.) Therefore, service interworking must include translations of these multiplexing fields as well.

Figure 2.25 shows that there does not exist an end-to-end connection in the case of service interworking. Instead, the two separate FR and ATM circuits are concatenated by the IWF. In addition, the IWF must implement the NLPID to LLC/SNAP conversion for this configuration to be useful to the majority of the CPE connected to FR and ATM networks today.

2.9 Summary

We have discussed a number of the prevalent WAN technologies deployed in multiprotocol enterprise networks today. These include private line TDM multiplexer networks, X.25 networks, frame relay, ATM, and IP networks. We have tried to give a sense of the history of these technologies and their different deployments. Each of these technologies has their own unique performance characteristics and pricing structures. Hence, different design issues come to bear when implementing and growing these respective networks. Finally, we have attempted to give a sense of the new developments in IP and ATM networking and have alluded to the activities, which indicate a convergence of these capabilities, for example, P-NNI, NHRP, and tag switching developments. These are truly interesting times we live in with respect to the rapid advances in data communications—and it is just these rapid advances that will keep network administrators and designers fully employed for many years to come.

In the next chapter, we begin to discuss the basics of delay and throughput performance analysis over wide-area networks. This discussion is at a level that is relatively independent of the specific WAN technologies discussed in this book. In later chapters we will combine our knowledge of the specific

WAN technologies from this chapter with our basic performance knowledge obtained from the next chapter.

References

[1] Tanenbaum, A., *Computer Networks: Towards Distributed Processing Systems*, Englewood Cliffs, NJ: Prentice Hall, 1981.

[2] Lin, S., and D. Costello, Jr., *Error Control Coding: Fundamentals and Applications*, Englewood Cliffs, NJ: Prentice Hall, 1983.

[3] Schwartz, M., *Telecommunications Networks: Protocols, Modeling and Analysis*, Reading, MA: Addison-Wesley, 1987.

[4] Bradley, T., and C. Brown, "Inverse Address Resolution Protocol," IETF RFC 1293, 1992.

[5] Turner, J., "New Directions in Communications (or Which Way to the Information Age?)," *IEEE Commun. Mag.*, October 1986.

[6] Handel, R., M. Huber, and S. Schroder, S., *ATM Networks: Concepts, Protocols, Applications*, Reading, MA: Addison-Wesley, 1994.

[7] Stevens, W. R., *TCP/IP Illustrated, Volume 1: The Protocols*, Reading, MA: Addison-Wesley, 1994.

[8] Comer, D., *Internetworking with TCP/IP*, Volumes 1 and 2, Englewood Cliffs, NJ: Prentice Hall, 1991.

[9] Dowd, K., *Getting Connected: The Internet at 56Kbps and Up*, Cambridge, MA: O'Reilly & Associates, 1996.

[10] Huitema, C., *IP Version 6: The New Internet Protocol*, Englewood Cliffs, NJ: Prentice Hall, 1996.

[11] Huitema, C., *Internet Routing Protocols*, Englewood Cliffs, NJ: Prentice Hall, 1995.

[12] Cole, R., D. Shur, and C. Villamizar, "IP over ATM: A Framework Document," IETF RFC 1932, 1996.

[13] deSouza, O., and M. Rodrigues, "Guidelines for Running OSPF Over Frame Relay Networks," IETF RFC 1586, 1994.

[14] Laubach, M., "Classical IP and ARP over ATM," IETF RFC 1577, 1994.

[15] Luciani, J., et al., "NBMA Next Hop Resolution Protocol (NHRP)," IETF RFC 2332, 1998.

[16] Cansever, D., "NHRP Protocol Applicability Statement," IETF RFC 2333, 1998.

3

Performance Analysis: Some Basic Tools

3.1 Introduction

This chapter provides the basic performance analysis background necessary to analyze complex scenarios commonly encountered in today's data network designs. The material in this section is by no means meant to be all inclusive, and the reader, if interested, is encouraged to pursue in more depth the materials identified in the references. Two texts come to mind that provide a more detailed and in-depth analysis: one by Bertsekas and Gallager [1] and the other by Schwartz [2].

Many of the ideas and techniques presented in this chapter are approximate when used to analyze practical applications. However, we believe that it is not important for the network analyst to provide an exact evaluation of the network performance, but only to develop an understanding of the network behavior in order to make correct design decisions to improve the network. An exact analysis would require an exact model of the traffic loads and patterns and exact understandings of the network components, for example, modems, transmission systems and paths, switches, network servers, and the applications. Because each of these inputs is impractical to characterize in the necessary detail, we instead strive for a reasonable and approximate understanding of each and rely heavily on approximate models. As will be discovered in the remainder of this book and based on our experience, we have found this approach to be successful in providing insights into multiprotocol network performance.

These are the two most often asked questions when assessing the performance of data networks:

1. What is the latency or delay for a packet to traverse the network?
2. What is the end-to-end throughput expected when transmitting a large data file across the network?

Network engineers should be able to answer these questions. In the event that the performance of the network does not satisfy the requirements of the end applications and users, the engineer must recommend modifications to the network design necessary to improve the performance. In general, the network engineer is responsible for designing the network to meet necessary performance requirements while minimizing the expense of maintaining the network.

The primary purpose of this chapter is to provide a fundamental overview of the considerations required to answer these two questions. We have organized this chapter as follows:

- *Network delays.* We begin with a discussion and identification of the delay components encountered in most data networks. Simply, this section discusses how to estimate the time for a single data packet to travel from the transmitting computer interface to the receiving computer interface across a typical data network. This analysis is fundamental to the capability to estimate transaction response times and file transfer throughputs of the end user applications.

- *Timing diagrams.* Timing diagrams present a pictorial representation of network delays. We outline the construction of timing diagrams and discuss their utility in developing an overall understanding of network level performance.

- *Pipelining.* Pipelining is a means for the network to achieve some level of parallelism in the delivery of data. This is accomplished by segmenting information into smaller packets for transport. We discuss the impact of pipelining on end-to-end performance.

- *Network throughput.* We build on the discussion of network delays by walking through examples of estimating file transfer throughputs across typical data network paths. We discuss the effects of network delays and protocol windows on file throughput.

3.2 Network Delays

Fundamental to the design and analysis of high-speed data networks is a thorough understanding of the impact of design decisions on end-to-end delays.

Data networks are characterized as communications systems that deliver data, in the form of packets, from one computer to another computer. The time necessary for the network to deliver these packets of information has many important ramifications. In general, we define *delay* as the *time to carry the packet of information between two defined points in the communications network path.*

The two points chosen to define the delay are dependent on the particular question to be addressed. For example, a common question asked of engineers is the delay the user of a computer would see when requesting a file from another computer attached to a common network. In this instance the points are (1) when the user initiates the request for the information on a distant computer, for example, the user enters a *get* request during a file transfer protocol (FTP) session, and (2) when the last byte of data is stored locally on the user's computer. In this example, the delay includes all data network components (discussed later in this chapter) as well as any delays associated with the two computers communicating with one another.

3.2.1 Delay and Latency

Within this book we attempt to define and maintain a distinction between *delay* and *latency*. We attribute to latency those delay contributors that are relatively fixed and beyond the control of the network engineer to control or modify. Examples of network latency are wire propagation delay or delays associated with switch or router forwarding within a common carrier network. All other delay contributors with which the network engineer has some amount of control over, we simply refer to as delay. Examples of delay are queuing delays in customer-premises equipment (CPE) or insertion delays on an access or egress link to a frame relay service.

We draw this distinction between latency and delay for a simple reason. It is extremely important to understand the design parameters, which can be modified in order to improve performance and to understand the impact of those design characteristics, which are immutable. Do not waste time trying to modify immutable characteristics.

We categorize network delay into the following components:

- *Propagation delay* is the time for a given signal to travel down the communication facility. This delay is associated with the physical distance that the communication facility spans and network electronic components such as digital cross-connect systems.

- *Transmission delay* is the time associated with inserting, or writing, the packet onto the communications facility.

- *Processing delay* is the time required for a processing unit to perform a required action on the packet as it traverses a network switching or translation device.
- *Queuing delay* is the time associated with a packet waiting in a buffer prior to a given action being performed on the packet, for example, processing or transmission.

Each of these delay components must be considered when analyzing a particular network, because each of these components can be significant. As an example, consider an IP data packet traveling across a router network as identified in Figure 3.1. Here an IP packet incoming to the router on the left is to be transmitted to the router on the right. We assume that the leftmost router is located in Los Angeles (LA) and the right router is located in New York City (NYC). They are connected via a long-haul serial line, roughly 3000 miles in length. The processor (CPU) in the LA router must decide where to forward the packet and performs a forwarding table lookup. Once this lookup is performed the CPU places the packet in the queue for the serial line interface card. The time for this table lookup and placing the packet in the interface card buffer is a *processing delay* for this packet. The packet finds multiple packets in queue ahead of it and must wait for these packets to be transmitted out of the interface card prior to its transmission. This waiting time is a *queuing delay* for this packet. Given that the packet is of finite length, for example, 1500 bytes, and the serial line is operating at a finite speed, there is a delay between the time the first bit of the packet is transmitted onto the serial line to the time the last bit is finally transmitted onto the serial line. This delay is referred to as a *transmission delay* for the packet. Finally, due to the finite propagation speed over the communication channel, there is a delay between the time when the LA router transmits

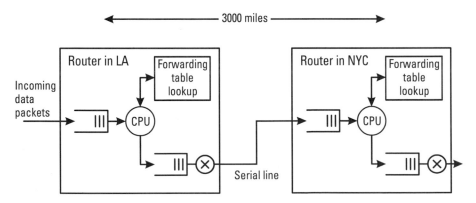

Figure 3.1 Delay components for a packet through a router path.

the first bit of the packet to the time when the NYC router receives the first bit of the packet. This network latency is referred to as the *propagation delay*.

In the subsections to follow, we discuss each of these components separately and in much more detail.

3.2.2 Propagation

We define *propagation delay* as the *time for the signal to travel from one end of the communications facility to the other end.* The propagation delay is not insignificant, and the speed at which a signal travels through a transmission system is less than, but comparable to, the speed of light. For reference, the speed of light in a vacuum is roughly 5.35 msec per 1000 miles, while communications facilities typically have propagation delays on the order of 8 msec per 1000 miles. Propagation delay can become relatively significant when analyzing the performance of high-speed data networks, for example, a DS-3 facility. For example, the time to insert a 1500-byte packet onto a 1000-mile DS-3 (or 45-Mbps) facility is only 0.3 msec. The propagation delay of 8 msec (best case) is significant in comparison.

For the transmission systems that are deployed in carriers' networks, other factors must be considered. Common fiber transmission channels in today's long-haul carrier networks perform numerous optical-to-electrical conversions in order to amplify the signal. (Newer fiber systems are deploying optical amplifiers, which eliminate the optical to electronic conversions.) These conversions add delay to the propagation times for the signal. Further, long-haul networks are comprised of point-to-point transmission systems that are interconnected via digital cross-connect systems. A corporation's private line circuit, the communications path between two networking components, is built from multiple point-to-point transmission systems interconnected via digital cross connects. These cross connects add processing and buffer delays as well. Field measurements of typical propagation delays for common transmission systems show that terrestrial facilities, such as microwave, coax, and fiber, all have propagation delays of roughly 8 msec per 1000 miles. In contrast, satellite transmission systems show propagation delays of roughly 5.5 msec per 1000 miles. The majority of satellite transmission systems are based on satellites in geosynchronous orbit. Therefore, the transmission path from transponder to satellite and back to another transponder is roughly 2 times 26,400 miles, resulting in a total, round trip propagation time of 280 msec.

The preceding calculations require the engineer to know the actual path length of the communications facility in question. This is rarely the case. What is known are the originating and terminating points for the transmission path. This is typically two known city or building locations. However, the transmission path is never a direct line between the two locations. A typical route between Albuquerque and Denver may travel through Los Angeles and San Francisco, depending on the fiber routes deployed. Long-haul carrier networks are built up from fiber routes that travel along "rights of way" belonging to the carriers. These rights-of-way typically follow highways, railways, and gas pipelines throughout the country. Figure 3.2 shows a typical facility map for a common carrier network. Further, depending on the availability of capacity on the transmission network, it may be in the carrier's interest to route the circuit along somewhat longer paths. This may be dependent on the time of day (when utilizing switched communications circuits) or it may change over a period of months as the carriers constantly re-groom their network circuits. It is not unreasonable to assume that the terrestrial circuit miles between two locations may be a factor of from 1.3 to 2 times the distance of a direct line between the locations.

Looking at Figure 3.2, one can see that relatively few fiber routes traverse the country. This impacts the propagation delays quoted in technical references of the long-haul carrier private line services. Even though the farthest distance between any two cities within the domestic United States is roughly 4000 miles (which should result in a propagation delay of less than 32 msec), one typically finds the carriers quoting propagation delays of less than or equal to 60 msec

Figure 3.2 A facility map for a common carrier network.

(roughly corresponding to a route distance of 7500 miles). This difference between city pair distance and route distance is exacerbated during times of network failures. During these periods, in the event that the network is self-healing, the route miles will probably increase. This will increase the measured propagation delays during these periods until the original route is restored.

Unlike other components of delay that we discuss in this section, there is not much the engineer can do about propagation delays. For this reason, we associate propagation delays with network latency. In this context, the following possibilities come to mind:

- Physically move the communicating locations closer together. This is probably not a practical solution.

- Redesign the protocol or application to be less "latency sensitive," or rewrite the application so that fewer message exchanges are required over the WAN to accomplish the same application task. (See the discussion of client/server applications in Chapter 9.)

- Move from satellite to terrestrial facilities. This is not always an option for many situations due to the lack of terrestrial facilities, for example, to remote mining locations or outer space.

- Review the facility routing of your network connections with your communications carrier. Attempt to have private line circuits routed on the most optimal (direct) routes between network end points.

- Switched facilities, for example, switched 56- or 384-Kbps circuits, will have different delay characteristics depending on the actual facility routing. Facility routing can vary based on network utilization or time-of-day routing patterns. This can have an effect on the performance of latency sensitive applications.

In summary, the propagation delay is the time for the signal to travel from one end of the communications facility to the other end. As discussed, this delay is a function of the nature of the communications facility, for example, satellite or terrestrial, and of the actual path length. The communications path can take some nonintuitive routes through carrier networks.

3.2.3 Transmission, or Insertion Delays, on Serial Lines

We define *transmission delay* (T_i) of a packet as the *time between the transmission of the first bit of the packet to the time of transmission of the last bit of the packet onto a communications facility*. Information is carried across data networks in the form of packets. These packets come in various forms and sizes, depending on the particular data networking technology. We measure the length of packets, *S*, in this book in terms of *bytes*. We represent the speed of a transmission link, *L*, in terms of *bits per second* (bps). For calculations in this book, we assume that a byte is equal to 8 bits.

There exists a delay associated with the transmission equipment writing the bits in the packet onto the communications facility. This delay is related to the packet length and the transmission channel speed as

$$T_s = 8 \times S / L \tag{3.1}$$

which is simply the *(number of bits in the packet)/(line speed in terms of bits per second)*. This seemingly simplistic expression carries with it some hidden complications. To begin with, the length of the packet is not always obvious. Usually the packet is defined in various other units, for example, bytes, characters, or octets. If the packet is given in terms of octets, then it is unambiguous to translate this into bits as an octet is defined as 8 bits. However, translating bytes or characters into bits can be problematic due to the fact that protocols are sometimes defined in terms of bytes or characters of differing lengths. For example, although most modern protocols are 8-bit protocols (such as HDLC, SDLC, bisync), there exist 6-bit protocols (such as ALC, a protocol popular in older airline reservation systems), 10-bit protocols (such as various asynchronous protocols with a stop and start bit for each byte), and we have even encountered a 9-bit protocol (the Universal Receiver Protocol used in the AT&T DATAKIT® fast packet switch is based on a 9-bit protocol). Therefore, some care should be taken in converting the byte count of a packet into the packet bit length. For calculations in this book, we assume that a byte is equal to 8 bits.

Also, the rate, or speed, of a communications channel may be unclear because (1) the channel speed may be defined in terms other than bits per second or (2) the channel speed may be variable or statistical in nature. It is common for modem channels to be defined in terms of the baud rate. The *baud rate* is defined as the *number of signal transitions per second* supported by the modem pair. A single signal transition is capable of conveying more than a single bit of information. Depending on the modem technology in question, the number of bits of information conveyed in a single signal transition varies from 1 to as high as 3 bits. Table 3.1 shows the relationship between baud rates and bit rates for some of the more popular modem standards defined.

With the advent of modem compression technologies, even when the bit rate of the transmission channel is known, the effective bit rate may be greater

Table 3.1

Relationship Between Baud Rates and Bit Rates for Some Popular Modems

Modem Standard	Baud Rate (transitions/s)	Bits per Transition	Bit Rates
v.34	9,600	3	28.8 Kbps
v.32 bis	4,800	3	14.4 Kbps
v.32	4,800	2	9.6 Kbps
v.22 bis	2,400	1	2.4 Kbps
Bell 212	1,200	1	1.2 Kbps
Bell 103	300	1	300 bps

when the bit streams can be compressed. The performance of the transmission channel will be a function of the type of compression schemes utilized and the nature of the data to be transmitted. In practice, for systems with compression enabled, it is necessary to run test measurements on the actual system and application in order to determine accurately the effective transmission speed of the communications facility. The effective bit rate of the transmission system may vary depending on transmission errors when coupled with forward error correction systems. Transmission systems are not perfect and sometimes the bit integrity of the packet is changed due to transmission errors. Error detection schemes have been developed that allow the receiving system to detect when the bit integrity of the packet has changed. More sophisticated schemes, called forward error correction systems, not only detect the change in the bit integrity of the packet but they also allow the receiving system to correct these bit errors (under many, but not all, circumstances). Some modem standards include forward error correction capabilities and, depending on the frequency of trans-mission errors, may modify over time the transmission baud rate, that is, may choose a lower baud rate to compensate for a higher transmission error rate. Also, dial-up modem connections must first initialize the communications between the modem pairs. The result of this initialization determines the trans-mission speed achieved for the communication path. This depends on the qual-ity of the systems subtending the path between the modem pairs. For example, v.34 modems are capable of achieving 28.8 Kbps. In practice, the transmission speeds can be less.

It is relatively simple to affect the transmission delay experienced by data packets. Referring to (3.1), we see that either one can decrease the length of the data packet or increase the speed of the communications facilities. The data packet length can be affected by *defining a smaller packet size.* Some protocols allow for the engineer to define the packet size. This can be useful depending on the nature of the application. For terminal-to-host, echoplex applications where a number of keystrokes are collected by a PAD and forwarded to the host, smaller packet size definitions can sometimes improve the performance of the network as perceived by the typist. However, smaller packet sizes imply greater protocol overhead, and can therefore diminish overall system throughput.

The speed of the communications facility can be affected by[1] *increasing the speed of the communications facility.* You can pretty much always increase communications facility speed (given that cost is not an issue). Analog modems can be replaced by faster modems or by *Integrated Services Digital Network* (ISDN) lines. Dedicated *Digital Data Service*® (DDS) lines can be replaced by DS-1 or DS-3 digital circuits.

In summary, the transmission delay is the time for a system to insert the packet onto a communications facility. This delay is proportional to the length of the data packet and inversely related to the transmission rate of the communication channel. However, due to the considerations discussed in this section, certain subtleties exist when estimating these delays. These subtleties include the bit nature of the protocol, the capability of the communications equipment to support compression, and the quality of the communications facilities, for example, the error rates of the line. For an overview of the practical aspects of data communications protocols, modems, error control, and so on, refer to McNamara [3].

3.2.4 Processing

We define *processing delay* as the *time required for a processing unit to perform a required action on the packet as it traverses a network switching or translation device.* As a packet traverses a network there exist switching and translation points where protocol processing or signal conversion processing is performed. This processing necessarily introduces an additional delay, dependent on the amount of processing required and the speed of the processor performing this task.

Consider the signal processing performed by a pair of v.34 modems. The modem converts a digital signal stream from a computer to an analog stream to

---

1. The apparent speed of a communications facility can be modified through the use of compression schemes. However, it is not always clear how this will affect the packet delay. In fact, packet delay is generally increased while overall data throughput is also increased.

transmit over the analog voice communications facility, and the other modem converts the analog signal back to a digital stream. If you were to connect two modems back to back and measure the time between the first bit of a packet entering the first modem to the time the first bit of a packet exits the second modem, you would observe a delay on the order of 5 to 70 msec. The majority of this delay is due to modem processing, although some of this time is due to a store-and-forward delay due to link level framing. These processing delays are typically quoted in the technical specifications of most modems on the market today. This measured delay depends on the particular modem implementation, the encoding scheme utilized by the modem pair, and the link level frame size (affecting the store-and-forwarding delay component). Typically, the higher the transmission speed of a given modem technology, the larger the modem processing delay incurred (i.e., you don't get something for nothing). You can expect pair-wise v.34 modem delays to be on the order of 70 msec. In addition, modems in v.42 mode also perform packet buffering, which adds to their observed delays.

In packet switching technologies, processing is required to perform the necessary link protocol terminations, forwarding table lookups, and internal transmissions across internal communications buses. This can be observed by measuring the time between the receipt to the switch of the last bit of a data packet to the transmission of the first bit of the packet onto the outgoing communications channel (when there exists no addition packet load on the switch). This measurement will provide a best case estimate for the switch delay. A given packet switch can contribute a nontrivial processing delay. However, due to the rapid increases in microprocessor speeds over the years the contribution of this delay component has been reduced dramatically.

Also new, lightweight protocols have been developed that rely on improvements in transmission systems and in turn reduce the amount of processing required. Examples include fast packet technologies such as frame relay and ATM and IP technology. As an indication of the amount of reduction in packet switching processing delays typical over the last 10 years, compare typical processing delays found in mid-1980s vintage X.25 switches versus mid-1990s vintage IP routers. Switching delays of 20 to 100 msec were not uncommon on unloaded X.25 switches. Today's routers, performing simple IP forwarding, can switch IP packets in less than 1 msec.

Equipment designers have several options to improve processing delays, including the following:

- *Faster processors.* Engineers should be sure that the problem is indeed a processing bottleneck. To determine this, they must understand the nature of the load on the processor. If a faster processor is implemented, it may introduce or create other bottlenecks/problems elsewhere in the network.

- *Process pipelining.* This divides the processing to be performed into separable tasks that can be processed in parallel. This effectively reduces overall processing delay when compared to serial processing.

- *More processors.* This is effectively the same as reducing the load on a given processor. This assumes that the processing can be performed in parallel for multiple packets.

- *Simplify the nature of the processing.* Sometimes the protocols or applications can be simplified to minimize the nature of the processing required. This technique has enabled fast packet switches to achieve extremely high packet processing rates.

In summary, processing delay is the time required for a processing unit to perform a required action on the packet as it traverses network switching or translation devices. This delay is a function of the networking technologies deployed, the vintage of the implementation, and the speed of the processing units performing the task. Modem pair-wise processing delays can be as high as 70 msec or more. Early X.25 switches can introduce delays on the order of 20 to 100 msec. Modern IP routers performing simple IP forwarding introduce processing delays of less than 1 msec.

3.2.5 Queuing

We define *queuing delay* as the *time a packet waits in a queue or buffer prior to receiving some type of service, for example, protocol processing, transmission onto a communications facility.*

3.2.5.1 Introduction to Queuing

In this section we discuss queues and their delays at a rather intuitive level. In Appendix A, we present a somewhat more mathematical treatment of queuing systems. Within the appendix, we present some representative formulas, which reinforce the intuitive concepts presented in this section. For an excellent overview of queuing theory, see Cooper [4].

Figure 3.3 shows the essential components of a queuing system. Packets enter the queue or buffer, and they wait for the server or processor to serve all the packets in queue ahead of our given packet. Once the processor initiates service on our given packet, the queuing delay ends and the service time begins. The type of service given is a function of the system.

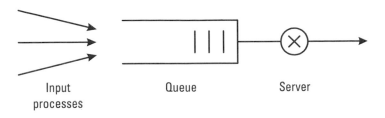

Figure 3.3 Components of a queuing system.

It turns out that the queuing delay is a function of the input process (i.e., the nature of the arrivals and the offered load), the output process (i.e., service time characteristics to process a customer), and the speed or rate of the server. The input process describes the nature of the packet arrivals into the buffer. The input process is described as *smooth* when the packets arrive at a fairly uniform rate, for example, one packet per second or 5 Kbps. The input process is described as *bursty* when the packets arrive in bunches, for example, 20 packets arrive in the first second and then no packets arrive for the next 20 sec and this process repeats.

The average rate of packet arrivals, referred to as the average load on the system, is required to further characterize the arrival process. The service process characterizes the nature of the service processing. The departure process is a function of both the arrival and the service process. Fixed sized packets being served by a fixed rate processor will have a uniform service time. If these packets had a *smooth* arrival process then they would depart the system in a *smooth* fashion as well. However, if these packets were variable sized, then their departure from the queue would be nonuniform.

It is fairly obvious that the speed of the server will affect the total queuing delay of a packet in the system. At first blush, it is not so obvious that the total queuing delay is a function of both the input and the output processes. Consider driving down the highway and arriving at a toll booth. If a large amount of other traffic is there and the toll taker is relatively slow, then the waiting time you experience to get your service at the toll booth will be quite large. Now, consider the impact of the arrival process on your waiting time. As the traffic load diminishes (and approaches zero) your waiting (or queuing) time goes to zero (i.e., the probability that on arrival at the toll booth you will find cars ahead of you is small). Also, if the time to process/serve each car increases (for example, the speed at which the toll taker collects tolls decreases), the typical waiting time each car experiences increases.

Not so obvious is the fact that your waiting time is also dependent on the nature of the arrivals and the nature of the service time statistics for a fixed offered load. If cars arrive at the toll taker's booth in bunches, then queuing delays increase on average reflecting the fact that the cars toward the end of the bunch have to wait for the cars in the front of the bunch. Whereas, if the cars are uniformly spread out in the time of their arrivals to the toll taker, then their waiting times will be smaller. If we fix the arrival process to be uniformly distributed, we can still affect the waiting time by varying the service time statistics (i.e., the variability is the time the toll taker takes to accept the toll). For a service time process that is the same for each car, as long as the offered load results in a stable system, the waiting time for each car is zero. However, for the same load, if we assume a variable service time, cars will typically experience a waiting time resulting from the fact that occasionally the service time will exceed the time between the uniform car arrivals. Hence, to determine the waiting times for queuing systems, both the nature of the arrival and the departure (or service time distribution) processes must be considered.

With this thought in hand, a rather complex mathematical model is usually required. More often than not, however, this level of detail is either not available or not required. Often simple rule-of-thumb estimates are sufficient. Systems have minimal queuing delays at low loads (for example, less than 10% utilization), their delays grow extremely large at high loads (for example, greater than 90% utilization), and for the most part they are roughly equal to the typical packet service time at moderate loads (for example, roughly 50% utilization).

In summary, queuing delay is the time a packet has to wait in a queue or buffer prior to receiving some type of service, for example, protocol processing or transmission onto a communications facility. The queuing delay is a function of the nature of the packet arrivals to the queue, and the nature of the service times for the various packets in the queue the number of servers.

One of the simpler and most useful results of queuing theory is the M/M/1 model. (See Appendix A for an explanation of notation and a further discussion on queuing results.) The M/M/1 model assumes Poisson arrivals and exponential services times. This model yields a simple expression for the queuing delay, D_q, as a function of the system load, U, and the mean service time, T_s, that is:

$$D_q = [U / (1 - U)] \times T_s \qquad (3.2)$$

Notice that these are extremely simple formulas and therefore are extremely useful aids to estimating ball-park queuing delays in network. One easy point to remember for simple estimations is that the queuing delay for a system at 50% load is simply equal to the service time. For example, if a communications facility is running at 50% utilization and the average transmission delay is 8 msec (the transmission delay for a 1500-byte packet on a 1.5-Mbps facility), then an estimate for the queuing delay is 8 msec. If you remember nothing else from this section on queuing delays, remember this.

Appendix A goes into a little more mathematical detail than is generally necessary. We offer it up for those readers who are interested.

3.2.6 Delay Synopsis

We have discussed four different delay and latency components found in communications networks. These are:

1. *Propagation* delay is the time for a given signal to travel down the communications facility. This latency is associated with the physical distance spanned by the communications facility.

2. *Transmission* delay is time associated with inserting, or writing, the packet onto the communications facility.

3. *Processing* delay is the time required for a processor unit to perform a required action on the packet as it traverses a network switching or translation device.

4. *Queuing* delay is the time associated with a packet waiting in a buffer prior to a given action being performed on the packet, for example, processing or transmission.

This ends our discussion of the fundamentals of a delay analysis in typical communications paths. This analysis will be relied on time and time again in the remainder of this book.

3.3 Timing Diagrams: A Data Communications Score Card

We now introduce the concept of a timing diagram. The timing diagram is an extremely useful tool and should be relied on when trying to develop an understanding of the delay points and packet flows through a data network. The timing diagram illustrates and records the details of the packet flow through network reference connections. We use the term *score card* because the timing diagram is an aid that keeps track of where frames are at any given time and how many packets or bytes remain to be transmitted at any given point in the reference connection at a given time.

The term *timing diagram* refers to a specific reference connection, which is often shown at the top of the timing diagram. Associated with processing and queuing points in the reference connection, for example, switching equipment, the timing diagrams contain a vertical line underneath these points. A time axis is associated with timing diagrams that runs from top to bottom, that is, time starts at the top and increases as we move down the timing diagram. Packet transmission times are represented by the width of the packet in traveling from one queuing and processing point to the next. A slope associated with the packet transmission indicates the propagation delay between the queuing and processing points. The greater the slope the greater the propagation distance between the two points.

At the queuing and processing points, the delay associated with the queuing and processing point is indicated by the vertical distance between the receipt of the last bit of the packet into the queuing point to the transmission of the first bit of the packet out of the queuing point. By summing up the individual delay components on a timing diagram, one is able to determine the total delay associated with a given transaction or transmission across the reference connection.

Figure 3.4 shows a relatively simple reference connection, similar to our LA to NYC example, and the associated timing diagram for the time to input a packet transmitted from a terminal to the distant computer on the far side of the reference connection. We will refer to this reference connection as reference connection #1 (RC#1).

The left-hand side of Figure 3.4 shows the time at which the user at terminal A hits the enter key (or submit button). This prompts the terminal to transmit the data packet toward minicomputer B. The packet queues at the terminal while waiting to be transmitted over the LAN segment. It is transmitted to the router, where it is processed and queued for transmission onto the WAN facility. The packet is then transmitted onto the WAN facility and propagates to the distant router at the right-hand side of the figure. Again the packet is processed and is queued for transmission onto LAN segment B. It is transmitted onto the LAN segment and into the transaction processing minicomputer B. The packet then must queue in the transaction processing computer's run queue to await processing by the application. The entire time for this packet to travel across the network and to start processing in the transaction processing computer is labeled the *input delay* in the figure.

In the timing diagram, time is assumed to flow linearly from the top to the bottom. Therefore, the longer the delay, the longer the gap between incoming and outgoing events. The longer the transmission time for a packet, the thicker the appearance of the packet width on the timing diagram. The greater

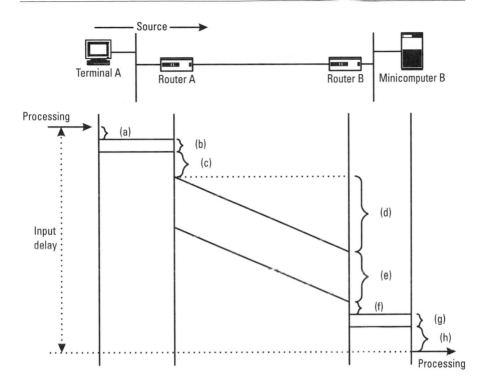

Figure 3.4 A simple reference connection and associated timing diagram.

the distance between end points, the greater the propagation delay, and the greater the downward slope of the packet traveling across a facility.

To illustrate these principles further, let us trace again the packet flow across the reference connection. This time we will concentrate on the label points (a) through (h) on the diagram:

(a) This gap represents the time that the packet sits in terminal A's network interface card (NIC) output queue awaiting permission to access the LAN segment. In this figure, we do not distinguish between processing delays and queuing delays; instead, we chose to combine these into a single delay. The longer the delay in the terminal, the larger the gap on the diagram, the further down the packet transmission is initiated on the diagram.

(b) This gap represents the transmission time of the packet onto the LAN segment. This shows up on the timing diagram as the packet thickness. The faster the transmission facility or LAN technology, the thinner the packet appears on the timing diagram. Also, this packet line appears essentially flat on the timing diagram because of the minimal propagation delay of the packet on the LAN segment. This is because of the geographic proximity of the terminal to the router on the common LAN segment.

(c) This gap, similar to gap (a), represents the processing and queuing delays experienced by the packet while within the router and waiting for transmission onto the private line facility. The longer the delay, for example, the greater the congestion on the WAN facility, the longer the queuing delay waiting for transmission, the greater the gap appears on the timing diagram.

(d) This gap accounts for the wire propagation delay over the private line facility that interconnects routers A and B. The farther apart the routers, the greater the propagation delay and the greater the slope of the packet line.

(e) Like gap (b), this represents the transmission time of the packet onto the private line facility. Because the private line facility in this example is slower than the LAN segment technology, the transmission time for the packet is greater. This is reflected in the fact that the thickness of the packet is greater than that for the LAN segment. For example, the time to transmit a 1500-byte packet onto an Ethernet segment is 1.2 msec, whereas the time to transmit this same frame onto a 1.5-Mbps T1 facility takes 7.7 msec, which is roughly 6 times slower. Therefore the thickness of the packet transmission on the WAN facility, in this example, should be roughly 6 times greater than for the same packet being transmitted onto the Ethernet LAN segment.

(f) This is another point in the reference connection where the packet will experience processing and queuing delays. This is represented on the timing diagram as another gap.

(g) This is identical to gap (b).

(h) This represents the waiting time for the transaction processor to begin processing the transaction.

The total input time is simply the sum of the individual delay components (a) through (h).

Figure 3.4 shows the timing diagram for a one-way "input time" for a packet to travel from the terminal to the host. Communication, however, is generally two way, and this is represented on a timing diagram by charting packet delays in both directions. This is illustrated in Figure 3.5, which

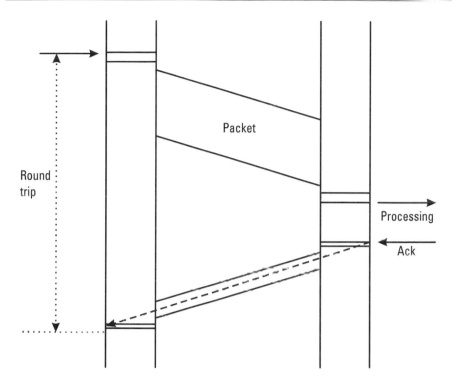

Figure 3.5 A timing diagram showing an explicit acknowledgment flowing in the reverse direction.

explicitly shows an acknowledgment flowing in the reverse direction in response to the receipt of the packet in the forward direction.

Figure 3.5 is typical of those used to analyze the behavior of windowing systems, as discussed in the sections following this one. It shows a large packet flowing from the left-hand side of the figure to the right-hand side of the figure. In response to the receipt of this packet, the right-hand station transmits an acknowledgment back to the left-hand side of the figure. Here we show the explicit details of the acknowledgment delays in the reverse direction. This level of detail is required when determining the round-trip delays in estimating response times or file transfer throughputs. However, for simplicity of presentation, we will often summarize the flow of an acknowledgment as a single arrowed line (this arrowed line is also shown in the figure). This will often be the case when discussing the various types of windowing systems and in qualitatively discussing their behavior.

Also, some protocols will attempt to "piggyback" acknowledgments on larger data packets that are flowing in the reverse direction. When this is the case, the complexity of the acknowledgment flow is increased. Essentially the ack must wait for a period of time for a packet flowing in the reverse direction. If no packet is queued for transmission within a given time-out period, then the ack is sent back to the transmitter on its own. However, if a packet is queued for transmission within the time-out period, then the ack is carried on this larger packet back to the transmitter. In this case, the delay characteristics for the ack assume the characteristics of the larger data packet. So whenever a single arrow is used to simplify the diagrams, remember that there is further detail hidden in this apparent simplicity.

The utility of the timing diagram is that it graphically highlights the delay contributions of each component in a communications path, shows the component's contribution to the overall delay a packet experiences, and establishes the correct relationship between these various components when determining how to sum all the components to develop an end-to-end delay estimate. In this sense, we refer to the timing diagram as a *data communications score card*. It also forces the engineer to understand the delay contributions in a given network connection and the relationship of the delays.

3.4 Pipelining

We now concentrate on the impact of protocol overhead and multiple hops on the overall insertion delays. The result of this discussion is the introduction of the concept of pipelining data across packet networks.

It is important to define more precisely what we mean by a transaction before we head into a discussion of transaction delays across data networks.

Loosely speaking, we would say that a transaction is a unit of work that an end user submits to a server. Typically, an end user submits a request for a limited amount of information from the server and the server responds by sending the information in the form of a single (or a few) data packets over the network. A transaction could be a single inquiry/response or it could be in the form of multiple inquiry/responses for some client/server applications. However, in general, it is assumed that a transaction has a relatively short duration, spanning only a few seconds.

In contrast to a transaction-based application, we will analyze a file transfer application. A file transfer typically involves a request for a large amount of information to be transmitted to the requesting user in the form of numerous data packets. This larger number of data packets far exceeds the end-to-end window of the transport layer over which the application runs, as discussed in following sections. Of course, there exists no clear boundary between a short transaction and a large file transfer, for example, some transactions request information fitting into tens of packets comparable to the end-to-end transport window.

We now consider several questions relating to the performance of a single transaction from terminal A to workstation B in the reference connection found in Figure 3.4. We refer to this reference connection as RC#1. The simplest question to ask is what is the end-to-end delay of this transaction across this reference connection.[2] As the single transaction is carried across the connection, it experiences three separate insertion delays, two while being inserted onto the LAN segments and one while being inserted onto the private line facility. This is shown in Figure 3.6.

The timing diagram also shows the additional insertion delays due to the packet protocol overhead. This protocol overhead can be nontrivial, for example, PPP, IP, and TCP protocol overhead can amount to 48 (20 bytes for TCP, 20 bytes for IP, and 8 bytes for PPP = 48) bytes for each telnet packet (and the data packets in a typical telnet session can average 10 to 20 bytes or less). The timing diagram on the left in Figure 3.6 shows the packet flow when the data packet is large enough to carry the entire transaction, while the timing diagram

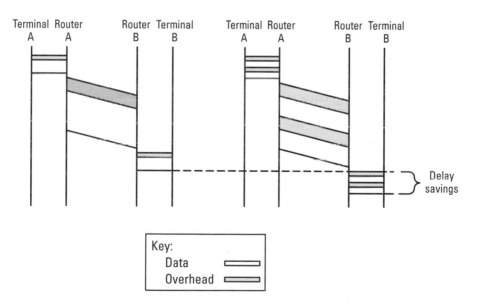

Figure 3.6 Transaction delay due to multiple packet insertions.

---

2. We identified numerous contributors to the delay experienced by a packet being transported across a data network, for example, insertion, queuing, and propagation. However, for the remainder of this section, we focus our attention primarily on insertion delays.

on the right shows the flow when the transaction is carried in two separate data packets. From a comparison of the two timing diagrams, it is apparent that the delay for the transaction is less when the total transaction is placed into a single data packet. This is due to the additional delays associated with transmitting multiple packet headers when breaking up the transaction into multiple packets. From this simple example, we would conclude that the optimal delay performance for the transaction occurs when choosing the packet size to be large enough to carry the entire transaction.

Consider now the case when our transaction is carried over the reference connection shown in Figure 3.7. We refer to this as reference connection #2 (RC#2).

The timing diagrams in Figure 3.8 show the effects of multiple WAN hops on the performance of the single transaction. The left-hand diagram

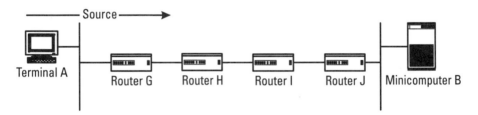

Figure 3.7 Reference connection #2.

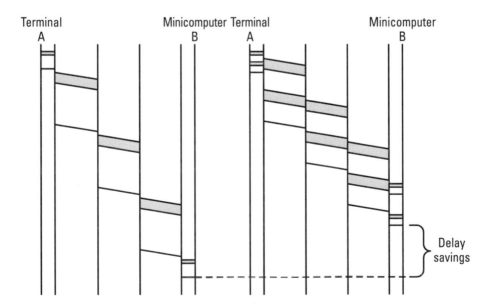

Figure 3.8 Timing diagrams illustrating the advantages of data pipelining.

shows the case where the packet size is chosen to carry the entire transaction, while the right-hand diagram shows the case where the packet size is chosen such that two packets are required to carry the entire transaction.

The total delay for the transaction to travel across the reference connection is less for the right-hand case where we have chosen a smaller packet size to carry the transaction. This is true even though the smaller packets force us to transmit a greater amount of protocol overhead onto the private line facilities. This is opposite of our conclusion when analyzing the single hop reference connection, RC#1.

The advantage of choosing smaller packets, for this reference connection, is that the network achieves a certain amount of parallelism in being able to simultaneously transmit portions of the transaction over multiple private line facilities. This effect, referred to as *packet pipelining*, is utilized in various high-speed packet networks, for example, frame relay and ATM, to lessen the negative impact that the packet network has on the end-to-end performance. The disadvantage of a large packet format is that the transaction is carried across the network in a store-and-forward fashion; the net result is an increase in the overall end-to-end delay for low-speed data networks.

Pipelining takes advantage of the trade-off of carrying greater overhead in order to reduce the cumulative effect of the store-and-forward delays inherent in data networks. For a given transaction size, number of network hops and protocol overhead size, there exists an optimal choice of packet size in order to minimize the end-to-end transaction delay. However, because these quantities are variable within a given network implementation, engineers must generally make a "best guess" at the optimal packet size, for example, the designers of ATM technology rather arbitrarily decided to fix their packet size to 48 bytes.[3]

Pipelining reduces delays if the underlying network is "slow." The relative advantage of pipelining when the underlying network is "fast" is lessened. For instance, consider transferring 1000-byte frames on an X.25 network with four serially connected 56-Kbps private lines versus transferring the same packet over frame relay with T1/T3 private line facilities. The transmission time of a 1000-byte packet in the X.25 network facility is (1000 bytes × 8 bits/byte) / 56 Kbps = 0.143 sec, and the time to transmit over four serially connected facilities is 4 × 0.143 = 0.572 sec (see Figure 3.9).

What if we were to reduce the packet size to 250 bytes? To estimate the transit time in this case look at the second timing diagram in Figure 3.9. The

---

3. OK, this decision was not arbitrary, it was more of a compromise between two opposing camps with different sets of requirements. The first wanted a 32-byte cell size to optimize performance for voice. The other camp wanted a cell size of 64 bytes, which was considered more appropriate for data applications.

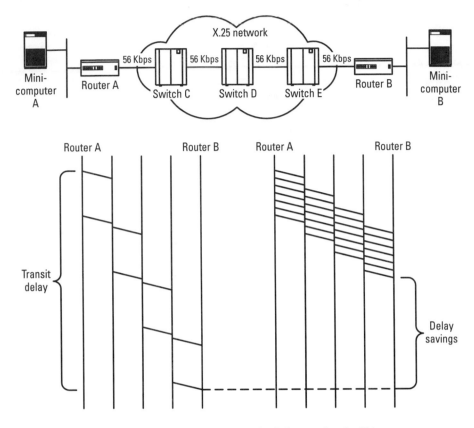

Figure 3.9 Improvement in transit delays due to pipelining on slow facilities.

total transit time can be estimated by following the first packet of the group (of four) across the first three 56-Kbps facilities. Due to the fact that this first packet is transmitted onto three separate facilities, its total transit time across those three facilities is roughly 3 × (250 bytes × 8 bits/byte) / 56 Kbps = 3 × 0.0357 sec or 0.107 sec. The entire four packets are transmitted onto the final slow facility (also indicated as highlighted packets). The time to transmit the four packets is 4 × 0.0357 sec or 0.143 sec. Then the time to transmit the 1000 bytes over the four facilities reduces to roughly 0.107 + 0.143 sec or 0.250 sec. This is to be compared with the 0.572 sec for the total transit time for the case of 1000-byte packets. This represents a reduction of the transmission delays by 56%.

In contrast, the comparable transmission time in the case of a T3 (45-Mbps) facility is (1000 bytes × 8 bits/byte) / 45 Mbps = 0.00018 sec for a single packet and across four facilities is 0.00072 sec. Since this delay is negligible compared to other delay components in networks (for example, propagation delays of 8 msec per 1000 miles), pipelining will provide little to no perceivable improvement in performance. To reiterate, the trade-off with pipelining is increasing packet overhead versus reducing packet transmission delays.

3.5 Throughput

In the previous sections, a fair amount of time was spent discussing the various contributors to delays that a packet traversing a communications network would experience. This is the base knowledge required to begin to address performance, engineering, and network design questions when analyzing high-speed data communications networks. We now begin to build on this base knowledge by discussing throughput computations. In doing so, we begin to reveal the higher level complexities of data communications networks.

Let us first define delay and throughput. For this discussion, the following definitions will apply:

- *Delay* is the time to complete a single, elemental function, for example, the time to transmit a single packet onto a communication facility or the time to serve a single customer in a queue.

- *Throughput* is the long-term rate of completion of numerous, elemental functions averaged over a period long compared to the delay to complete a single, elemental function, for example, the time to send a large data file comprised of numerous data packets.

At first thought, one might assume that there exists a simple relationship between the delay and the throughput of a system. In fact, this relationship should be of the form $X = T_s^{-1}$, where X is the system throughput and T_s, as defined here, is the system delay. This relationship does hold for some, but not all systems. We refer to systems for which this relationship between throughput and delay holds as *simple throughput systems*. An example of a simple throughput system is a transmission facility on which data packets are transmitted. If the delay to transmit a single packet is 10 msec, then the system throughput is 1/(10 msec) or 100 packets per second. In contrast, a *complex throughput system* is defined as one for which the simple inverse relationship between delay and throughput does not hold. A common example of a complex throughput system in data communications is a windowing protocol system.

Often, when a computer attempts to transmit a file consisting of numerous data packets across a data network, it does not simply transmit the packets continuously one after another. Instead, to avoid overloading the communications network or the receiving computer, it sends some number of packets and then stops until the receiving computer acknowledges the receipt of some or all of the packets.

The number of outstanding data packets sent before waiting for the receiving computer to acknowledge is referred to as the *window size*. Two of the different types of windowing systems are as follows:

- *Simplex windowing systems.* Here the transmitting system is allowed to send N_w packets before ceasing transmission. Once it receives a single acknowledgment for the N_w packets, it sends the next N_w packets and so on. N_w may be sent to one, two, or a larger number of packets.

- *Sliding windowing systems.* Here the transmitting system can have at most N_w packets outstanding, that is, have transmitted without receiving an acknowledgment, before stopping transmission. A sliding window system associates a single acknowledgment with each data packet or group of data packets and can transmit another packet for each acknowledgment received. This allows the window to "slide" without having to stop and wait.

Windowing systems are further categorized as dynamic versus fixed, and go-back-*n* versus selective-repeat. A *dynamic windowing system* is one that incorporates a feedback mechanism to readjust the size of the window, N_w, that it maintains. A *fixed windowing system* does not adjust the size of its window. The distinction between go-back-*n* and selective-repeat systems results from the possibility of packet losses in data communications networks. Packet losses can occur due to a number of effects, including misdelivered packets, transmission errors, time-outs, and buffer overflows.

Besides providing feedback to the windowing systems, acknowledgments serve to notify the transmitter of the successful reception of the data packet by the intended receiver. In the event of a packet loss, the receiver will not receive an acknowledgment for the lost packet, and it will have to retransmit the packet. For a go-back-*n* windowing system, the transmitter will retransmit the lost packet and all subsequent packets (whether lost or not). For a selective-repeat windowing system, the transmitter will only retransmit the lost packets.

Using this categorization of windowing systems, we can classify some of the more common transport windowing protocols as follows:

- The IP-based Transmission Control Protocol (TCP) employs a dynamic, go-back-n, sliding window protocol. TCP actually employs a byte-streaming protocol, where it maintains a count of the bytes transmitted and the acknowledgments specify the highest count byte consecutively received. However, the underlying networking layer, IP, employs a packet-based transport and hence the performance of TCP windowing can be assumed to be packet based as well. The actual window to transmit is set by the TCP receiver. Also, the adjustment of the window size is dynamic with respect to the characteristics of the underlying network (delay, loss, etc.). TCP is continually evolving. Selective repeat capabilities are being incorporated into various implementations.

- The Novell NetWare protocols initially utilized a fixed, simplex windowing system. Prior to their software release 3.11, NetWare transmitters would send a single packet and wait for the acknowledgment before sending the next packet. This proved to severely limit system throughput (as discussed later). Since then, Novell has enhanced the capabilities of their transport protocols in several ways. (See Chapter 8 on Novell networking.)

- The IBM Systems Network Architecture (SNA) employs a session level windowing mechanism, referred to as *session level pacing*. This windowing system is a simplex, fixed windowing scheme with the twist that the first packet of the packet stream is acknowledged (not the last, which is typically assumed). Upon receipt of this acknowledgment, the session level pacing transmitter is allowed to transmit the next packet stream. In addition to session level windowing, SNA also employs windows at the virtual route level and the link-level, that is, SDLC/LLC2.

- The AppleTalk Data Stream Protocol (DSP) employs a static, go-back-n, sliding window protocol. It is similar to TCP in that it is also a byte-streaming protocol.

As demonstrated by the examples given, it is typical to find that the so-called Internet-based protocols, for example, TCP/IP and NetWare, usually implement transport level windowing schemes, whereas others, most notably SNA, employ windows at the session, routing, and link layers.

Another approach taken to limit buffer overflow in some more recent implementations of data protocols is a rate-based flow control. Here, the transmitter explicitly spaces the transmission of successive data packets to control the system throughput. Typically the spacing is controlled by a system clock in the transmitter. Adaptive schemes are possible by providing the transmitter feedback on the utilization of the end-to-end system. Based on this feedback, the transmitter can increase (slow down) or decrease (speed up) the spacing between data packet transmissions. Examples include the ACF in frame relay implementations and the interpacket gap in NetWare.

We break up the remainder of our discussion into a description of windowing and rate-based systems in the next section. We follow this with an analysis of how to determine the optimal window sizes in various WAN environments and present simple formulas for throughput and bulk transfer times.

3.5.1 Windowing Systems

For most purposes, understanding the performance and behavior of throughput systems in lossless environments is sufficient. We provide a discussion of throughput systems in lossy environments as a digression in Appendix B. In this section, we will derive simple expressions, which relate the throughput of windowing systems to the packet delays across the network and to the number of bytes in the window. This derivation is not rigorous, but instead will rely on a few simple examples. We begin with an analysis of a simple windowing system, and then generalize this discussion to a more complex sliding windowing system. Before proceeding, we want to emphasize that the following discussion focuses only on the windowing system being the bottleneck to the overall system throughput. In some instances this is the case; in others it is not. We will conclude the overall discussion of end-to-end throughput by investigating all the potential throughput limiting factors in a path. The system throughput is then related to the resource with the minimum throughput, be it the windowing system, a given network transmission facility, or even a processor.

3.5.1.1 Simplex Windowing Systems

Our first example is based on a simplex, fixed windowing system. This is a relatively simple system to analyze. Assume for this example that the hosts have direct frame relay interfaces, that is, that there are no intermediate routers. The reference connection is shown in Figure 3.10. Here, the connection between the transmitting and receiving hosts is a frame relay WAN accessed at 56 Kbps. The frame relay network consists of three frame switches and two 1.5-Mbps trunks.

In Figure 3.11, we present a timing diagram that explicitly demonstrates the delay components when sending a packet from the transmitter (e.g., a transmitting computer) to the receiver (e.g., a receiving computer).

The timing diagram shows the delay components in computing the round-trip time for sending a packet from the transmitting host in our

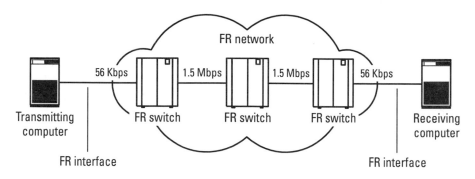

Figure 3.10 An example reference connection over a wide-area frame relay network.

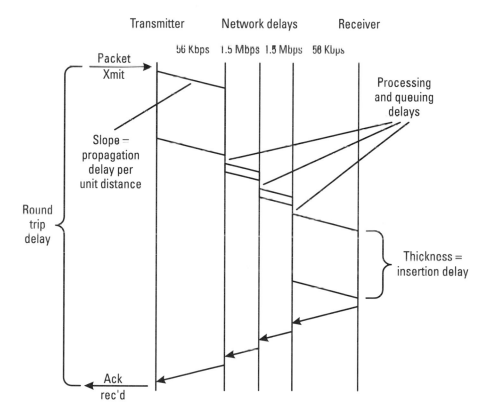

Figure 3.11 Timing diagram for a simplex windowing system and the network connection shown in Figure 3.10.

reference connection to the receiving host, and the receiving host transmitting an acknowledgment back to the transmitting host. Time begins when the transmitting host queues the packet for transmission onto the 56-Kbps access line to the frame relay network. The first bit of the packet arrives at the first frame relay switch after a propagation delay (which is a function of the distance to the first switch and the propagation delay per unit distance). The switch then accumulates the remainder of the packet. This is a function of the access line rate and the number of bits in the packet, which is reflected in the diagram as the thickness of the packet transmission. The first switch then processes and queues the packet for transmission onto the internal 1.5 Mbps network facility.

Again, the time to transmit the packet is reflected in the thickness of the packet transmission time (which is now smaller due to the high-speed network trunk facility). This process continues until the packet is received at the receiving host. The receiving host then generates an acknowledgment, and queues this for transmission back to the transmitting host. The acknowledgment (which is considerably shorter than the original packet) travels the reverse path back to the transmitting host. We define the *round-trip delay* as the *total delay between when the transmitting host first queued the packet for transmission onto the access facility to the time the transmitting host receives the last bit of the acknowledgment and finishes the acknowledgment processing.*

Now, consider the definition for the behavior of a simplex, fixed windowing system. Here the transmitter is allowed to send a single packet, and then it must wait for the reception of the acknowledgment prior to sending the next packet, and so on, until the transmitting host has sent all of the data it wishes to send to the receiving host. We idealize the previous timing diagram by hiding the complexity of the frame relay network delays into a single delay component, referred to as the *network delay.*[4]

Figure 3.12 shows multiple rotations of the simplex, fixed windowing system. This process will continue until the transmitting systems has sent its entire data file. As seen from the previous timing diagram, for each round-trip delay period, the transmitter is able to transmit a single packet (which in the case of a simplex, fixed windowing system is equal to the entire window's worth of data bits). Therefore, the end-to-end throughput is calculated simply as the ratio of the number of bits in the packet (we are assuming a fixed sized packet) divided by the round-trip delay:

$$X = W / R_d \qquad (3.3)$$

---

4. This is useful for simplifying the example, but is also necessary in practice. It is often the case that the frame relay network provider will not divulge the details of the VC to the end user.

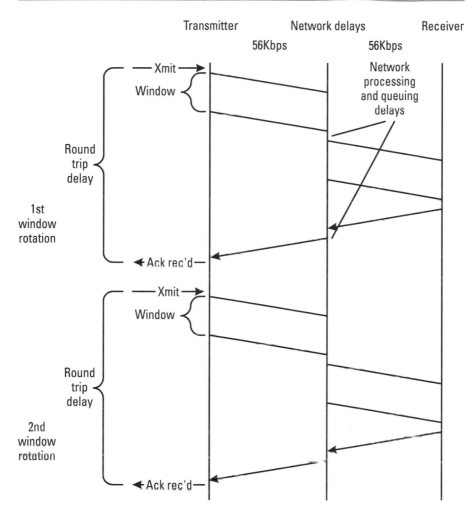

Figure 3.12 Timing diagram showing multiple window rotations for a simplex, throughput system.

where X is the end-to-end throughput,[5] R_d is the round-trip delay for a packet, and W is the window size (which in this example is simply the number of bits in each packet transmitted). Note that the computation of R_d will reflect the lower level protocol overhead while the W does not.

---

5. Notice that the throughput is different than if we computed it assuming that the transmitter can transmit unimpeded, that is, that the throughput is simply the inverse of the packet transmission delay onto the frame relay access facility. Therefore, we refer to this windowing system as a complex throughput system.

Clearly, this formula holds for any simplex, fixed windowing system, independent of the window size. This is true independent of whether the transmitting host must stop sending after a single packet or multiple packets, as long as the receiver collects all the packets in the window, and it returns a single acknowledgment for the entire window. A simple way to visualize this is to assume that the packet in the preceding timing diagram is really two packets, each half the size of the original. The only change to the above diagram, for the most part, is that the transmitter sends two consecutive packets, then stops to wait for the acknowledgment as shown in Figure 3.13.

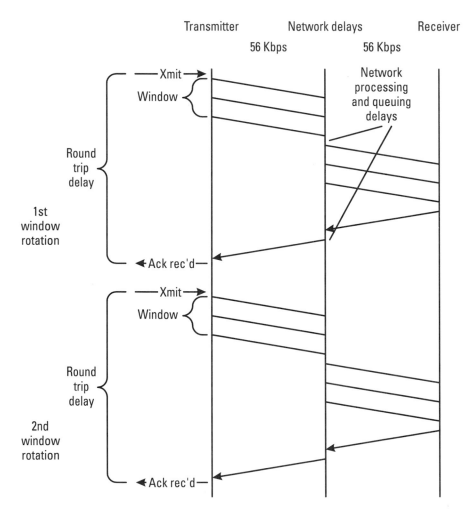

Figure 3.13 A simplex, fixed window system with multiple packets per window.

We have made several assumptions to estimate the end-to-end throughputs for these simplex, fixed windowing systems. Two of these assumptions are as follows:

(1) *The assumption of static loads on the queuing delay components in the end-to-end path.* In reality the loads on the end-to-end queuing components vary over time. It is often the case that estimates of network loads have to be made, or more often, estimates of the network delays are made based on guidelines and bounds generated by the network providers, administrators, and so on.

(2) *The assumption that the packets of a given window see identical delays across the network.* In reality, the consecutive packets within the same window should see increasing network delays due to the effective load the previous packets add to the network components. This assumption is valid when the ratio of the access transmission times to the network trunks transmission times and network processing times are large, for example, for a 56-Kbps access line and 1.5-Mbps network trunk speeds as in Figure 3.10.

We will make liberal use of assumptions in order to develop simple expressions of network performance. However, the use of these expressions is not meant to replace actual network measurements. Instead their primary value is in aiding the network analyst, engineer, and administrators in developing an understanding of the essential aspects of the networks and protocol interactions. This will aid the engineer in developing useful models of network performance in order to predict behavior of new network applications and to develop useful capacity management guidelines in order to grow their networks in a graceful fashion.

Let us get back to the discussion of simplex, fixed windowing systems. Looking at Figure 3.13, we observe that the use of the transmitter and access line resources is very inefficient; that is, after transmitting the window, the transmitter must wait for the acknowledgment, causing the access line to go idle. This effect causes the calculated throughput for these windowing systems to be smaller than the maximum attainable end-to-end throughput for the system. The maximum attainable throughput is estimated by identifying the single, slowest component in the end-to-end path, for example, the slow access line to the WAN or the speed of the transmitter output device driver, and computing the bit rate at which that single component can transmit or process data. This slowest component is the bottleneck to the system throughput. In Figure 3.13, the maximum rate is reasonably assumed to be 56 Kbps, which is the WAN access line speed. Assuming a round-trip delay of 200 msec and a window size of 4000 bits, we get a windowing throughput of 20 Kbps (which is significantly less than 56 Kbps). In this case, we say that the system's throughput is *window limited*. One can define the system efficiency as the ratio of actual to the maximum achievable throughput, which in this example is 20/56 or 0.42. Systems employing simplex windowing systems are always *window limited*. As a result, sliding windowing systems were developed.

3.5.1.2 Sliding Window Systems

Sliding window systems allow the transmitting host to continue sending packets even though the acknowledgment for the first packet has not yet been received by the transmitter. This helps to eliminate the inefficiency seen in the simplex windowing systems discussed previously. Figure 3.14 shows a timing

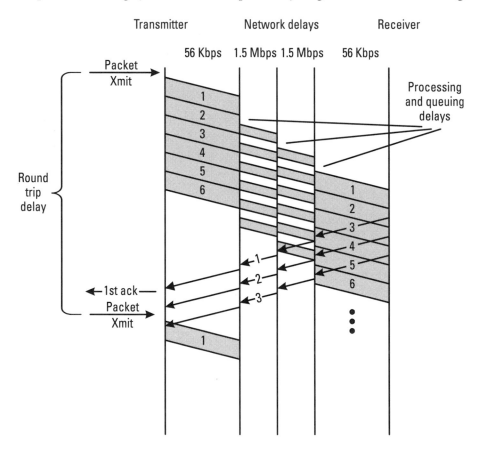

Figure 3.14 Timing diagram for a sliding windowing system and the network connection shown in Figure 3.10.

diagram for a sliding windowing system running across the reference connection shown in Figure 3.10. Here, the window size is six packets. Notice that the transmitter continues to transmit packets up to its window prior to receiving the acknowledgment on the first packet. Once it receives the acknowledgment for the first packet it is allowed to send one more packet. In fact, it is allowed, from this point on, to send a packet for each acknowledgment received.

Allowing the transmitter to send additional packets before receiving the acknowledgment of the first (and successive) packet improves the end-to-end system efficiency. However, the estimate for the throughput is identical in form to the expression for the throughput of the simplex system in (3.3). We estimate the throughput by observing that the transmitter is allowed to have up to N packets in transit within the round-trip delay cycle. Therefore, we have

$$X = W / R_d \tag{3.4}$$

where X is the system throughput, R_d is the round-trip delay of a given transmitted packet (and its associated acknowledgment), and $W = 8 \times N_w \times S$ is the system window size. This is the product of the number of packets in the window, N_w, and the (average) length of the packets, S (converted to bits by the factor of 8). This approximate expression for the throughput holds only as long as the window is not too large. If the window becomes too large then other factors tend to limit the system throughput, for example, facility bandwidth, and processor throughput. Strictly speaking, this expression represents the system throughput in the event that the window is the throughput limiting factor in the system.

As an example of the utility of this expression, consider the question of tuning an end-system window to ensure a given throughput between two computers of 1.5 Mbps (but no more). Assume that one computer is in LA and the other is in NYC. First, let us estimate the round-trip delay. The distance, as the crow flies, between LA and NYC is roughly 3000 miles. Given that the facility route miles are expected to be roughly 1.3 times longer, we will assume that the facility route mileage between LA and NYC is roughly 4000 miles. This equates to a propagation delay of 32 msec one way, and 64 msec round-trip. Rounding this up to 100 msec to account for additional networking delays, and inverting (3.4) to solve for the window size as a function of round-trip delay and throughput, we calculate that $W = X \times R_d = 1.5$ Mbps $\times$ 100 msec = 152 Kbits = 18,750 bytes. This is a significant window size in comparison with typically sized windows in the more popular networking protocols.

3.5.1.3   Rate-Based Throughput Systems

Several years ago rate-based throughput systems were proposed, most notably the "leaky bucket" mechanism [5]. Rate-based throughput systems control not the number of outstanding packets in transit, as in windowing systems, but the explicit rate, for example, packets per second, at which the packets are transmitted. These systems are implemented as access control filters in frame relay and ATM access ports, as discussed in Chapter 2.

Figure 3.15 shows a schematic of a rate-based throughput system in which asynchronously generated data packets are queued for transmission in a data buffer (strictly speaking, this represents a delaying ACF). In a typical frame relay implementation, the rate-based control exists in the access port, the asynchronous packet arrival represents packets arriving from the access line, and the time smoothed packet stream on the right-hand side of this figure represents the packet stream allowed access to the frame relay network. Queued packets

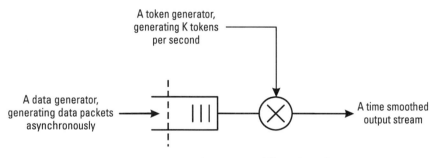

Access control filter schematic

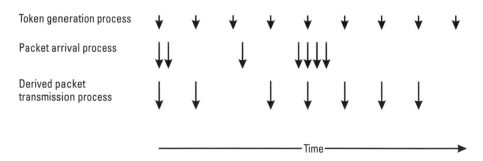

Figure 3.15   A simple schematic of a rate-based throughput system (top) and its effect on an example arrival process (bottom).

are transmitted, one at a time, when a token is available. Tokens are generated at a constant rate, for example, one token every 10 msec. When no data packets are queued, no traffic is admitted into the network.

Also shown in Figure 3.15 is an example of a derived packet transmission process, based on the regular token generation process and the asynchronous packet arrival process. Notice that the packet transmissions are smoothed relative to the initial arrivals, and there exists a minimum spacing between the initiation of packet transmissions, which is simply the inverse of the token generation rate.

Rate-based schemes were proposed as a mechanism to reduce the probability of overload in packet networks. This is accomplished by spreading out the packet arrivals to a congested network resource, thus reducing the queuing delays and reducing the number of packets in the congested resource's buffer. We know from the preceding discussion of queuing delays and the effects of various arrival processes on the waiting time that by smoothing the arrival process we reduce the waiting times of the number of packets in queue.

Several variations of this rate-based scheme have been proposed and developed. These include (1) the leaky bucket mechanism, which adds a finite sized buffer for the tokens; and (2) a closed-loop feedback mechanism that adjusts the rate at which the tokens arrive. These variations were discussed in Section 2.4.

Adding a finite sized token buffer to rate-based schemes has the effect of collecting tokens during idle periods up to the finite token buffer size. When new packets arrive, they are immediately transmitted until the tokens in the buffer are exhausted. From this point on, the packets are queued and the system behaves as if the token buffer did not exist. Therefore, the behavior of this leaky bucket mechanism is intermediate between a system having no rate-based mechanism and that of a rate-based system with no token buffer. This is shown in Figure 3.16.

At low loads and small transaction sizes, the token buffer scheme has little effect on the transmission of the arriving packets. This is because, with a high probability, tokens exist in the token buffer that allow for immediate transmission of the arriving data packets. At high loads and/or large file transfers, a small probability exists that there are enough stored tokens queued to allow for immediate transmission of the data packets, and the data must be queued until the necessary number of tokens appear. Overall, this has the advantage that for small transaction-oriented applications, few additional delays are added to end-to-end performance due to the leaky bucket algorithm. However, for large file transfer applications, the leaky bucket algorithm tends to smooth the loads by queuing the data packets and pacing them out onto the transmission facility; hence, reducing the probability of buffer congestion.

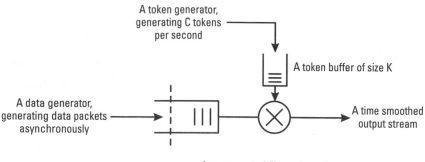

Access control filter schematic

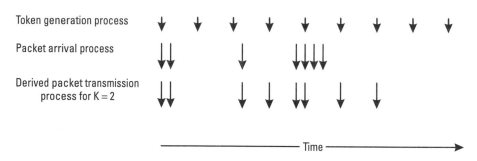

Figure 3.16 The leaky bucket algorithm components adding a finite token buffer ($K = 2$ tokens) to the rate-based scheme in Figure 3.15.

Although not discussed in the section on queuing models, an expression that estimates the mean queuing delay in the leaky bucket scheme [6] is

$$E(W) = [U / (1 - U)] T_s \times \exp[-(1 - U)K/S] \quad\quad (3.5)$$

where U is the system load, S is the system average packet size, and K is a measure of the size of the token buffer (in the same units used to measure the average packet size). This expression is derived based on several simplifying assumptions, for example, Poisson arrivals, exponential service times, and infinitesimal token sizes. Even so, this expression verifies the discussion in the previous paragraph. That is, for transaction-oriented traffic, the existence of the token buffer reduces the mean waiting time. If the token buffer does not exist, $K = 0$, then the mean waiting time reverts to the familiar M/M/1 expression. However, for finite token buffers, the mean waiting time is dramatically reduced. This is illustrated in Figure 3.17.

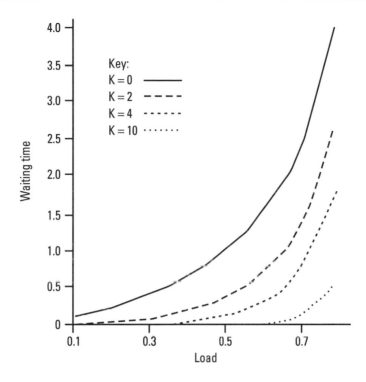

Figure 3.17 A plot showing the effects of increased token buffer size on the expected waiting time.

Although the token buffer size has a dramatic impact on the expected transaction waiting time in the token buffer scheme, this has little or no impact on the long-term throughput of this system. The long-term throughput in this scheme is determined solely by the rate at which the tokens accumulate.

3.5.2 Lossless Throughput Systems Summary

We discussed three different throughput systems in this section:

1. A simplex windowing system, where the transmitter had at any instant in time a single packet in transit to the receiver. As we saw, this tended to severely limit the overall system throughput performance.

2. A sliding windowing system, where the transmitter was allowed to have multiple packets in transit toward the receiver at any instant in time. This tended to improve system performance. This also raised

the issue of how to set or tune the system's sliding window. This tuning represents a trade-off between large system throughputs and low packet losses.

3. A rate-based throughput system, in which the transmitter explicitly spaced the transmission of packets based on an internal clock. This had the advantage of reducing congestion by smoothing out the packet arrival process.

Fundamentally, all three systems are implemented to reduce the probability of overflowing buffers. This occurs in situations where large speed mismatches exist between the maximum transmitter rate and the minimum network transmission rate or the receiver rate. Without some bound on the number of outstanding data packets (for example, a sliding window) or a bound on the transmission rate (for example, a rate-based transmitter), buffers would eventually overflow, causing severe network congestion and packet losses. For the interested reader, the impact of packet loss on throughput system performance is presented in Appendix B.

3.5.3 Optimal Window Sizes and Bulk Data Transfer Times

We wish to spend a few moments talking about *optimal* window sizes. We feel it is important to understand the dynamics of windowing systems and the effects of varying the size of the transport protocol window. We realize that tuning transport window sizes is often out of one's control. However, that should not give us cause to ignore the topic. By understanding the potential effects and by predicting their existence, other tools may be considered to improve overall network performance. (We discuss various tools and capabilities in Sections 4.5 and 5.4 in later chapters.)

Consider, RC#1 when carrying file transfer traffic (instead of transaction-based traffic). Suppose that terminal A has a large file, for example, 10 MB, to transmit to minicomputer B. Further assume that the transport protocol uses a sliding window scheme. Given this, what is the optimal window size to carry the file transfer across RC#1?

Figure 3.18 shows the startup events for a file transfer over our RC#1 in the case of three progressively larger window sizes. The timing diagram on the far left represents a window size of two data packets, the middle diagram represents a window size of three data packets, and the timing diagram on the far right represents a window size of four data packets. In each case roughly the same events occur.

On startup, the transmitting workstation can transmit its entire window over the relatively high-speed LAN network to the router connecting the LAN

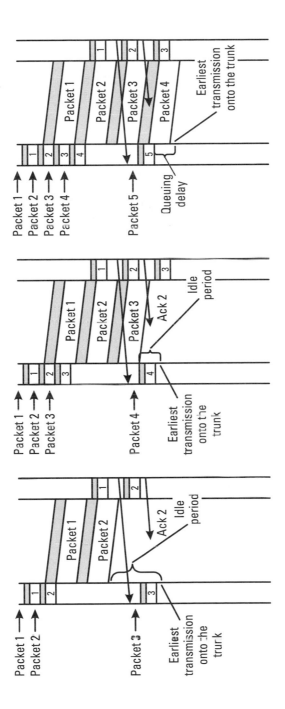

Figure 3.18 File transfer startup for progressively larger window sizes.

to the WAN. The router is then forced to buffer the data packets until it is able to insert the packets onto the relatively low-speed WAN private line facilities. Eventually, the first data packet arrives to the receiving host, which acknowledges it. The receipt of the first acknowledgment by the source opens the transmitter's window, allowing it to transmit the next data packet. Again, due to the high-speed mismatch between the fast LAN and the slower WAN private line facility, this packet may have to queue in the router to await transmission.

Now, notice the differences in the waiting time for the data packet that is transmitted in response to the first acknowledgment in all three cases, as discussed next.

For the case of a sliding window of two, the packet experiences no waiting time at all to be transmitted onto the WAN private line facility. This is because the private line facility has been idle for a period of time waiting for the arrival of the next data packet. This represents an inefficient utilization of the slow private line facility. During the file transfer, in order to minimize the transmission time of the file, we would like the slowest link in the reference connection, that is, the WAN facility in RC#1, to be running at close to 100%. Otherwise, this particular end user will observe a network with rather sluggish performance. This may seem somewhat nonintuitive because we would also expect that the line utilization should be kept at or below 70% in order to maintain good end user performance. However, these are two different concepts.

In the first case, during the period of the file transfer, we would like the transfer to be able to perform well in the absence of other traffic and fully utilize the network resources. On the other hand, when measuring the average utilization on a given resource in the network over some period of time, for example, 15-min intervals, we would like this time average utilization to be 70% or less.

For the case of the sliding window of three, the packet also experiences no waiting time at all to be transmitted onto the WAN facility. This is because the WAN facility is idle at the time of arrival of this next data packet. However, unlike before, the WAN facility was not idle for an extended period of time. In fact, it only recently became idle. In this case, the WAN facility will be running at near 100% utilization.

For the case of the sliding window of four, the packet experiences a waiting time at the router before being transmitted onto the WAN facility. This is because the WAN facility is slow enough that it takes longer for it to transmit the entire sliding window than for the round-trip time associated with transmitting the first data packet across the reference connection and receiving the first acknowledgment at the transmitter. Because the WAN private line facility in RC#1 is the slowest link in the example, it is not possible to achieve a higher throughput than supported by this link. The only effect that increasing the sliding window has at this point is to cause a larger and larger queue to develop in the router. Increasing the window will not give the end application an improvement in performance. Therefore, why increase the window beyond this point?[6]

In this particular set of examples, we see that the window size of three achieves good end-system performance while minimizing the size of the queue at the slow link.

The preceding description of the dynamics of the windows tended to be very atomic in nature. Another way to think about choosing optimal window sizes for sliding window protocols is more flow based. For sliding window protocols like TCP, a very simple formula is used to compute the optimal window size:

$$\text{Optimal window size } W^* - \text{Bandwidth of slowest link in the path} \times \text{Round-trip delay for a packet}$$

$$(3.6)$$

This formula is often referred to as the *delay bandwidth product*. By drawing timing diagrams, you can see very clearly that if you set the window size W to this optimal value, then acks will be received at the exact moment that you complete inserting the window on the slowest link. If $W > W^*$, then window is too large because acks will be received before window closes. If $W < W^*$, then acks will be received after window closes, meaning that the system can take more data.

In most cases, the sliding window size must be somewhat larger than the number of the slowest links in the reference connection. How much larger depends on the specifics of the reference connection, for example, the nature of the propagation delays and the number of additional higher speed links.

Because of the dependence of the "optimal" window size on the specifics of the reference connection, it is not possible to pick the right window size to accommodate all applications and all paths through a given enterprise network. Instead, the analyst must optimize for the average or the worst case reference connection. Once the reference connection is chosen, then the analysis can be performed and an appropriate window size can be determined. Unfortunately, many transport level windowing implementations do not let you pick window sizes. For these, the best you can do is determine the default value of the transport level windows for the protocols supported in the network and understand the impact on end user performance that these windows will have. Also, for

---

6. You may ask, why not? We will discuss why not in Section 4.4.

many protocol implementations the sliding window size is dynamic, and the end systems can adjust the window size based on lost packets and calculated delays based on receipt of acks.

3.5.4 Throughput Summary

Let us now wrap up our discussion of the throughput analysis by building a simple formula for estimating bulk data transfer times. We first need to find the system throughput and then estimate the bulk transfer time.

We have identified several considerations when determining end-to-end system throughput, including processor speeds, facility rates, rate-based access controls, and windowing throughputs. In analyzing the throughput of given data applications, each must be considered separately, and then combined to fully understand the interactions in the end-to-end path. The observed system throughput, X_{syst}, is determined as follows:

$$X_{syst} = \min\left\{ X_{cpu}, X_{facility}, X_{access}, W_{window} \right\} \tag{3.7}$$

were X_{cpu} is simply the inverse of the time for a given processor in the path to process information related to a single data packet (converted to bits per second), $X_{facility}$ is the bandwidth of the transmission facilities in question, X_{access} is the throughput of the access control mechanisms (as determined by the token generation rate), and X_{window} is the throughput limit as determined by the nature and size of the windowing schemes involved. (Often several windows are layered on top of one another and each must be considered and the minimum throughput determined.) All of these have the potential for being the throughput bottleneck, so each must be individually considered and the minimum must be identified.

For example, consider what is required in determining the throughput for a system such as that shown in Figure 3.10. Let us assume an end-to-end sliding window. Also, assume rate-based access mechanism controlling packet access exists to the frame relay WAN with an effective token arrival rate of 16 Kbps. Given these assumptions, we are required to perform multiple levels of analysis.

First, an estimate of the windowing throughput should be made that ignores for the moment the access control. If the windowing throughput is determined to exceed 16 Kbps, then the access control is the throughput bottleneck limiting the system throughput to 16 Kbps. If the windowing throughput is determined to be less than 16 Kbps, then it is the throughput bottleneck.

Of course, other throughput limitations may exist, for example, processor limitations. It is the job of the network engineer to identify all of the potential throughput bottlenecks in their networks and to understand the impact they may have on end users applications.

Once the system throughput, X_{syst}, is identified, it is a simple matter to estimate the data transfer time for the file transfer in question. If the file to be transferred is $F \times 8$ bits in length and the system throughput (or throughput bottleneck) is X_{syst} bps, then the file transfer time, T, is

$$T = F \times 8 / X_{syst} \qquad (3.8)$$

As an example, if $F \times 8 = 800$ Kbits and the system throughput is $X_{syst} = 16$ Kbps, then the file transfer time is $T = 800,000 / 16,000 = 50$ sec.

One final word of caution. Care should be exercised when using this expression to ensure that common units and consistent measurables are being used. For example, it is a common mistake to forget to include protocol overhead when estimating transfer times. In this case, either (1) the X_{syst} must be decreased to account for packet overhead (i.e., to describe the throughput in terms of user data bits per second) and use the application level file size figures, or (2) expand the file size to include the packet overhead. Because the protocol overhead is a function of the specific protocol suite, we discuss these overhead issues in the later chapters of this book.

3.6 Summary

In this chapter, we have tried to give the network engineer an overall appreciation of the complexities of analyzing application performance over data networks. We began with a detailed discussion of the various factors affecting the delay in transmitting a single packet across a data network. We used this information to show how to estimate the system throughput. We discussed various types of flow control mechanisms that have been implemented in networking protocols that ensure that network and end-system buffers do not overflow due to the transmission of too much data in too short a time period. These mechanisms included windowing systems and rate-based flow control systems. Finally, we ended the chapter with a section on optimal window sizes and file transfer times.

We hope this chapter was sufficient to provide the reader with an overall appreciation of various performance issues. In the later chapters of this book we will revisit many of these same issues and performance questions within the context of specific network protocol implementations and protocol interactions.

References

[1]   Bertsekas, D., and R. Gallager, *Data Networks*, Englewood Cliffs, NJ: Prentice Hall, 1987.

[2]   Schwartz, M., *Telecommunications Networks, Protocols, Modeling and Analysis*, Reading, MA: Addison-Wesley, 1987.

[3]   McNamara, J. E., *Technical Aspects of Data Communications*, Bedford, MA: Digital Press, 1982.

[4]   Cooper, R. B., *Introduction to Queuing Theory*, New York, NY: Elsevier, 1981.

[5]   Turner, J. S., "New Directions in Communications (or Which Way to the Information Age?)," *IEEE Commun. Mag.,* October 1986.

[6]   Cansever, D., and R. G. Cole, private communications, 1992.

4

Techniques for Performance Engineering

4.1 Introduction

In this chapter we present general techniques for performance engineering data networks. We begin by discussing a collection of techniques: load engineering, application characterization, application discrimination methods and collection and traffic monitoring. We end this chapter with a section that ties these various techniques and tools together. The context we choose is a discussion of the process to follow when implementing a new application in an existing data network. The discussion of load engineering in the first section of this chapter builds on our knowledge of queuing systems, their stability, and general load versus delay characterization. This knowledge is leveraged to define measurable component thresholds in order to maintain satisfactory application level performance over the enterprise network.

We next classify applications into two broad categories: *latency sensitive* and *bandwidth sensitive*. Even this broad classification of application types is useful when engineering data networks for application-level performance objectives. This discussion is followed by an extensive taxonomy of techniques to minimize cross-application interference effects in multiprotocol, enterprise networks. Techniques discussed here include type of service routing, priority queuing, processor sharing, adaptive controls, and selective discards.

This is followed up by a section on network management and data collection tools. These are useful in several areas including (1) data collection and application-level characterization and (2) traffic monitoring and forecasting. The general techniques we consider in this chapter are used in one way or another for performance engineering data networks. To reinforce this concept, we conclude the chapter with a step-by-step identification of the process to follow when deploying a new application over an existing corporate network.

4.2 Load Engineering

In Chapter 3 we developed various queuing models relating component delays to their utilization. We also know that queuing systems are stable, that is, they have finite queuing time and queue lengths, when their load remains less than 100%. The goal of load engineering is to ensure that the load on the system remains less than an identified threshold, above which performance degradation occurs. This is accomplished through (1) defining appropriate target component utilization on critical components within the network, (2) constant monitoring of the load on system components, and (3) making necessary adjustments to component load in the event that it becomes too high or too low, based on a developed action plan.

Within this section, we concentrate on a simple queuing model akin to a packet buffer system feeding a private line facility. In Section 5.4 in the following chapter, we discuss load engineering of a more complex frame relay interface and its associated virtual circuits.

As discussed in Chapter 3, the queuing time is a function of T_s(average), the average service time, and U, the utilization. The component utilization is a product of the arrival rate, L, and T_s(average), that is, $U = L \times T_s$(average). In fact, the queuing delay increases when either (or both) the service time or the arrival rate on the system increases. As a rule of thumb, queuing delays in a system should not overly dominate the service times; say, queuing time should never be more than twice the service time.

Looking at the curve for the queuing time in an M/M/1 system, shown in Figure 4.1 (and discussed in Chapter 3), this implies that the load on the system should remain less than roughly 70%. In the figure, the queuing time is measured in units of T_s(average). This point can be thought of as the location in the curve above which small increases in load will result in large increases in queuing times. If instead we wanted to keep the queuing delays less than or roughly equal to the insertion times, then from the figure we see that the utilization should be maintained at less than or equal to 50%. Again, these estimates are based on the M/M/1 queuing model and results will vary depending on the specific queuing model used.

From Chapter 3, we know that the queuing delay, Q, as a function of utilization, U, is:

$$Q = [U / (1 - U)] \times T_s\text{(average)}$$

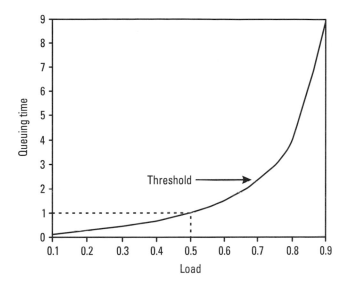

Figure 4.1 The expected queuing time as a function of load for the M/M/1 queuing.

For the queuing delay to be equal to T_s(average), we see that the utilization should be equal to 50%, that is,

$$Q = [0.5 / (1 - 0.5)] \times T_s(\text{average}) = T_s(\text{average})$$

For the queuing delay to be equal to $2 \times T_s$(average), we see that the utilization is equal to 66%:

$$Q = [0.66 / (1 - 0.66)] \times T_s(\text{average}) = 2 \times T_s(\text{average})$$

This is consistent with our reading of the delay versus load curve shown in Figure 4.1.

A more general queuing model discussed in Appendix A is the M/G/1 model. This model incorporates the variability of the packet sizes in its expression for the queuing delay. The queuing delay expression from the M/G/1 model is:

$$Q = [(U \times V) / (1 - U)] \times T_s(\text{average})$$

where V is a measure of the variability of the packet transmission time, that is, $V = (1 + C_V**2) / 2$, and C_V is the ratio of the standard deviation of the packet transmission time to the average packet transmission time. As we discuss in Appendix A, $V = 1$ for the M/M/1 queuing model. The M/G/1 expression simply states that the greater the variability in the packet sizes, the greater the queuing delay experienced on the transmission facility. So, the higher the variability in packet sizes, the more conservative one must be about loading the connection. To make this a little more quantitative, say we choose to try to maintain queuing delays to be no more than twice the average packet insertion time on the transmission facilities. Then the M/G/1 queuing model yields:

$$Q = 2 \times T_s(\text{average}) = [(U \times V) / (1 - U)] \times T_s(\text{average})$$

Eliminating T_s(average) on both sides of this equation yields

$$2 = [(U \times V) / (1 - U)]$$

Rearranging this expression and solving for U we get:

$$U = 2 / (2 + V)$$

If $V = 1$, as it is assumed to be in the M/M/1 model, we get that $U = 66\%$ as a target threshold (which is roughly our 70% threshold in the previous paragraph). However, if the variability in the packet sizes is higher, say, $V = 2$, then we get $U = 50\%$ as a target threshold. This validates our claim that the higher the variability in packet sizes, the more conservative one must be about loading the connection.

One last consideration in defining thresholds is the period of time over which the loads are measured. Or, in other words, should the thresholds (and their corresponding averaging) apply to 1-, 5-, or 15-min intervals? And should the engineer focus on the average results over a number of sampling intervals or the peak result? Further, at what point should the engineer decide to make capacity management decisions to increase the necessary bandwidth? If the chosen threshold is exceeded once a day for one month, does this constitute the point at which more capacity is ordered? Or should more capacity be ordered only when 20% of all measurements during the 8-hour work day exceed the chosen threshold?

Unfortunately there are no clear and definitive answers to these questions. Much depends on the nature and the time sensitivity of the applications running over the network and on the amount of data the engineer wishes to store and analyze. For example:

- *Fifteen-minute averages.* In most situations it is probably reasonable to tract 15-min intervals and monitor the average utilization threshold over this period. Fifteen-minute intervals are short enough to catch the hourly variations in the daily behavior of corporate employee activities. However, it is not so short as to cause the data collection and monitoring systems to have to store large amounts of data.

- *Five-minute averages.* In some situations, for example, reservation systems where end customers of the corporation are immediately affected by the system delays, network engineers should consider shorter time periods over which to tract thresholds. This would give the engineer greater visibility into the nature of the variations in the network loading, while not requiring the collection and storage of too much data.

- *Five-minute averages with hourly peaks.* In some situations it is more important to keep track of the extremes in the variation of the utilization. In these cases, taking relatively short averaging periods, for example, 5-min, and monitoring the hourly peak 5-min period is appropriate. Here, the thresholds are set for these 5-min, hourly peaks and capacity management decisions are based on these peak values.

Of course, there are an infinite number of possibilities and combinations. Note that the shorter the time interval over which the component values are averaged, the greater the amount of data to be collected and the greater the storage requirement on the data collection system.

Further, for a fixed threshold, for example, 70% utilization on a private line facility, the shorter the time measurement interval, the greater the bandwidth requirement to maintain the chosen component levels. This is due to the fact that short measurement intervals will show a greater variation in the averaged values and in turn increase the likelihood of seeing component thresholds exceeded.

At one level it would be nice to be able to answer all of these questions by relating all of this back to end-to-end performance objectives. But this is rarely feasible. At another level, just by going through the thought process of setting thresholds and measurement periods (and collecting and analyzing them), the network engineer has accomplished the majority of the work discussed in this section, independent of the specific periods, peaks, thresholds, and so on.

Once the component thresholds are identified and the measurement intervals chosen, then methods to monitor the component metrics must be determined. Monitoring tools is one of the topics discussed in Section 4.5. However, before moving on, let us first mention several ways to modify the load on a queuing system in the event that it is determined through monitoring that the component threshold is consistently being exceeded. These include:

- Changing the number of servers in the system; for example, we can add servers in a client/server environment when the server is being overloaded, or we can add parallel communications facilities between two locations in the event that the transmission line is being overloaded.

- Changing the rate at which the servers operate; for example, we can increase the speed of the transmission facility that is being overloaded.

- Providing preferential treatment to the more important applications, hence reducing their effective utilization (see Section 4.4).

To summarize, the keys to successful load engineering are:

- To clearly define thresholds for each component that is critical to the end-to-end performance of the network;

- To ensure that the load on the particular components in question is constantly monitored; and

- To develop a strategy to actively adjust the load in the event that the specified threshold is either being exceeded or the load is too small in comparison to the threshold.

To accomplish these tasks, measurement and monitoring equipment need to be utilized. These are discussed in Section 4.5. But first we wish to discuss some techniques to characterize applications and to discriminate between application types in network deployments.

4.3 Latency-Sensitive and Bandwidth-Sensitive Applications

The first section of this chapter outlined a methodology for load engineering components within a data network. The methodology consisted of component level thresholds from load engineering models and monitoring of the components to maintain loads less than or equal to the desired thresholds. Measurement systems are deployed in order to maintain the component level thresholds and to support capacity management tools and methods. But how do these component thresholds relate to the end-to-end performance desired by the end user or application running over the data network? To answer this question, it is necessary to understand the applications running over the data network and their characteristics.

This section discusses various application types or profiles. We use these profiles to better understand how to engineer the underlying network components in order to achieve desirable end-to-end performance. We identify two categories of applications, *latency-sensitive* and *bandwidth-sensitive* applications.[1] In this section we identify and define these application types. In the next section, we discuss issues associated with supporting both application types within a common packet network infrastructure.

Latency-sensitive applications are those whose performance critically depends on the delay of the underlying network. In Chapter 3, we distinguished between latency and delay as a way to reinforce the notion that some delays are immutable, that is, cannot be changed by modifying the underlying network infrastructure. For instance, propagation delay is ultimately bounded by a fundamental law of physics, that is, the finite speed of light.

Latency-sensitive applications do not perform well over a network with significant delays. These applications are often very "chatty" in nature. A single user transaction, for example, a request for a piece of information, may be composed of numerous, elemental exchanges between the requesting client and the responding server. This is illustrated in Figure 4.2. Here a user makes a request for some information from the remote server. The application invokes numerous elemental exchanges (for example, multiple SQL queries to a database server) of information between the client and the server before sending the user the specific information requested. In this instance, the user-perceived delay between the issuance of the request and the receipt of information is proportional to $\sim N \times$ (Network round-trip delay). Therefore, the network round-trip delay is magnified by a factor of N in the eyes of the end user. The factor of N can be as large as several hundred for some applications (see Chapters 9 and 11 for examples of such applications).

This tends not to be a problem if the underlying network is a LAN with round-trip delays on the order of a few milliseconds. However, if the application runs over a WAN where the round-trip delays are typically in the hundreds of milliseconds, the impact on the end user performance would be devastating.

Applications such as these abound in corporate data networks. Often, the application developers do not give WAN performance issues adequate attention until after full-scale deployment. When deployed over the WAN, the

---

1. Many applications have the combined characteristics of latency and bandwidth sensitivity. These are treated in more detail in Chapter 9.

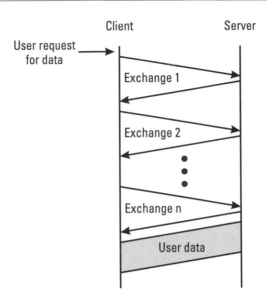

Figure 4.2 An example trace of a latency-sensitive application.

frailties of these latency-sensitive applications are exposed. Several approaches can be taken to alleviate this situation, including migrating to three-tier architectures or relying on the use of thin-client solutions. This topic is discussed in Chapter 9.

In contrast to latency-sensitive applications are bandwidth-sensitive applications. Bandwidth-sensitive applications typically involve the transfer of high volumes of data with relatively few elemental exchanges between the client and the remote server machines. This is illustrated in Figure 4.3. Here the user perception of application delay is dominated by the connection bandwidth, and is proportional to ~(Volume of data) / (Connection throughput). Bandwidth-sensitive applications show little to no dependence on the round-trip delay. These applications are written to require few elemental exchanges, in contrast to latency-sensitive applications.

4.4 Methods to Discriminate Traffic in a Multiprotocol Network

Some latency-sensitive applications tend to exchange a high number of relatively small packet transactions, such as many client/server applications. Other latency-sensitive applications, such as Winframe and telnet, rely on the network to essentially echoplex packets off of remote servers before presenting the

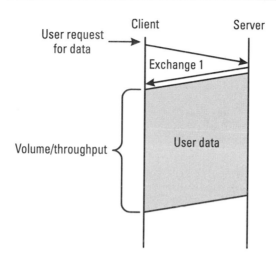

Figure 4.3 An example trace of a bandwidth-sensitive application.

typed characters or the mouse actions at the user's terminal screen. For good user performance, these packet exchanges must not encounter excessive delays.

In contrast, bandwidth-sensitive applications tend to transmit large volumes of traffic within a short period of time. If care is not taken in engineering, these applications may cause temporary network congestion. If the network is simultaneously carrying both types of applications, then the presence of the bandwidth-sensitive applications may adversely degrade the user perceived performance of the latency-sensitive applications. The several techniques available to mitigate these cross-application effects are the topic of this section. In the following chapter, we revisit this discussion within the context of frame relay networking.

Consider the case where we are mixing the two types of applications simultaneously carried over the reference connection in Figure 4.4. We first consider the impact the underlying window size associated with the bandwidth-sensitive application has on the performance of the latency-sensitive application. The result of this inspection will lead us to the conclusion that we should not rely on network buffers to store our end system's data. We follow this analysis with a discussion of the techniques available to network designers and engineers to discriminate between traffic types within the network.

4.4.1 Window Size Tuning

In this section, we analyze the effect the bandwidth-sensitive application window size has on latency-sensitive applications. The result of this analysis is that

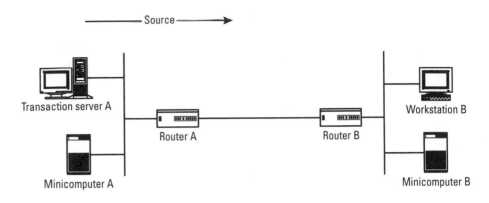

Figure 4.4 Reference connection #1.

windows should be tuned large enough that the file transfers can reasonably fill the facilities along the data path, but not be so large that they cause a large amount of queuing within the data buffers of the intermediate network routers and switches.

Although we find this statement to be true, it is often the case that the engineer has little control over the specifics of the transport window sizes. We discuss this in more detail later.

Now that we know the conclusion, let's see how this conclusion comes about. Consider the performance of reference connection #1 (RC#1, shown in Figure 4.4), when the transaction and the file transfer servers are simultaneously active. What is the impact of the file transfer traffic on the observed performance of the database transaction application? Assume for the moment that the file transfer has been ongoing for some time and look at the timing diagrams in Figure 4.5.

Figure 4.5 shows the timing diagrams for the file transfer traffic in "steady state" for two different window sizes, windows of size three and seven. These timing diagrams are similar to those we have shown earlier with respect to file transfers, only they are somewhat more cluttered. However, this clutter is necessary to observe the effect we want to analyze.

Notice that for the case of the smaller window size (that is, the timing diagram on the left-hand side of the figure), the private line facility is fairly highly utilized yet there is little or no queuing delay experienced by the file transfer data in the WAN router. The upper part of this timing diagram roughly represents a steady-state behavior of the file traffic over RC#1. The window is smoothly rotating and the WAN facility shows a fairly high utilization. Then, about halfway down the timing diagram, the transaction application kicks off and transmits a transaction packet onto the LAN segment. This transaction

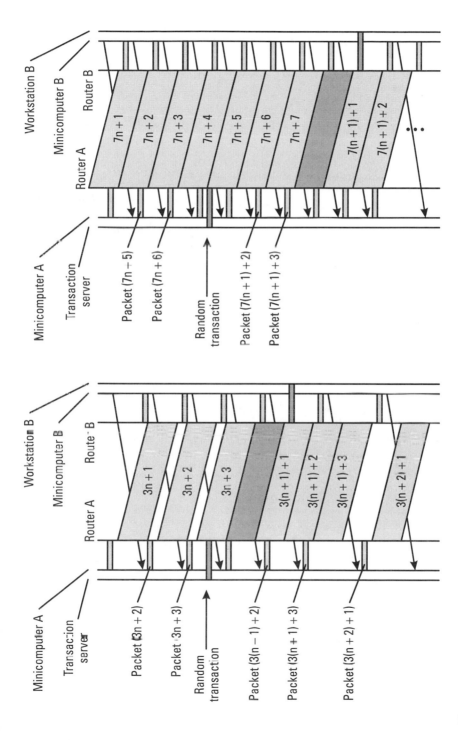

Figure 4.5 Timing diagrams showing cross-application effects.

packet gets interleaved between packets of the ongoing file transfer. Due to the relatively high speed of the LAN segment, gaps exist between the ongoing file transfer packets, and the transaction packet has little effect on the file packets. Once the transaction packet enters the router, the interesting question is how long does the transaction packet have to wait before it gets transmitted onto the relatively slower WAN private line segment. In the case of the timing diagram on the left-hand side of the figure for the case of the smaller file transfer window size, the wait for the transaction packet is small. The window is tuned to keep the WAN facility at a little less than 100% utilization (as is evident by the existence of small idle periods between file transfer packet transmissions). Hence, the transaction packet has a queuing delay of (on average) one-half the insertion delay of a file transfer packet (onto the private line facility).

As an example, for a WAN private line facility running at 56 Kbps and a file transfer packet size of 560 bytes (roughly the size of an off-LAN IP packet carrying FTP data), one-half the insertion delay is roughly 40 msec (560 × 8/56,000 sec divided by 2). This would not be noticeable to the end user of the transaction application.

The timing diagram on the right-hand side of Figure 4.5 shows comparable behavior for the case in which the window size of the file transfer application is larger, that is, seven packets. As we now know, the main effect of increasing the window size in these example reference connections is to place more of a burden on the buffers in the intermediate routers to queue the file traffic packets. Increasing the window beyond a given size does not improve the performance of the file transfer application.

This point is again evident in Figure 4.5. In this example, the router is roughly buffering from two to three file transfer data packets while queuing for transmission onto the relatively slow WAN facility. So now what do we expect to happen when the transaction application packet is transmitted? As before, it is interleaved with the file transfer packets on the LAN segment to the router. It must now wait its turn for transmission onto the WAN facility and hence it is queued. But instead of only having to wait on average for half a packet insertion time, it has to wait for roughly two and a half file transfer packet insertion delays.

Using the same example conditions as in the previous paragraphs, the transaction packet will experience a queuing delay of roughly 200 msec. This is within the perception of an end user and can be exacerbated by larger window sizes or a greater number of simultaneous file transfers or a greater number of transaction packets.

For the conditions assumed in drawing the timing diagram in Figure 4.5, anything larger than roughly a window size of three will cause larger and larger delays for the transaction applications. Further, increasing the window from one to two to three will improve the end-to-end performance of the file transfer traffic. But increasing the window size greater than three will not improve the end-to-end performance of the file application and it will only degrade the performance of the transaction application. From this example, we can then conclude that the "optimal" window size for the file transfer is three. This example demonstrates the necessity of choosing a window size large enough to support sufficient file transfer throughput but not so large as to degrade the performance of other applications sharing the same network resources. However, while it is important to understand the dynamics of this situation, window size tuning may be appropriate only on special occasions. It cannot be recommended as a general method for traffic discrimination because (1) dissimilar TCP stacks deployed within an enterprise will have different configuration capabilities and characteristics, (2) it is hard to implement/configure options on all clients in an enterprise, and (3) it is not always clear what to configure the options to because of the heterogeneous nature of the enterprise network. Therefore, other methods of traffic discrimination are usually relied on.

We are not quite ready to leave the example in Figure 4.5. We now want to ask another question: What can be tuned or implemented in this example to improve the relative performance of the file and transaction traffic when mixed on a common network segment? We have already discussed in some detail one way to improve the relative performance of these interacting applications, that is, optimal tuning of the file transfer window size. A number of other techniques have been discussed in the industry and have been implemented in various data products, in networking protocols, and in data services, including the following:

- Type of service routing;
- Priority queuing;
- Bandwidth or processor sharing (also referred to as weighted fair queuing);
- Adaptive controls; and
- Selective discards.

We now discuss each of these other techniques in turn as they apply to our first example of mixing various types of applications onto a common multiprotocol data network.

4.4.2 Type of Service Routing

Within most data networks, there are often several different routes to a given destination. This is by design, usually to provide a high level of reliability within the overall network. These different paths to the same destination may ride over similar types of facilities, but these facilities may have different engineering rules applied to their capacity management or have different bandwidths associated with each, or they may ride over different types of facilities. For example, one route may travel over satellite facilities while another route over terrestrial facilities.

In any event, these different routes may have different performance characteristics associated with them. The path utilizing satellite facilities may be characterized as a high-throughput yet high-delay route. The path utilizing terrestrial facilities may be characterized as a low-throughput, low-delay route. In this case, our transaction traffic would be perceived to perform better by the end user if it were carried over the path utilizing the terrestrial facilities. The file transfer application would be perceived to perform better by the end user if it were carried (or routed) over the path utilizing the satellite facilities. This is the essence of type of service (TOS) routing.

Because different applications impose different performance requirements on the underlying network, it seems reasonable to route the various applications over the "optimal path" within the network for the particular application in question. In practice, this capability is described by first defining the set of performance metrics necessary to specify all reasonable application requirements. The metrics typically include delay, delay variation, throughput, reliability, and cost. The requested values for each of the metrics and allowable groupings are defined. Then their encoding is specified. Often only a rather crude set of values is specified: high, medium, or low delay; high, medium, and low throughput; and so on. Typical groupings may be high throughput with high-delay service or low throughput with low-delay service. These groupings are termed the type of service.

To implement TOS routing, several capabilities are necessary. For virtual circuit networks the requested TOS values for the connection are indicated in the call setup message. In datagram networks, the requested TOS values must be carried in every datagram header. For TOS routing, separate routing tables must be maintained for each possible TOS supported within the given network. Also, the network should have developed separate capacity management capabilities and guidelines for each TOS supported in the given network. When strict TOS routing is desired, some negotiation capabilities should be supported. In practice, all of these are extremely complex to develop, implement, and maintain.

The most common implementations of TOS routing support throughput metrics (often referred to as *throughput classes*). These implementations are found in X.25 (referred to as throughput class), frame relay (referred to as *committed information rate*), and ATM (referred to as *sustainable cell rate*) networks supporting SVC signaling capabilities or PVC manual provisioning capabilities. All of these technologies support the routing of a VC, which can sustain the throughput defined by these terms, for example, throughput class, CIR, and SCR.

Finally, to be truly useful in our examples, the end applications need some capability of communicating the necessary TOS to the routing entity in the network. This is rarely available in layered networks due to the lack of consistent standards in this area.

So how would TOS routing improve the relative performance of our file transfer and transaction-based application traffic in our reference network? Figure 4.6 shows an expansion of RC#1 with multiple routes providing different TOS paths. The satellite path provides a high-throughput and high-delay TOS, and the terrestrial path provides a low-throughput and low-delay TOS. One strategy would be to route the file transfer traffic over the satellite path and the transaction traffic over the terrestrial path; this essentially isolates the cross-application effects. Specifically, the transaction traffic does not get caught in a large queue behind the file transfer packets, while queuing for insertion onto the same communications facility.

4.4.3 Priority Queuing

In the examples given in the last section, the file transfer and transaction application data essentially collide at the gateway buffer onto the WAN facility.

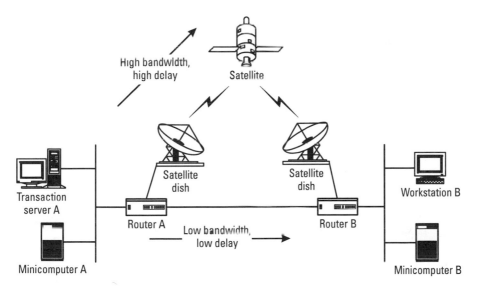

Figure 4.6 Reference connection showing multiple paths with different TOS.

Window size tuning minimizes to some extent the impact of the collision. TOS routing avoids the collision altogether by routing the different applications' data onto different buffers and WAN facilities out of the LAN gateway. As discussed, the TOS routing capability requires the routers to identify the different TOS requests and routes these requests appropriately. Another mechanism, which has the potential to greatly minimize the collision of the various applications traffic, is *priority queuing*.

Like TOS routing, suppose that a mechanism exists for communicating to the gateway the type of application data being carried within a given data packet. It would then be possible for the router to favor one application's data over the other by moving it "to the head of the line" so to speak. If the transaction application data were placed at the front of the queue holding the file transfer traffic, then the transaction data would only have to wait for the insertion of the current file transfer packet onto the facility to complete before it was transmitted onto the same WAN facility.

In this case (that of a single transaction packet), the maximum queuing delay would be a single file transfer packet insertion delay. And the average queuing delay would be one-half the file transfer packet insertion delay.[2] This mechanism is termed *priority queuing*.

As an example, for a WAN facility running at 56 Kbps and a file transfer packet size of 560 bytes, one-half the insertion delay of the file transfer packet is roughly 40 msec. This would not be noticeable to the end user of the transaction application.

Strictly speaking, it is not necessary to communicate the nature or type of application data being carried within the data packet. It is only necessary to communicate the *priority level* of the data. One could easily imagine a need for several priority levels. One example would be to have high-, medium-, and low-delay TOS priority levels. Packet level headers would then only be required to indicate the priority level of the packet.

A typical implementation would allocate separate data buffers for each interface on a gateway, router, or switch; there would be one queue per priority level. When the router has to insert the next data packet onto the facility, it takes the first packet from the highest priority queue containing data. This type

---

2. If the transaction packet arrives to the router buffer just prior to the completion of the current file transfer packet insertion onto the facility, its queuing time will be essentially zero. If it arrives just following the start of the current file transfer insertion onto the facility, its queuing time will be essentially a full file transfer packet insertion time. Then, assuming that the transaction packet arrives independently of the state of the file transfer packet insertion, its average queuing time will be one-half the file transfer packet insertion time.

of priority queuing mechanism is referred to as *nonpreemptive priority queuing.* This is shown in Figure 4.7.

Conversely, a *preemptive priority queuing* implementation allows for the server to halt (or suspend) the insertion of the lower priority packet on the arrival of a higher priority packet. Once the higher priority packet (and all other subsequent higher priority packets) is serviced, then the service of the preempted packet can start over (or resume).

In practice, the majority of the priority queuing implementations are of the nonpreemptive priority case. However, there are cases, especially on low-speed serial links where routers will support a form of preemptive priority, based on a packet fragmentation scheme. This is necessary when the high-priority traffic has very stringent delays and/or delay variation requirements and is being run over relatively slow links. One such example would be when attempting to carry packet voice over an enterprise data network.

As for the case of implementing a TOS routing capability, priority queuing schemes require that information on the nature (e.g., the priority level) of the application be passed down to the entity building the packet level data header. It is the packet header that must carry the indication of the priority level of the application. In a single, end-to-end, networking protocol

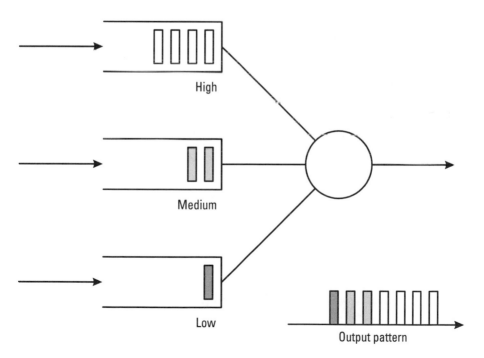

Figure 4.7 A three-level priority queuing system.

environment this could be indicated through an application programming interface (API) and then down to the appropriate protocol level.

However, in most of the implementations of priority queuing in multi-protocol networks, the priority level of the data packets is usually inferred based on some higher level indication of the application. This is usually satisfactory, but it sometimes may not be granular enough for the needs of the application and end users.

The priority queuing schemes generally base the decision of the priority level of the data packets on one of the indicators discussed next (the specific implementation depends on the particular type and manufacturer of the networking equipment).

Priority Level ID

Some packet level protocol headers carry explicit priority level bits. For example, the IPv4 packet headers carry a TOS field indicating special packet handling. The first three bits were designated as precedence bits and the next five indicated the type of service requested by the hosts (low delay, high throughput, and so on).

Until recently, this field has usually been ignored, however its use is beginning to gain popularity within the differential services architecture being proposed within the IETF. In this architecture, the TOS field is referred to as the DS octet and the first five bits are used to indicate priority packet handling at the network edge, with the sixth bit indicating whether this packet is conforming to a traffic contract or not. The rest of the bits are not used at this time. This, however, still requires that an entity (in this case the IP packet header processing entity) infer or be told the priority level to set. This can be done by other methods discussed later or could have been indicated through an API.

This method, where the priority level is directly indicated in the packet header, offers simplicity of protocol processing for the intermediate routers along the data path. Some of the schemes identified later require that the queuing entity search far into the higher level protocol headers to infer a priority level, and this can have a detrimental impact on the protocol processing entity (increase the processing delays at this location in the network).

Protocol ID

Packet level headers contain a field to indicate the higher level protocol carried within it. For example, in an IP header this is the protocol ID indicating TCP, UDP, ICMP, and so on. This works for our purpose in some situations, but it relies on a rather low-level protocol type in order to base priority queuing decisions for particular applications. For example, if it is desirable to give telnet terminal traffic a higher priority than file transfer protocol (FTP) traffic, then this scheme will not work because both of these higher level protocols are carried over a TCP transport-level connection. Hence, both telnet and FTP would receive the same priority level treatment under this method.

High-Level Protocol Identifier

Switches and routers can be programmed to look deeper into the protocol headers within the packet. One popular approach is to configure a router view the TCP or UDP port numbers, which identify the protocols riding over these transport-level protocols. Recently, this has been referred to as level 4 switching, to indicate the protocol level at which the routers are required to process in order to make switching or filtering decisions.

TCP port numbers indicate whether the data are from a telnet application or an FTP application. The router can then rely on the port numbers as an indication of the type of priority to be given to the data packet. This is valuable in that the higher level protocol indications are "closer" in some sense to the application and thus they give a better capability to discriminate file applications from transaction applications.

The price for this type of approach is that it is relatively more difficult for most high-speed switches or routers to search this deep into the protocol headers. This consumes more processing capabilities in the router and will, in turn, negatively impact the overall packet throughput on the router.

Protocol Encapsulation Type

Devices that operate at the link level, for example, FRADs or ATM switches, could conceivably rely on the protocol encapsulation type indication within the link level encapsulation protocols. An example is the encapsulation for frame relay networks defined in RFC 1490 (and discussed in Chapter 2). This would base the prioritization on the NLPID. Basing priority levels on the protocol encapsulation type would allow switches to prioritize based on the type of protocol. This is useful when, for example, you want to give all of your Novell NetWare Inter-Packet eXchange (IPX) traffic priority over TCP/IP applications (primarily carrying e-mail). The downside of this method is that you cannot prioritize, for example, telnet traffic over FTP transfers. Although this technique is occasionally discussed, it is not implemented in practice. One strong criticism of this approach is that it would violate the requirement that virtual circuits do not reorder the packets on the individual circuits.

Incoming Port or Interface

Suppose a router has two LAN interfaces, such as an Ethernet and a token ring. The token ring LAN carries primarily SNA traffic, and the Ethernet carries primarily native, non-real-time TCP/IP applications such as e-mail. Then, it would be useful for the router to give all packets incoming from the token ring LAN a high priority level and all packets incoming from the Ethernet LAN a low priority level.

However, like the case of basing priority levels on encapsulation types, this is a relatively nondiscriminating method for assigning priority levels. Also, unless there is some method of indicating the priority level on the packet, this method provides only local priority service; this does not extend across to the rest of the network.

Source/Destination Address Pair

Often traffic from a given source or traffic between a given source/destination pair is to be given priority handling. This can be accomplished in the routers by examining, for example, the IP addresses of the source and the destination indicated on the IP packet header. This is a fairly flexible form of identifying high-priority traffic in router networks.

Circuit ID

Virtual circuit switches base switching decisions on a circuit ID. They could just as well base a priority-level decision on a circuit ID. If the access device (for example, a FRAD) applied one of the other methods for assigning priority levels to packets and ran multiple virtual circuits to all other end points, then it could place the higher priority packets on one virtual circuit and the lower priority packets on another virtual circuit. This would extend the priority treatment of the data packet beyond the access facility and across the entire virtual circuit network.

Packet Length

Some systems have been known to assign priority levels based on packet size. The theory here is that terminal traffic is comprised of generally small packets while file transfer traffic is comprised of larger packets. This is not always true. Further, extreme care must be taken to ensure that the wrong packets are not reordered along the network connection. This could cause high levels of end-to-end retransmissions and greatly degrade the perceived performance of the network. In general, avoid these techniques for "guessing" at the priority level of the traffic.

The methods discussed for determining priority can be categorized by the type of discrimination, that is, explicit versus implicit, and the level in the protocol stack at which the discrimination occurs, for example, layer 2, layer 3, and so on. We summarize these in the Table 4.1.

As a rule of thumb, priority queuing methods are useful when accessing relatively low-speed facilities, for example, analog lines or 56-Kbps to fractional-T1 digital lines. At high speeds the relative benefits of priority

Table 4.1
Categories of the Priority Discrimination Schemes

Priority Determination	Discrimination Type	Discrimination Layer	Comments
Packet length	Implicit	Not applicable	Rarely implemented (thank goodness)
Port or interface	Implicit	Layer 1	Can prioritize one token ring versus Ethernet, or one department versus another
Source/destination address pair	Explicit	Layer 1	Prioritize all traffic from a given source or between a source/destination pair
Encapsulation indicator	Explicit	Layer 2	Can prioritize, for example, SNA over IP, however causes packet reordering on a VC, not implemented (thank goodness)
Protocol ID	Explicit	Layer 3	Can prioritize, for example, IPX over IP
High-level protocol ID	Explicit	Layer 4 (plus)	Can prioritize, for example, telnet over FTP

queuing are diminished (and sometimes are detrimental if the processing overhead incurred by enabling priority queuing greatly reduces the overall throughput of the networking device). Also, attempts should be made to ensure that the total amount of high-priority traffic is small relative to the amount of low-priority traffic. A good rule is to keep the high-priority traffic to less than 10% of the low-priority traffic. After all, if all the traffic were treated as high priority then the system would degenerate to a simple, single queue system.

Low-priority queue "starvation" can also occur if the higher priority traffic load becomes too great. This may cause retransmit timers associated with low-priority traffic to expire and greatly degrade the performance of these applications.

4.4.4 Processor Sharing

Suppose that routers were to put packets into different buckets (depending on some criteria), and then transmit one packet at a time from each bucket onto the transmission facility. If a given bucket were empty, then it would simply jump to the next bucket containing a data packet. We refer to this type of packet transmission system as a *processor sharing* system. (A related system is

referred to as a *weighted fair queuing* algorithm; see the discussion in the following paragraphs.)

A processor sharing system is fundamentally different from a priority
queuing system in that a processor sharing system treats all types of packets (or
traffic) equally in some sense. A priority queuing system, however, explicitly
favors one type of packet (or traffic) over other, lower priority traffic type.[3] This
is shown in Figure 4.8. Notice the predicted output pattern for this processor
sharing system and compare it to the predicted output pattern in Figure 4.7 for
the priority queuing system.

Here, the processor sharing server effectively interleaves the packets from
the various packet buffers. In this sense it behaves differently from a priority
queuing system. The processor sharing system treats all queues equally by
transmitting a packet from each occupied queue in a round-robin fashion. The

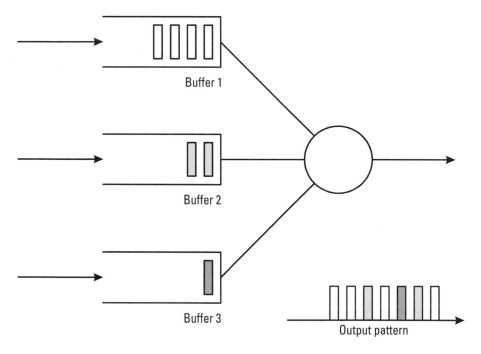

Figure 4.8 A three-bucket, packet-level processor sharing system.

3.   Other types of processor sharing systems exist where the data from each bucket are essen-
     tially transmitted a byte at a time. These are referred to as byte-interleaving, processor shar-
     ing schemes, as opposed to the packet-interleaving, processor sharing schemes discussed
     earlier.

priority queuing system explicitly favors those packets in the queue designated high priority.

Some processor sharing schemes allow for a weight to be assigned to specific flows. This gives a larger proportion of the bandwidth to some flows over others. For example, flow A can be assigned twice the bandwidth as the other flows by serving two packets from flow A's queue versus one packet for each visit to the other queues. Various schemes like these exist, for example, proportional round-robin, weighted processor sharing, or weighted fair queuing algorithms, which are variations on the processor sharing algorithm.

Assume that a single transaction packet arrives at a processor sharing system and is placed into an empty bucket.[4] In this case (that of a single transaction packet), the maximum queuing delay would be a file transfer packet insertion delay of $N-1$ packets, where N is the number of active, separate queues in the processor system.

In contrast, the maximal delay in the priority queuing system would be a single file transfer packet insertion delay. The factor of $N-1$ arises because the processor sharing system treats all queues equally. This maximal delay for active queues would occur if all the other queues were occupied and the transaction packet is queued just following its "turn" in the processor cycle. If the transaction packet arrived to its queue just prior to its turn, then its delay would be roughly zero. Hence, the average delay would be one-half the maximum delay: $(N-1) \times$ (Insertion delay) / 2.

As an example, for a WAN facility running at 56 Kbps and a single active file transfer with a packet size of 560 bytes, the average delay, under processor sharing, for a packet belonging to an interactive traffic stream that is queued separately is roughly 40 msec (that is, half of 80 msec, the insertion delay of a 560-byte packet on a 56-Kbps link). This would not be noticeable to the end user of the transaction application. For two simultaneously active file transfers this delay would double; for three, it would triple; and so on. On the other hand, if interactive traffic is explicitly given highest priority, then the maximum queuing delay (due to packets from other traffic streams) for an interactive packet will be half the full insertion delay, *no matter how many queues are maintained and active.*

Although processor sharing is not as effective at reducing the transaction delays as a priority queuing scheme, it does have several advantages. Like priority queuing schemes, a processor sharing scheme can be implemented based on very low-level indications, for example, virtual circuit identifiers, packet level

---

4.   In the case where the transaction arrives at a nonempty queue, the delay bounds discussed in this paragraph are increased by the number of transaction packets in queue ahead of the transaction in question times $(N-1)$ packet insertion delays.

flow identifiers, or packet destination or source/destination pairs. Processor sharing is effective at sharing bandwidth across all active circuits or flows, and avoids "starvation" effects, which can occur in priority queuing systems. Also, by assigning unequal weights to the different buffers, a form of favoritism can be assigned to "higher priority" traffic.

4.4.5 Adaptive Controls

Adaptive controls attempt to minimize the length of the queues at the resource in contention by monitoring, either implicitly or explicitly, the state of the buffers. In our example, where a large file transfer and a small transaction are contending for a common communication facility, a large queue develops when the file source is sending too much data into the network at any given time. These data end up sitting in the buffers for access to the common network facility, which causes an excessive delay for a small, delay-sensitive transaction. By minimizing the length of this queue, an adaptive control scheme can improve the transaction delays.

There are basically two mechanisms that adaptive controls can apply to manage queue buildup: dynamically adjusting the rate at which the source transmits into the network (commonly referred to as *traffic shaping*) or dynamically adjusting the transmission window size of the source. When the queue is determined to be building up too large, then the rate or the window can be decreased. In the event that the queue is determined to be extremely small, then the rate or the window of the transmitter can be increased. By constantly monitoring and adjusting the rate or window size, the system can achieve a reasonable trade-off between high throughputs and low delays.

The adaptive source requires some form of feedback on the state of the network path in order to make the appropriate changes to the transmitter. The form of the feedback can vary, being either explicit or implicit. This is shown in Figure 4.9.

The lower path shows an explicit congestion indication being sent to the source. Receipt of this message would then allow the transmitter to either decrease its transmission rate (either by spacing out its packets or decreasing its transport window) or increase it depending on the nature of the message.

Several methods exist to implement this explicit notification. One method would have the congested resource generate a signaling message and transmit it to the source of the traffic. Another method would have the congested resource setting a bit (or bits) in the protocol header of the data packets as they traverse the congested resource. The receiver would then pass this back to the transmitter through a similar indication on the packet headers.

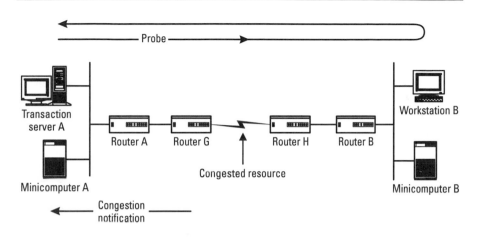

Figure 4.9 Implicit and explicit feedback on network congestion.

The upper loop refers to an implicit method of sensing network congestion. One example of this method is to base adaptive decisions on a round-trip time measurement. Here the "probe" could be a single data packet in the forward direction and the acknowledgment in the reverse direction. Then, the onset of congestion in the data path would cause delays in queuing and increase the measured round-trip time. The transmitter could then slow down its transmission rate or decrease the size of its transport window. This would help to alleviate the congested resource. Another method of implicitly sensing network problems is to rely on packet loss to indicate congestion.

An example is the TCP slow start adaptive algorithm (see the discussion in [1]). Here, the transmitter would sense a lost packet through the windowing acknowledgments (or lack thereof) and would infer that the packet was dropped due to congestion. The transmitter would then decrease its transmission rate (through one of several methods already discussed). This is not a direct measure of network congestion because packet loss can occur in networks due to other reasons, for example, bit errors on transmission facilities or misrouted packets.

By adapting the transmission rate based on some form of feedback from the network on its congestion state, transmitters attempt to keep packet buffers from growing too large. As we have discussed, this will help in reducing the network delays for all data including transaction-oriented application data. This will help to reduce the negative, cross impact of file transfer traffic on transaction data. We have already discussed similar algorithms in Chapter 2 on frame relay congestion control methods.

4.4.6   Selective Discards

Selective discard mechanisms discriminate among various sources of traffic during periods of extreme resource congestion by discarding packets from select sources while not discarding from other sources. It is argued that this is a method of reserving critical network resources during congestion periods for time-sensitive or high-priority applications.

This reservation mechanism implicitly reserves resources for critical applications by explicitly denying access to noncritical applications. This scheme relies on the same methods to identify those applications deemed noncritical as the other schemes discussed in the previous sections. Often noncritical applications are considered to be those utilizing certain TCP port numbers, for example, FTP is deemed noncritical, while telnet is deemed critical.

Some implementations of selective discard rely on the specific switching equipment to look deep into the packet headers to determine the nature of the applications and to determine which packets to discard during resource overload. Other schemes have end systems tagging the packets as discard eligible through some network protocol-specific indication. Others tag packets based on a bandwidth contract as measured at the access control point into a network. One such implementation, known as *discard eligibility*, is found within the frame relay network standards as discussed in Chapter 2.

The selective discard strategy is fundamentally different than the other methods discussed within this chapter. The trigger for the selective discard action to begin is usually a buffer congestion threshold. These thresholds are usually set to a significant fraction of the total buffer space allocated to the resource. Otherwise, the majority of this buffer space is wasted. Once the threshold is reached, the selective discard mechanism is initiated until the buffer utilization drops below a lower level congestion threshold (in order to prevent thrashing). What is different with this scheme is that it is responding to large buffer utilization, instead of responding to delay and bandwidth considerations. Long before the time the buffers have reached their threshold, the delays for the time-sensitive applications have exceeded reasonable limits. Therefore, this should not be considered a useful mechanism to ensure good transaction application performance in multiprotocol networks.

Often heard in discussions of packet discard techniques is the random early detection (RED) strategy implemented in many IP-based router networks. This is a technique in which routers attempt to mitigate the onset of router congestion by initiating a limited level of random packet discards based on buffer threshold settings. By performing a low level of packet discards, the router is relying on throttling back TCP slow start implementations and damping the load somewhat on the router. This has been shown to be a very effective technique in maintaining high network utilization while mitigating the effects of congestion. From an Internet provider perspective, RED provides significant benefits to their overall customer base. However, from a critical end-applications perspective (which is the focus of this section and, in fact, this book), RED does not significantly benefit particular applications by providing significant differentiated service. These are fundamentally different goals.

4.5 Data Collection

We have discussed application characteristics, load engineering and capacity management, and methods to give preferential treatment of one application type over another when designing and implementing data networks. However, to deliver on a capacity management plan and to maintain the desired performance engineering, one needs to collect data on various aspects of the network. Data collection techniques are discussed in this section. Here we identify various types of data collection tools.

To help accomplish the tasks that arise only occasionally, specialized monitoring equipment, such as LAN/WAN analyzers or protocol traces, can be utilized. However, for ongoing tasks, such as capacity management, one needs to rely primarily on the existing measurement capabilities of the networking equipment and network management systems employed. We briefly discuss three types of collection tools: LAN/WAN analyzers, SNMP management tools, and RMON2 probes. We end this section with a brief classification of commercially available tools that are useful in this context.

4.5.1 LAN/WAN Analyzers

LAN/WAN analyzers are devices that can tap into the LAN/WAN and capture the data being transmitted over the LAN/WAN segment. These devices store some aspects of the data onto disk to be retrieved at a later date. Sniffers capture all the data, but usually one applies filters to look at some particular aspect of the information of use in characterizing the traffic load and in developing an understanding of the protocol models required for trouble shooting. LAN/WAN analyzers typically store the link level (and can be configured to capture higher level protocol information such as the IP or IPX) packet formats, the byte count of the data fields, whether the packets were received in error, and time stamps showing the time at which the packet data were captured.

From this, the analyst can compute the interesting statistics associated with the data traffic, for example, average and standard deviation of the packet sizes, the traffic loads, and even the point-to-point requirements by mapping end-system addresses in the packet headers to the destination locations.

Another important use of LAN/WAN analyzers is traffic characterization of specific applications, particularly client/server applications. This is of great utility as discussed in Section 4.3.

This equipment tends to be rather specialized and therefore relatively expensive. However, it is not necessary to widely deploy this equipment, and it is mostly used on an individual case basis for specific reasons. These are typically used to help resolve problems, to help build an understanding of traffic load over a given, finite period of time, or to characterize specific applications prior to generally deploying the application on the enterprise network.

4.5.2 Network Management Systems and RMON Probes

Here we focus on those aspects of management systems that capture and store traffic data in networks on an ongoing basis. This is a distributed functionality, in that individual network components are required to collect local information, then periodically transmit this information to a centralized network management system that can process this information into quantities of utility to the network analysts and administrators.

Local information typically collected on the individual network components includes the following:

- *Serial trunk interfaces:* the number of bytes and frames transmitted and received, the number of errored frames transmitted and received, the number of frames discarded due to buffer overflows, and so on;

- *Multiplexed interfaces* (such as frame relay or ATM): the number of bytes and frames transmitted and received, the number of errored frames transmitted and received, the number of frames discarded due to buffer overflow on an individual virtual connection basis, the number of established switched connections on an interface basis, and so on;

- *Switch processors:* the number of frames or cells forwarded over the switch fabric, the utilization of call processors, the size of the run queue in the processor, and so on;

- *Routers:* the number of packets forwarded as a function of the protocol type, for example, IP, IPX, AppleTalk, the utilization of the routing processor, the size of the routing tables, and so on; and

- *Network servers:* the utilization of the processor and disk, the number of simultaneous sessions established, and so on.

These parameters are stored locally and are periodically (in periods ranging from 5 min to 1 h) transmitted to a centralized management system when prompted by the management station. The management system will process and summarize the information and archive the results for trend analyses by the network analysts. Useful trend analyses might include the weekly variations in the utilization of a trunk interface into the WAN or the monthly increase on the peak utilization or the variation in the utilization of a network server, as discussed in Section 4.2. Useful troubleshooting information includes the number of frames or packets received in error or the number of frames discarded due to buffer overflows on a particular hour of a given day.

The trend today is to try to standardize the type of information to be collected and stored by the network management systems. The type of information to be collected is found in an SNMP management information base (MIB) for the particular device or technology in question. Many of the MIB definitions are standard, but many vendors have developed proprietary extensions that are specific to their particular equipment.

The types of information identified in the preceding list are local to particular devices comprising the network. As mentioned in the preceding paragraph, this type of information is captured within the MIB for the device. A standard MIB common across all SNMP manageable devices is defined and referred to as MIB II [2]. Because this type of information is local to a specific device, it is up to a central management system to correlate these individual pictures into a view across a specific subnet or LAN segment.

To eliminate this burden on a central management system, and to provide a richer set of subnet-based statistics, the RMON MIB was developed. The IETF has developed a set of recommendations for a standard remote monitoring capability. Hardware devices that implement these recommendations for a standard set of subnetwork monitors are referred to as *RMON probes*. RMON2 extends the monitoring capabilities of the RMON above the MAC layer.

We prefer to think of these additional RMON2 capabilities in terms of three different levels of traffic analysis. Level 1 of the traffic analysis is overall byte count and average packet sizes. Level 2 of the traffic analysis is a breakdown via layer 3 protocols, for example, IP versus IPX, DECnet, AppleTalk. Level 3 of the traffic analysis is a detailed analysis of IP traffic into TCP and UDP components along with further breakdown of TCP traffic according to port numbers and so on. This information, collected from networks and stored within the probes can be communicated to a central management system through the SNMP [3].

For these reasons, RMON2 probes offer a rich set of collection capabilities, which are extremely useful in capacity engineering. For more information on RMON2 probes see [2].

It is the job of the analyst to develop a capacity management strategy that relies primarily on the information types and statistics collected as identified in the MIBs available in the components deployed within their data network. As such, the MIBs must necessarily contain the required local information that the analyst uses to develop an understanding of the important end-to-end performance requirements of the applications riding over the multiprotocol network.

The capabilities provided by these activities in the IETF and afforded to a network management tool by RMON2 (or similar function) probes have led to the development of a host of monitoring, trending, and performance management tools. These tools can help the analyst by delivering much of the capabilities required in their capacity management needs. These are discussed next.

For an excellent discussion of IP management standards and MIBs, refer to [3], and [2] for RMON.

4.5.3 A Taxonomy of Commercially Available Tools for Performance Engineering Data Networks

As mentioned in this section on data collection, several commercially available tools cover all aspects of data networking: configuration management, equipment inventory, capacity planning, application analysis, performance modeling, troubleshooting, and traffic generation.

Although all of these tools are important in their own right, we focus below on a set of tools that we think is indispensable in performance engineering data networks, especially WANs. The objective is to provide the reader with a high-level view of the tool's capabilities and strengths. In this process, it is likely that some important aspects of the tools may not be mentioned. For more details, the reader is referred to vendors' Web sites.

We classify tools for performance engineering as follows:

- Sniffers;
- Capacity management tools;
- Application analysis tools;
- Predictive modeling tools.

We discuss each type separately next and give some examples of each.

Sniffers
Sniffers are passive devices that can be placed on a LAN or WAN segment to collect traffic statistics and packet-level information on the segment. They can be set up with various filters to selectively collect data. For instance, one

can isolate specific conversations between a client and a server by filtering on their MAC or IP addresses.

Sniffers are indispensable for performance engineering for two reasons: troubleshooting and application characterization. Many examples in the later chapters of this book contain descriptions of how sniffer protocol traces were used to troubleshoot performance problems and to characterize application behavior over WANs. Some tools even have the capability of directly reading sniffer protocol traces for performance modeling purposes.

The flagship tools in this category are the Network Associates Sniffer® and Wandel & Goltermann's Domino®.

Capacity Management Tools

Tools in this category are primarily software based. They rely on network components (routers, bridges, hubs, CSU/DSUs, servers, and so on) to collect measurement data in accordance with the IETF's SNMP/RMON/RMON2 standards and MIBs. Centralized network management stations poll the network components for information on the MIB variables and report on all aspects of network capacity, network congestion, application usage, and so on.

Examples of such tools are Concord Communications' Network/Router Health/Traffic Accountant®, INS's E-Pro®, Cabletron's Spectrum®, and Net-Scout Systems RMON2 probes and NetScout Manager.

Some tools from vendors such as Visual Networks and Paradyne can be used for capacity planning and troubleshooting. However, they are hardware and software based. The hardware is in the form of a "smart" DSU interfacing with the WAN.

Application Analysis Tools

These tools use self-generated measurements and probes to estimate application-level performance. They rely on various application profiles that capture the essence of the application-level traffic flows to measure expected application performance.

These tools require a centralized management server as well as the deployment of distributed software clients at the remote locations from which performance measurements are desired. Examples of this type of tool include Ganymede's Pegasus® tool.

Predictive Modeling Tools

These tools are either simulation tools or analytical modeling tools. Simulation tools build a logical model of the network and the applications in question, and run a simulation collecting the performance metrics of interest. Examples of these tools include MIL 3's OPNET® and IT Decision Guru®. Analytical modeling builds reference connections in the network and, along with an application profile input, provides answers to "what if" scenarios such as increased bandwidth and protocol tuning. A prime example of analytical tools is the Application Expert from Optimal® Networks.

Other predictive tools have or build application profiles and deploy these profiles on remote software clients. These profiles are then run from the remote client over the existing network and performance metrics of interest are collected. Ganymede's Chariot® tool is a good example of such a tool.

4.6   An Example: Deploying New Applications

We end this chapter by providing a high-level discussion of the necessary steps that one should take before deploying a new critical application over an existing WAN. This discussion ties together all of the techniques presented within this chapter.

The analysis and design of data networks and their performance evaluations are only as good as the inputs to the design and the ongoing monitoring of the implemented network. Similarly, when developing the engineering rules for the data network to support the rollout of a new application, the effectiveness of the engineering rules is only as good as the input to the engineering models.

We recommend the following four steps in the context of deploying new applications over a WAN.

Application Characterization

Characterize the new application to be carried over the network and its performance requirements, for example, delay and throughputs. When entering into the process of redesigning a network to support a new application, it is important to have first built up an understanding of the nature of the specific application in question.

We are strong proponents of a top-down approach, as far as possible, to design, engineer, and maintain data networks. This begins with time spent in defining the performance requirements of this application. Networks are built to support a variety of applications and application types.

Applications range from highly transactional, or latency-sensitive, applications where delay is critical, to bulk data transfer, or bandwidth-sensitive, applications where throughput is critical, or hybrid applications, which share the characteristics of sensitivity to latency and bandwidth. Various tools are available to the engineer to help in characterizing the behavior of the application. These include sniffers, product literature, and vendor support.

Ideally, at the conclusion of this first phase, the network engineer should be able to draw a timing diagram for the important network applications. This level of detail will aid in (1) the redesign phase in determining if the design will meet the application-level performance requirements and (2) the capacity management phase in developing a component-level performance allocation. However, it is not always possible to develop this level of detail, depending on the specific application in question.

Traffic Matrix

Identify the traffic loads and develop a point-to-point traffic matrix related to the new application. Once the application characterization is complete, the next stage is the characterization of the traffic, including its arrival patterns, its point-to-point flows, and its offered load.

This information should be presented in terms of the point-to-point traffic loads between the data sources and the data sinks. This is best built up from the volume of traffic a given application transaction or file transfer produces and the frequency of the transmissions of these transactions between all sources and destinations during the busiest period of the day. Often these traffic volumes must be estimated based on issues such as (1) assumptions regarding number of users at a given location, (2) assumptions regarding usage concurrence, and (3) best case and worst case bounds.

The sum total of this information is the traffic matrix. The traffic matrix is used to determine the additional network connectivity and bandwidth requirements during the network redesign phase.

Network Redesign

Redesign the network and identify the affected network components to support the new application characteristics and traffic. The first step in the network redesign phase is to determine the specific network components that will be affected by the deployment of the new application. At a minimum, the engineer must identify the components whose capacity must be increased to support the bandwidth and connectivity requirements specified in the new traffic matrix.

At this point, the engineer should have a network layout, including the expected loads on each of the network components, for example, links and PVCs. The engineer can now develop the expected delays and throughputs for the dominant applications to be carried over the design. This is accomplished by developing the appropriate timing diagrams for these applications over typical reference connections within the network layout. From the timing diagrams, one can determine the realizable end-to-end delays and throughputs. If these realizable delay and throughput estimates meet the initial set of performance requirements for the applications, then the initial design is finalized. If the realizable performance estimates do not meet the initial requirements, then the design must be modified until the performance objectives are met.

If the new applications are characterized as latency sensitive, and must compete for network resources with various bandwidth-sensitive applications, then the engineer should consider the various methods of traffic discrimination, which were discussed in Section 4.4.

Capacity Management

Modify the existing capacity management tactics to maintain acceptable network performance for the existing and new applications. Redesigning a network in this a priori fashion is useful but is never sufficient. New users and servers are constantly added to existing networks. This has the effect of changing the nature of the traffic flows over the network and hence changing the network performance characteristics. Also often it is impossible to fully characterize the new application being deployed to the level of detail suggested earlier. Therefore the network analyst/network administrator must design a strategy to continuously monitor the network and plan its growth in order to maintain the desired level of performance as changes to traffic occur or as an understanding of the nature of the traffic improves.

Ideally, the capacity management strategy is devised by reverse engineering the desired end-to-end delays or throughputs into the individual network component metrics that can be monitored effectively. This is accomplished in two parts.

First, the end-to-end delays must be mapped into the individual component delays. We refer to this as *component allocation*. Next, the component delays must be mapped to component metrics, which are measurable. It is generally not possible to directly measure component delays. Therefore, these must be mapped to directly measurable quantities, such as the component utilization. This relies on load engineering and associated load engineering models.

Consider these four steps as a recipe for performance engineering when deploying a new application. Each step is as important as the next and none should be skipped when designing an enterprise network from scratch or when deploying a new application onto an existing network. However, it must be remembered that these guidelines cannot be fully realized for all applications being deployed. Consider the rollout of a company-wide Microsoft Exchange® e-mail platform, or an IBM/Lotus Notes® deployment, or an Intranet Web-based application. For these applications, it will be very hard, if not impossible, to obtain traffic matrices, usage patterns, and performance requirements prior to their deployment. For other applications, for example, the deployment of client/server applications, this approach may be more fully realized.

4.7 Summary

In this chapter, we covered several topics regarding general performance engineering techniques. The first topic was load engineering where we discussed the definition of directly measurable component thresholds and their relationship to maintaining acceptable network performance. We next discussed latency- and bandwidth-sensitive applications and the problems of simultaneously supporting both on a common enterprise network. This led us to list the various traffic discrimination techniques available to minimize negative cross-application effects. Following this we presented a section on tools for maintaining data networks. This touched on tools for monitoring, capacity management, network planning, and troubleshooting. We ended this section with a brief discussion of a method to follow for the deployment of a new application over an existing data network.

References

[1] Stevens, W. R., *TCP/IP Illustrated, Volume 1: The Protocols*, Reading, MA: Addison-Wesley, 1994.

[2] Stallings, W., *SNMP, SNMPv2 and RMON*, Reading, MA: Addison-Wesley, 1996.

[3] Rose, M. T., *The Simple Book: An Introduction to Management of TCP/IP-Based Internets*, Englewood Cliffs, NJ: Prentice Hall, 1989.

5

Frame Relay Performance Issues

5.1 Introduction

In this chapter, we discuss the performance impact of carrying data applications over frame relay networks. This extends the discussion of WAN performance characteristics in the previous chapters. While we use frame relay as the specific packet technology to illustrate design and performance engineering considerations, much of the discussion in this chapter applies to other packet technologies as well, for example, ATM.

Frame relay is the technology example we chose to illustrate the design and performance engineering considerations because of its dominant growth during the late 1990s. More and more enterprise networks are moving from a private line environment to a frame relay environment. We will present the advantages and the perils in making this migration from a performance perspective.

Our focus in this chapter, and in fact in the majority of this book, is public frame relay networking. The majority of frame relay-based enterprise networks are implemented over public frame relay services. In rare instances, the frame relay network infrastructure is privately held by the enterprise.

Although the technologies and vendor products may be similar, there are differences between these two alternatives, primarily in the amount of control and information over the internals of the frame relay network implementations. With private implementations of frame relay, corporations have complete control and a view into their frame relay implementations. With public frame relay implementations, the internals of the frame relay network are under the control of the frame relay service provider. These service providers may not choose to disclose to end users all of the internals of their network service.

Corporate data networks are often migrated from private line environments to frame relay with the expectation that end user response times will improve. However, results are not so predictable and depend on a variety of factors, including bandwidth, traffic volumes, message sizes, traffic mix, and network design. Frame relay has different latency and throughput characteristics when compared to private lines. Therefore, a successful frame relay network implementation must take into consideration these various factors.

The material in this chapter will guide the network engineer though the various performance engineering considerations when planning, implementing, performance troubleshooting, or redesigning a frame relay network. The material herein is kept at an end protocol generic level, in that we do not discuss the specifics of TCP/IP, Novell's NetWare, and IBM's SNA protocols.

We first compare and contrast private line and frame relay performance for the simplest of reference connections. We then examine global frame relay network internals and issues. We follow this with a section on bandwidth engineering and traffic/application discrimination methods. We conclude this chapter with a brief discussion of various topological considerations in frame relay network designs.

5.2   Private Line Versus Frame Relay

Let us begin by considering the results of a straightforward private line replacement. Consider the private line reference connection shown in Figure 5.1. This was reference connection #1 (RC#1) in the previous chapter. Here two routers are interconnected with a direct private line facility. Private line connections are

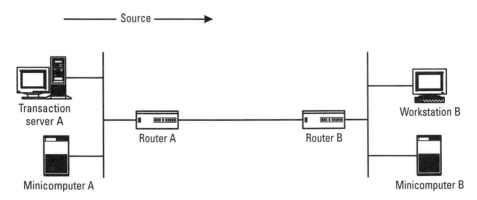

Figure 5.1 A simple private line reference connection.

characterized by a constant propagation delay and fixed bandwidth. The timing diagram associated with this reference connection was discussed in Chapter 4, and we repeat it here for convenience (see Figure 5.2).

To begin our discussion on the impact of relying on a shared virtual private network such as frame relay, we introduce a second reference connection (RC #2) for our analysis. In Figure 5.3, the same premises configurations as in Figure 5.1 are maintained, but the direct private line facility between the two routers is replaced with a shared frame relay network. This figure shows some of the internals of the frame relay connection, that is, three frame relay switches in the path. In general, the specifics of the frame relay connection, such as the number of hops and the specific route a VC takes, are hidden from the end user.

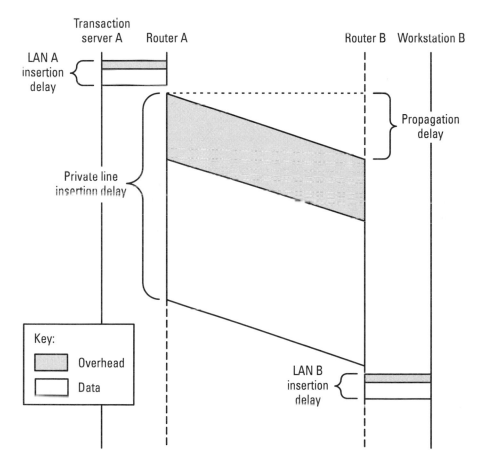

Figure 5.2 Timing diagram associated with the reference connection in Figure 5.1.

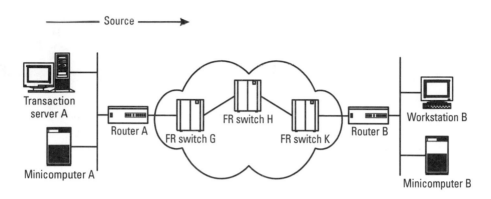

Figure 5.3 A comparable frame relay reference connection.

5.2.1 Dual Insertion Delay

What are the performance characteristics of this frame relay reference connection in relation to the comparable private line connection? Although it is true that frame relay can provide significant performance gains, there are many instances when performance will degrade when migrating from a private line environment if proper care in engineering is not taken.

For the moment, let us assume that the capacity of the internal frame relay switches and the facilities are infinitely fast; that is, they contribute zero delay to the path other than the usual wire propagation delay. Then the timing diagram for this fictional frame relay reference connection—to be compared with the timing diagram in Figure 5.2—is shown in Figure 5.4.

By now, we have worked enough with the concept of timing diagrams to interpret the diagram shown in Figure 5.4. Because we have assumed that the frame relay network is infinitely fast in all aspects, the packet width is infinitesimal while traversing the frame relay network. Even so, it appears that the presence of the frame relay network in this path has essentially doubled the insertion delay of the packet over the WAN (assuming the insertion delay dominates the router-to-router propagation delay). This is because the access frame relay switch—that is, frame relay switch G in Figure 5.4—must accumulate the entire packet from router A before it can encapsulate it into the internal networking protocol and transmit it to the next frame relay switch in the path.

This holds true across the frame relay network. Frame relay switch H must accumulate the entire packet prior to transmitting it onto frame relay switch I, and so on. This assumes that the frame relay network is not performing any type of pipelining, that is, packet segmentation. Later on in this section we will discuss the impact of network pipelining.

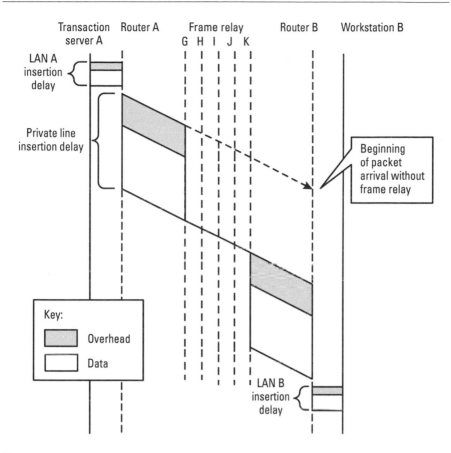

Figure 5.4 Impact of an infinitely fast frame relay network.

Finally, the last frame relay switch in the path must receive the entire packet before it can begin transmission of the packet onto the slower egress facility to router B. So, we observe that even for an infinitely fast frame relay network, the "router-to-router" delay a packet experiences traveling from router A to router B will roughly double in comparison to the comparable private line case.

As an example, assume a packet size of 1500 bytes, a private line speed of 56 Kbps, and a propagation delay of 10 msec. In this case, the time to move the packet from router A to router B is simply the insertion time of the packet, for example, 1500 bytes × 8/56,000 bps = 214 msec, plus the propagation delay of 10 msec, for a total delay of 224 msec. If we simply assume that the corresponding frame relay configuration has an egress and access port speed of 56 Kbps and the frame relay network internals are infinitely fast, then the time to move the packet from router A to router B is the sum of inserting the packet onto the access line plus the propagation delay plus the time to insert the packet onto the egress line. Because both the access and egress is 56 Kbps, we estimate the total delay is 214 msec + 10 msec + 214 msec, or 438 msec. This is roughly twice as large as the corresponding private line delay.

In reality, rarely is the corresponding frame relay connection implemented in this fashion. In fact, several methods are available to improve the relative performance of the frame relay connection in comparison to private line scenarios. We now discuss these options.

It is hard to imagine how to improve on an "infinitely fast" frame relay network. However, in the preceding example, we have implicitly assumed that a given packet is carried within a single frame relay network internal packet. This is not always the case. Could performance improvements be achieved by using several smaller network internal packets and *pipelining* the data across the frame relay cloud? (Refer back to Section 3.4 where we discussed the advantages of pipelining packets over multihop data networks.)

Consider the case where access frame relay switch G encapsulates the packet into three separate internal frames. Then after about one-third of the packet is accumulated into switch G, it can transmit this portion of the data packet onto network switch H, thereby speeding along the data across the frame relay cloud. This is illustrated in Figure 5.5.

In this case, network pipelining has done nothing to speed the delivery of the packet to router B. The reason for this is apparent in Figure 5.5. While network pipelining was able to deliver the first two internal frames to the far end frame relay switch K, this switch still had to wait for the last internal frame to arrive prior to transmission of the beginning of the full packet onto the router B.

You might ask why the last frame relay switch had to wait for the arrival of the last internal frame to arrive prior to transmission. Why could it not have begun the transmission of the first frame of the packet onto the egress line?

We refer to this capability as *egress pipelining*, when the egress frame relay switch, for example, switch K in Figure 5.6, begins transmission onto the egress facility prior to fully receiving the entire frame from the internal network. When the internal network uses multiple internal frames to carry a single data packet and these small internal packets travel separately across the internal network, we refer to this as *internal pipelining*. As shown in Figure 5.6, egress pipelining can improve the end-to-end performance of the data packet when traveling from router A to router B. However, this performance improvement is an illusion due to our assumption of an infinitely fast frame relay network, as we see next.

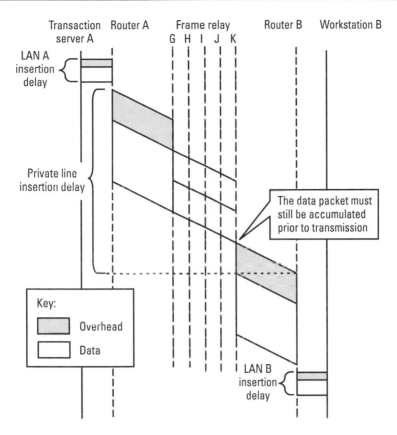

Figure 5.5 Timing diagram showing the impact of network pipelining on end-to-end packet delay.

5.2.2 Delay Variation

In order for the egress pipelining strategy we have just described to work, the spacing between the arrival of the internal protocol packets to the last frame relay switch in the path must be constant and equal to the transmission time of the data portion of the internal protocol packets onto the egress facility. This is illustrated in Figure 5.7.

The left-hand side of Figure 5.7 shows a timing diagram for the case where no delay variation exists in the frame relay network. This is evident in the figure because each internal frame experiences exactly the same delay through the frame relay network. The right-hand side of the figure shows an identical timing diagram except for the fact that the internal frame relay network is assumed to be finite in speed. This will result in a different delay experienced by each of the internal network frames when traversing the frame relay network.

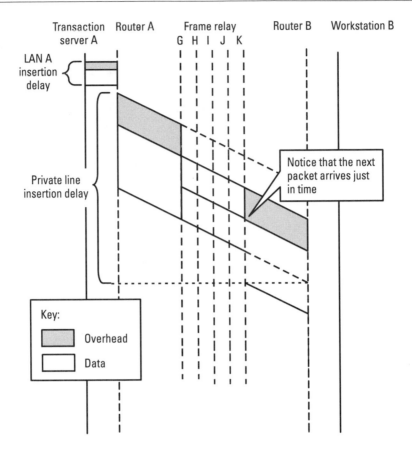

Figure 5.6 Timing diagram showing the effect of egress pipelining.

In our example, the third internal frame experiences a greater delay than the first two frames. This causes a gap to occur in the transmission of the overall frame relay packet onto the egress line. Router B would see a bad frame error check on that frame and would then discard the packet.

To ensure that gaps like this do not occur in the delivery of frame relay packets to end systems, frame relay networks do not perform egress pipelining.[1] Hence, the mere presence of a frame relay network will increase the apparent

---

1. It is possible for the egress frame relay switch to delay the initial insertion of the frame onto the egress line for a period of time, and then begin the insertion. Hopefully, the initial delay will compensate for the greatest delay experienced by the remaining packets traversing the frame relay network. Then the apparent gap in the frame would not occur on the egress line. However, this is a statistical game that is not implemented in current frame relay switches. So we ignore this possibility in the remainder of this book.

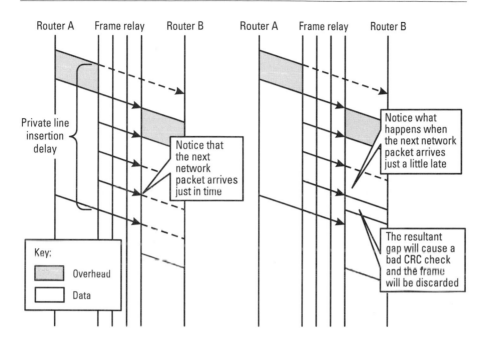

Figure 5.7 Impact of network delay variation on egress pipelining.

insertion delay of a packet when compared to a comparable private line connection.

In addition to the store-and-forward delays in frame relay, additional delay is incurred due to the finite speed of their components, for example, switches, trunks, and routing. The delays associated with switching are typically small, on the order of 1 msec or less. The delays associated with insertion delays on internal trunks are also typically small, for example, on the order of 1 msec or less. This is due to (1) frame relay networks performing internal pipelining that results in the frame relay network carrying small internal packets, and (2) internal trunks that are typically 1.5 Mbps or higher.

Finally, an additional consideration in comparing frame relay network delays to private line delays is that the physical paths over which the connections travel will be different. Remember that propagation delays are dependent primarily on the length of the physical path followed by the connection. Therefore, different routing in the frame relay and the private line cases will cause different delays. Given that the network infrastructure supporting private line services in most common carriers is more extensive than the corresponding frame relay infrastructure, this will cause further increases in the realized delay in the frame relay case.

Delays within the frame relay networks are often small, around tens of milliseconds total under most operating conditions. (We discuss the effects of network congestion in Section 5.4.) However, the additional store-and-forward delay caused by the frame relay network can be on the order of 100 to 200 msec. For example, the insertion time of a 1500-octet frame relay packet onto a 56-Kbps serial line is 214 msec. This assumes that a separate 56-Kbps port is introduced when moving to frame relay. Typically, the extra port added is higher speed, say, 256 Kbps or higher. So, one option to reduce the apparent store-and-forward delays is to increase the speed of one or both the serial access/egress lines into the frame relay network. This is illustrated in Figure 5.8.

By increasing the serial line speed on the left side of the frame relay network from 56 to 256 Kbps, these diagrams illustrate the improvement in delay. Further improvements can obviously be accomplished by increasing the serial access line speeds on both sides of the network. Although these options do improve the end-to-end performance, they do so at an additional cost.

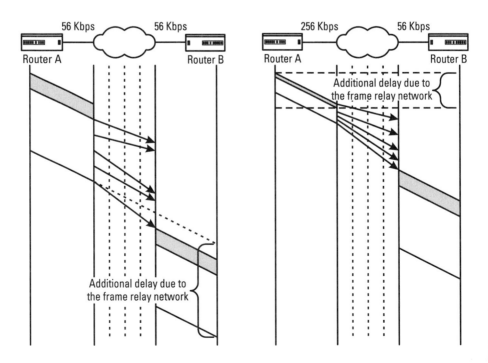

Figure 5.8 Timing diagrams illustrating the effect of increased serial line speed.

5.2.3 Application Impact of Migration to Frame Relay

For some applications, this additional cost is well worth the expense. For chatty applications—some client/server applications—or "non-WAN friendly" protocols, it is important to minimize the additional delays incurred by migrating from a private line environment to frame relay. Because these applications tend to be highly interactive or conversant, the additional delays incurred due to the frame relay network are magnified from the end user/application perspective. We have referred to these applications as latency-sensitive applications (see Section 4.4 in the previous chapter). In contrast, those applications, which we characterized as bandwidth sensitive, show few effects when migrating from private line to frame relay circuits.

We provide Figure 5.9 to aid in the identification of those applications that require careful attention when migrating from private line to frame relay networks. This figure outlines tolerance zones for various application types within the context of a *delay versus bandwidth* plot. Applications found at the bottom of the plot require low delay connections to perform adequately. Those found at the far right-hand side of the plot require high bandwidth connections

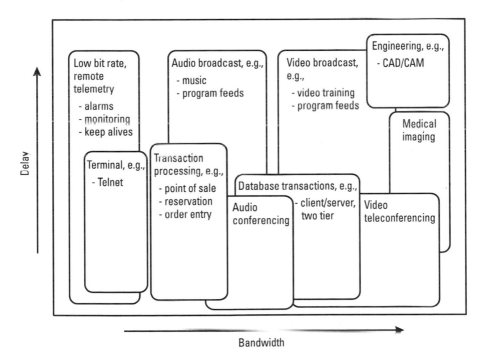

Figure 5.9 Application profiles within the delay versus bandwidth plot.

to perform adequately. Those applications that tend to be highly interactive, for example, telnet, transaction processing, teleconferencing, videoconferencing, are those that require low delay and require some care in their migration to frame relay.

5.3   Global Frame Relay Connections

It is often the case that the corporate data network is required to support international locations. This presents additional considerations for the network engineer when relying on international frame relay circuits. Specifically, consider the global frame relay reference connection shown in Figure 5.10.

This figure shows the path from the client—located in one country—to server—located in another country—to be a concatenation of multiple autonomous frame relay networks. In particular, this figure shows a larger backbone frame relay network, denoted as network 2, surrounded by two in-country frame relay networks, denoted networks 1 and 3.

Typically the networks are built on different frame relay products, for example, Cisco IPX® switches, Northern Telecom's DPN switches, or Lucent's BNS-2000® switches. Further, these networks may be managed, engineered, and provisioned by different frame relay service providers, for example, AT&T (United States), Stentor (Canada), Embratel (Brazil), Telintar (Argentina), and IDC (Japan). These various frame relay services are physically interconnected via frame relay network-to-network interfaces (NNIs).[2] Like the frame relay user-to-network interface (UNI) definitions, the NNI defines how two frame relay networks interwork to support end-to-end user virtual circuits. Beyond the capabilities defined within the NNI standards, the various service providers must develop bilateral agreements to address issues related to provisioning and maintenance coordination, interface points, the nature of the interconnect,

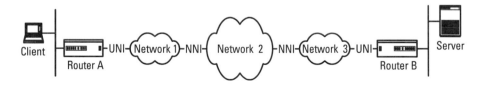

Figure 5.10 A typical global frame relay reference connection.

---

2. Depending on the equipment in use, some interconnects between frame relay providers may not be standard NNI based.

sales, and pricing. However, because the products and methods differ between the parties involved in the bilateral agreements, the end-to-end management capabilities afforded to the customer are often less than that received from a single service provider. In this case, it is difficult to establish service level agreements for the end-to-end connections.

Given that the typical global frame relay connection is a concatenation of multiple frame relay networks interconnected over NNIs, the capabilities of the NNI are critical. Several NNI issues are worth consideration:

Store-and-Forward Delays

The NNI is a frame relay interface, not an internal trunk protocol. For data frames to be transmitted over the NNI, the frame relay packets must be reconstituted prior to transmission over the NNI links. This adds additional delay due to store-and-forward insertion and additional protocol processing. For a 1500-byte frame, the additional round-trip delay per NNI is approximately 20 msec, assuming T1 rate NNIs. Some NNIs run at higher rates, and hence the added delay will be proportionally less.

Link Status and Failure Recovery

Some link level management capabilities extend across the interface through the NNI link management protocols. One NNI interface, of an NNI interface pair, will periodically poll the other interface with status_enquiry messages. Asynchronous link management interface (LMI) messages are also supported to indicate the status of, for instance, far-end equipment.

The LMI helps to determine the status of the NNI and the status of the separate VC segments carried over the NNI on the multiple autonomous frame relay networks. However, in the event of an NNI failure, no automatic rerouting is defined. For this reason, it is typical to find that the frame relay switches on either side of the NNI are colocated in order to minimize the probability of NNI link failure, although this is not always the case.

Congestion/Bandwidth Management Mismatch

There is a possible "impedance" mismatch between the two frame relay networks on either side of the NNI in the way they handle bandwidth management and congestion control. For example, network 1 may provide an open-loop congestion management scheme, whereas network 2 may provide a closed-loop scheme. Also, the separate service providers may implement different capacity management algorithms and internal overbooking factors. In these cases, it is difficult to predict the performance of the concatenated connection. In cases where the two service providers use the switches from the same frame relay equipment provider, it is possible that the equipment vendor provides a proprietary enhancement to the NNI. This enhancement may provide an extended, end-to-end bandwidth management over the NNI.[3] To exacerbate the concern over congestion management, global frame relay networks often rely on less than T1.5 trunks. Hence, congestion management becomes relatively more important in global frame relay implementations.

Performance Objectives and Measurement Capabilities

The end-to-end performance objectives are a function of the multiple service providers' engineering and capacity management and other operations, administration, maintenance, and provisioning (OAM&P) processes. These, of course, will differ across multiple service providers. When performance problems do arise, it is more difficult for customers to determine the root cause. When excessive delays are a concern, the individual providers can send test_delay messages over their portion of the customer VC to directly measure the round-trip delays. However, as shown in Figure 5.11, these do not provide complete coverage across the end-to-end VC.

Missing are the delays across the UNIs and, more problematic for companies, the delays across the various NNIs. Furthermore, the providers may run the test_delay measurements at different times and hence the results may not be useful to the problem at hand. A simpler approach is for the customer to send a "ping" packet across the VC and then subtract the ingress and egress delays, taking into account expected load on the access and egress ports. The result will be an estimate of the round-trip delay on the multiprovider frame relay network.

5.4   Bandwidth Sizing

We now focus on the issues associated with designing frame relay networks to support desired performance levels. The overall process of network design

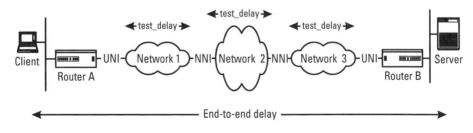

Figure 5.11   End-to-end delay versus individual network test-delay measurements.

---

3. One example of this is the Cisco Foresight® enhancements to the IPX frame relay NNI. In this case Cisco relies on consolidated link layer management (CLLM) messages to pass closed-loop congestion notifications over the NNI.

involves numerous steps beginning with building an understanding of bandwidth allocation to the various network-based applications. The process ends with details of statistically multiplexing traffic onto a common frame relay infrastructure. Bandwidth allocation requires network managers to understand the volume of traffic associated with each instance of a network application and their concurrency.

Although some application traffic may be easily obtained, other applications may not be fully deployed. It is difficult to predict concurrency among users. This forces the network designer to make an educated guess at the traffic volumes associated with these applications, very often using worst case conditions. The traffic estimates are used to build a traffic matrix, which characterizes the point-to-point traffic flows between the various end points comprising the enterprise network. The network designer then uses the traffic matrix to create a frame relay design. This frame relay design will identify the access points into the frame relay network, the size of the access ports and the number, size of point-to-point topologies of the various virtual circuits between the end systems connecting to the frame relay network, and an estimate of the expected performance for the known applications. As new applications become widely deployed, it is imperative to actively monitor the port/VC utilization to trend expected application performance.

Due to the statistical multiplexing inherent in frame relay technology, several critical areas exist where overbooking of traffic can occur. The network analyst has little control over the nature of the engineering of the internal frame relay service; this is under the control of the frame relay service provider. However, it is up to the network analyst to design and engineer the frame relay access and egress facilities and the VCs that interconnect these ports and to determine the CIR values for each VC in the network. This is the primary topic of the next section on sizing ports and PVCs. Following this section, we address the issues associated with network congestion in frame relay networks. Here we identify methods to provide differential service to diverse applications in situations where congestion would potentially occur in frame relay network deployments.

5.4.1 Sizing Ports and PVCs

We take the following approach to sizing frame relay ports and PVCs. We first lay out an initial frame relay design, taking into consideration both topology and bandwidth issues. When these considerations are addressed, the result is an initial frame relay design. This initial design is essentially an equivalent private line replacement, giving little or no consideration to frame relay specific concerns other than simple overbooking rules. We then consider methods to fine-tune the initial frame relay designs. Three considerations are discussed: load engineering to a reasonable traffic load, interactive traffic performance, and file application throughputs.

5.4.1.1 Initial Frame Relay Design Consideration

Network design and performance engineering go hand in hand. Network design is extremely important in maintaining and supporting multiprotocol environments. It is generally the first step in laying out the network and in building up a conceptual model of the network and its traffic flows. Once developed, this model should be continuously updated and refined through the appropriate network monitoring.

A frame relay design is characterized by the topology of the network and the configurations of the attached CPE, for example, routers. This section deals with the rationale to be followed when developing an initial frame relay network topology.

The frame relay network topology is characterized by the number and location of the frame relay ports, the size of the individual ports, the interconnection of those ports, and the size of the interconnections between the ports. At a minimum, the frame relay topology must be capable of carrying the traffic load, which is described in the *traffic matrix*. The traffic matrix defines the point-to-point traffic flow between all locations attached to the network. Figure 5.12, and its associated table, Table 5.1, give an example traffic matrix for a four-site network.

Table 5.1 gives the traffic flows from the location in the first column to the location in the top row. This traffic flow is often taken to be the peak traffic flow averaged over a small period of time. For example, sample traffic over a month and average the samples into 15-min periods. Then take the busiest 15-min period over the month, or take the average of the busiest 15-min period for each business day during the month, or some similar averaging methodology.

Table 5.1
Traffic Matrix for a Four-Location Design (in Kbps)

	City A	City B	City C	City D
City A	—	128	96	128
City B	48	—	8	8
City C	48	8	—	6
City D	56	16	8	—

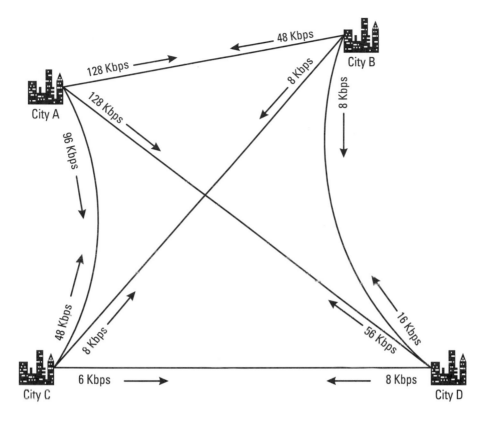

Figure 5.12 A traffic matrix for a four-location design.

Use this information to build up a traffic matrix such as the one shown in Table 5.1. Typical of most networks, one location appears to have the majority of the traffic into and out of its location. This location is city A according to this traffic matrix. The other locations have a lesser amount of traffic between themselves. This location is typically the corporate data center and has Internet connectivity and the majority of the business partner connectivity as well.

An initial pass at a frame relay design would have full mesh topology where each city pair has a virtual circuit between them. The size of the individual virtual circuits (VCs) is equal to the peak traffic flows between the city pairs. This is shown in Figure 5.13.

The access ports must be sized as well. An initial estimate is to size the access ports to either no less than one-half of the sum of the CIRs of the individual VCs configured on that port, or the single largest CIR on that port. This represents an overbooking factor of no more than 2-to-1. A quick look at the traffic matrix shows that 352 Kbps flows out from city A (total), 152 Kbps

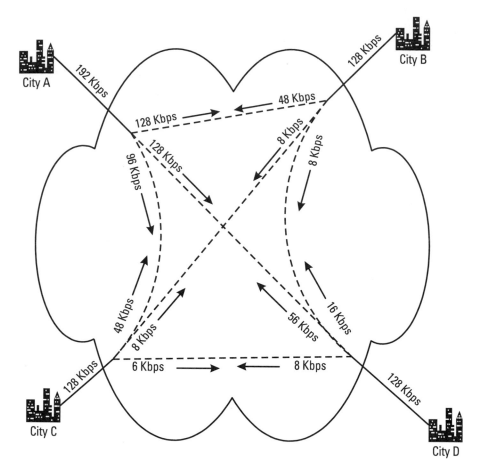

Figure 5.13 Full mesh VC topology frame relay design.

flows into city B (total), 112 Kbps flows into city C (total) and 142 Kbps flows into city D (total). Going with an overbooking factor not to exceed 2-to-1 and taking into consideration that port speeds are limited to increments of $n \times 64$ Kbps, we get the following port speeds for the various cities: City A's port speed is 192 Kbps, city B's port speed is 128 Kbps, city C's port speed is 128 Kbps (because the CIR of the VC from A to C is 96 Kbps) and city D's port speed is 128 Kbps.

For small network designs this is a reasonable approach. However, problems quickly develop when trying to apply this methodology onto larger frame relay designs. For one thing, whenever a new location is added, the number of VCs that must be added to the network is proportional to the total number of sites on the network. This soon becomes unmanageable. Full mesh designs soon exhaust the VC capacity of routers and of the frame relay switches in the network (see the discussion in Section 5.7). Therefore, as the total number of locations grows, a more sparse VC topology must be adopted.

As an example, given that the traffic flow in our four-location network is primarily into and out of city A, perhaps a star topology can be adopted. In this case only three VCs totals are required, one from each of the smaller locations (that is, cities B, C, and D) to the larger location, city A. This topology scales linearly with the number of locations; that is, the number of VCs is proportional to the number of locations, whereas the full mesh topology scales as the square of the number of locations, that is, the number of VCs required is proportional to the number of locations squared.

In this sparse VC topology, the traffic between any of the satellite locations, for example, cities B, C, or D, must travel two hops in order to reach their destination. For example, traffic between cities B and C must flow through city A. To account for this, the size of the VCs between city A and the satellite cities must be increased to account for this additional flow. The total traffic flowing over the VC from city A to city B is now 128 Kbps + 8 Kbps (for C to B traffic) + 16 Kbps (for D to B traffic) = 152 Kbps. In order to accommodate this traffic a CIR of 192 Kbps must be used. Similarly, for the total traffic flowing on the A-to-C circuit we get 112 Kbps, which must be accommodated on a CIR of 128 Kbps. For the traffic flowing on the A-to-D circuit we get 142 Kbps, which must be accommodated on a CIR of 192 Kbps. Given the additional flow of traffic into and out of the city A hub location, we must readjust the port speed for this location. Using the not to exceed a 2-to-1 overbooking rule for port A, we get for the sum of the CIRs a total of 512 Kbps. Dividing this by two we get a frame relay port speed of 256 Kbps. Similiar considerations may warrant increasing the speeds of ports B and D. This sparse VC topology design is shown in Figure 5.14.

We have taken this design about as far as possible without modeling, in more detail, the nature of the applications running over the network and the nature of the frame relay admission control and congestion management schemes. We discuss these issues next.

Finally, when choosing the specific VC topology in the frame relay design, consideration of the interaction between higher level addressing and subnet architectures must be made. This was discussed in the section on IP technology in Chapter 2.

5.4.1.2 Port and VC Load Engineering

The rules we discussed previously were simple. They stated (1) the total traffic on a VC should be less than the CIR and (2) the port should not be overbooked by more than 2-to-1. Let us elaborate further on these issues.

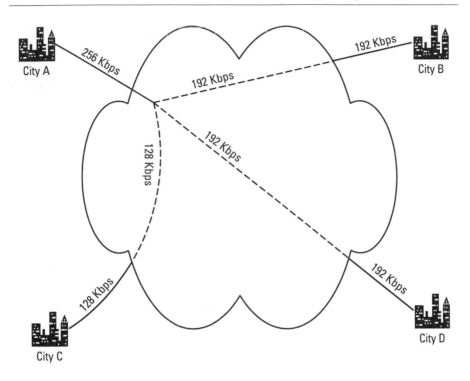

Figure 5.14 A sparse VC topology frame relay design.

Regarding the sizing of the CIR, it is generally true that public frame relay networks support average throughputs on VCs in excess of the CIR. Therefore, sizing CIRs can be done in a fashion somewhat more aggressive than the corresponding sizing of private line facilities. The CIR should generally be larger than the total traffic flow of the mission critical applications on the circuit. This is independent of the nature of the congestion avoidance mechanism in place in the frame relay network—whether it relies on an open-loop or a closed-loop mechanism. In either case, traffic that is sent onto the VC in excess of the CIR may be subject to potential performance degradation. In the open-loop case, this CIR excess traffic will be subject to higher discards during periods of network congestion. In the closed-loop case, the traffic sent in above the CIR may cause excessive network delays; this is due to its being buffered in the frame relay access module. If this is a concern, then a more conservative approach may be warranted. One approach is to keep the CIR utilization below 70%.

However, you may want to implement different sizing rules for non-mission-critical applications. Some applications run well over "best effort" delivery networks. Examples of such applications may be e-mail running over SMTP on TCP. (SMTP, the Simple Mail Transfer Protocol, is discussed in

[1].) For these applications, it is reasonable to consider sizing the CIR such that their utilization exceeds 100%. In fact, some frame relay service providers support "best effort" or 0 K CIR PVCs for these types of applications.

In general, we recommend a sizing approach that tries to find a middle ground in these two different philosophies. To take advantage of the excess bursting capabilities afforded in frame relay services while maintaining good performance for mission-critical applications, the following approach to sizing CIRs might be considered: *Our guidelines have been to carefully monitor the PVCs if their utilization exceeds 70%, but actually UPGRADE only if the utilization exceeds 100% for, say, 10% to 20% of the time.*

Regarding the port overbooking factor, a standard recommendation is that the overbooking factor on a port should not exceed 2-to-1. However, this makes some assumptions regarding the noncoincident behavior of the data traffic on the separate VCs. For instance, if all of the VCs on a port, which is engineered to the 2-to-1 rule, were simultaneously running at 100% utilization relative to their CIRs, then the port would be attempting to run at 200%. This would clearly lead to problems if this were a frequent and persistent occurrence. On the other extreme, if it were known that the traffic on the various VCs comprising the port was noncoincident, then a higher overbooking factor could be supported, for example, 3-to-1 or even 4-to-1. Also, the appropriate overbooking factor will depend on the number of VCs and the relative speed mismatch between the individual CIRs and the port speed. In fact, a larger overbooking factor can be used if all the CIRs are small relative to the port speed. A smaller overbooking factor should be used in the event that the CIRs are larger. This is due to the law of large numbers [2]. This theorem basically states that the total traffic flow is smoother, that is, shows relatively less variation about its mean, if it is composed of a large number of lower volume flows, than if it is composed of a small number of higher volume flows. The smoother the total traffic flows, the larger the overbooking factor.

5.4.1.3 File Application Throughputs

We now take our design considerations to a finer level of analysis by considering the impact of our design decisions on the performance of file transfer of size F, a packet size of S, and a window size of W. *We focus on the case where the access controls are based on a leaky bucket algorithm with access buffering. This system, as discussed earlier in Chapter 2, is characterized by the size of the token buffer K, the replenishment rate of the tokens or CIR C, and the port speed P.* The leaky bucket algorithm allows a data packet entry to the frame relay network only if the token buffer contains enough tokens to match the size of the data packet. For this reason, K should always be set larger than the maximum data packet size on the frame relay access line.

The CIR places an upper bound on the system throughput assuming no bursting. Here, we study the case where the CIR is not the limitation to the system throughput, but the end-to-end window is the limit to the system throughput. In this case, the file transfer packets arrive at the access control filter at a rate less than the arrival rate for the tokens. Therefore, it is reasonable to assume that whenever a file transfer packet arrives from the frame relay access line into the access control filter, there is an ample supply of tokens in the token buffer to immediately allow the data packet access to the frame relay network. In other words, we argue that in this limit the contribution of the access control filter to the data packet delay is negligible.

So our approach is as follows: We compute the round-trip packet delay, ignoring the access control filter and the limitation to the VC throughput as determined by the CIR. Once we have computed the round-trip packet delay (R_d), we then estimate system throughput (X) by using the formula $X = W/R_d$, where W is the number of bits in the end-to-end window. Our throughput estimate is then the lesser of the CIR, the slowest link in the path, or X.

Let us work through an example. Assume that the system window size is two packets, and each packet is 512 bytes in length. We will ignore all higher level and frame relay protocol overhead in this example. Assume the frame relay access and egress lines are running at 56 Kbps, the total wire propagation delay is 20 msec (roughly equivalent to 2000 miles), the subnets behind the routers are 10-Mbps Ethernets, and the packet acknowledgment is 50 bytes in length.

Further assume the frame relay network trunks are running at 1.5 Mbps, there are four trunks as indicated in Figure 5.3, and the internal frame relay packet sizes are 53 bytes. The various delays associated with this reference connection are given in Table 5.2. Also refer to Figure 5.15 for the corresponding timing diagram.

Table 5.2
Delay Worksheet for the Reference Connection in Figure 5.15

Delay Component	Computation	Value (msec)	Multiplier	Total (msec)
Packet insertion delay on the Ethernets	512 × 8/10 Mbps	0.4	2	0.8
ACK insertion delay on the Ethernets	50 × 8/10 Mbps	0.04	2	0.08
Packet insertion delay on the 56-Kbps lines	512 × 8/56 Kbps	73.0	2	146.0
ACK insertion delay	50 × 8/56 Kbps	0.7	2	1.4

Table 5.2 (continued)

Delay Component	Computation	Value (msec)	Multiplier	Total (msec)
Insertion delay on the FR internal trunks	53 × 8/1.5 Mbps	0.3	4	1.2
ACK insertion delay	50 × 8/1.5 Mbps	0.3	4	1.2
Wire propagation delay	10 msec/1000 mi	20.0	2	40.0
Processing delay through each router	Assumed	1.0	4	4.0
Processing delay through a FR switch	Assumed	1.0	10	10.0
ACK processing delay in the receiver	Assumed	1.0	1	1.0
			Round-trip	205.7

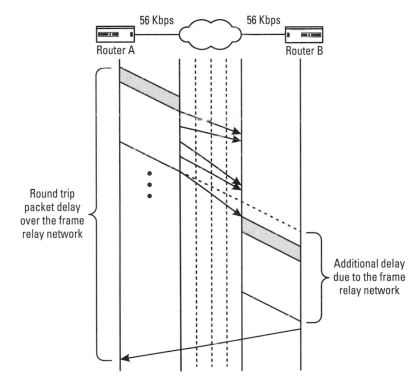

Figure 5.15 A timing diagram for the frame relay reference connection.

As an example of how this worksheet shown in Table 5.2 was developed, let us take a look at the row for the delay component labeled "Processing delay through a FR switch." The second column associated with this row indicates the method of computing this value. In this case, the quantity is either assumed or derived. The third column gives the value associated with a single instance of this delay, that is, 1.0-msec delay per each FR switch. The fourth column—labeled "Multiplier"—indicates the number of FR switch processing delays that contribute to the round-trip delay. (Remember that the round-trip includes both the packet transit and the return of the ack.) In this case the multiplier is 10. There are five FR switches in the reference connection and, therefore, a total of 10 FR switches are encountered in the round-trip path. The final column is the total contribution of this delay contributor to the round-trip delay. Finally, all of these delay contributors are summed to yield the total round-trip delay of 205.7 msec. Note, this is best case. That is, we assume there are no other traffic sources contending for the network.

Given a round-trip delay of R_d = 205.7 msec and W = (2 × 512 bytes) (8 bits/bytes) or 8192 bits, our estimate for the system throughput is

$$T = W/RT = 8192 \text{ bits} / 0.2057 \text{ sec} = 40 \text{ Kbps}$$

Referring to the connection in Figure 5.3, we find that the slowest link in the path is the 56-Kbps frame relay access and egress lines. Thus the maximum throughput obtainable in this example is 56 Kbps, as determined by the slowest facility. We just computed the impact of the end-to-end window on the system throughput and found it will limit the throughput to no more than 40 Kbps. The one determinant we have yet to consider is the effect of the access control filter. If the CIR of the connection is less than 40 Kbps, then it will be the bottleneck to system throughput. For example, if the C = 32 Kbps, then the system will not be able to obtain a throughput higher than 32 Kbps. However, if the CIR is between 40 and 56 Kbps, then the end-to-end window will be the determinant of the system throughput. Therefore, to achieve maximum throughput for this reference connection, the CIR should be set to a minimum of 40 Kbps; otherwise, the CIR will be the throughput-limiting factor.

In short, our recommended approach to determining the throughput for a reference connection over a frame relay network is as follows. First determine the reference connection and its relevant parameters—link speeds, number of routers and switches, and path length. Compute the round-trip delay for a packet and its corresponding acknowledgment. This ignores the frame relay access control parameters and impact. Compute the end-to-end throughput, taking into account the system window using the formula $X = W/R_d$. Now, compare the window-limited throughput that is obtained from the previous formula with the slowest link speed and with the throughput limit as determined by the CIR. The realized system throughput is the lesser of the window limit, the slowest link limit, and the CIR limit. In Chapters 7 and 8, we will discuss in greater detail file transfer and delay performance of TCP/IP and Novell applications, respectively.

5.4.1.4 Interactive Traffic Performance

We now consider the impact of our design decisions on the performance of isolated interactive traffic. Unlike the throughput analysis just discussed, this analysis will depend on the specific details of the access control mechanism implemented in the frame relay access modules. We focus on the same case as before, where the access controls are based on a leaky bucket algorithm with access buffering. Whereas, in the case of computing the throughput the relevant access control parameter is the CIR, for estimating interactive delay performance the relevant ACF parameters are both the size of the token buffer and the CIR.

Consider the example reference connection discussed in the previous section. Let us estimate the transaction delay for a single transaction consisting of 106 octets of input data that generates 1024 octets of output data, similar to an SNA 3270 or an SAP R3 application. We will ignore the protocol overhead for this exercise. Assume this represents a single isolated transaction; that is, the frame relay access and egress lines have been idle for a period of time.[4] During this idle period, the token buffer has been fully replenished. Finally, assume the token pool size is 2048 octets.

Given that we have chosen the token pool size to exceed both the input and the output transaction sizes, it will have no impact on delaying the data at the access control filters. This also implies that the CIR will have no bearing on the result of our transaction delay computation, given these transaction sizes, transaction loads, and token buffer pools. We will rely on a delay worksheet shown in Table 5.3. In this table, we identify the delay components for the 106-octet input traveling from the client to the server across the reference connection and the delay components for the 1024-octet output traveling in the opposite direction. Then the sum of all the delay components yields the final transaction delay. Again, this is best case.

Table 5.3 shows an input delay of 58.9 msec, an output delay of 248.4 msec, and a total transaction delay of 307.3 msec. It is not intuitively

---

4. In cases where this assumption is not true, we presented a formula in Section 3.5.1, (3.5), that captures the load/queuing behavior of the leaky bucket access control filter.

Table 5.3
Delay Worksheet for the Isolated Transaction Delay

Delay Component	Computation	Value (msec)	Multiplier	Total (msec)
Insertion delay on the Ethernets	106 × 8/10 Mbps	0.08	2	0.16
Insertion delay on the 56-Kbps lines	106 × 8/56 Kbps	15.1	2	30.2
Insertion delay on the FR internal trunks	106 × 8/1.5 Mbps	0.6	1	0.6
Pipelining delay on the FR internal trunks	53 × 8/1.5 Mbps	0.3	3	0.9
Wire propagation delay	10 msec/1000 mi	20.0	1	20.0
Processing delay through each router	Assumed	1.0	2	2.0
Processing delay through a FR switch	Assumed	1.0	5	5.0
			Input delay	58.9
Insertion delay on the Ethernets	512 × 8/10 Mbps	0.4	2	0.8
Insertion delay on the 56-Kbps lines	1024 × 8/56 Kbps	146.3	1	146.3
Pipelining delay on FR egress line	512 × 8/56 Kbps	73.1	1	73.1
Insertion delay on the FR internal trunks	53 × 8/1.5 Mbps	0.3	1	0.3
Pipelining delay on the FR internal trunks	53 × 8/1.5 Mbps	0.3	3	0.9
Wire propagation delay	10 msec/1000 mi	20.0	1	20.0
Processing delay through each router	Assumed	1.0	2	2.0
Processing delay through a FR switch	Assumed	1.0	5	5.0
			Output delay	248.4
			Round-trip	307.3

obvious how these delay estimates were generated from looking at the table. Remember that this level of detail must be extracted from the corresponding timing diagram, which is shown in Figure 5.16.

The simplest way to map the information in the timing diagram to the corresponding worksheet is to follow the last byte of the input and output messages across the timing diagram. For example, the entire packet must be inserted onto the LAN segment—not shown in the timing diagram—prior to the last byte being received at the WAN attached router. This time is simply the insertion time of the packets onto the 10-Mbps Ethernet LAN segment. This is the first entry in Table 5.3.

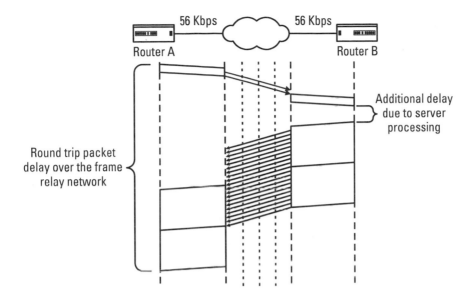

Figure 5.16 Timing diagram for the isolated transaction.

Next the message experiences a processing delay in transit through the router; we assumed 1 msec for this processing delay. Next, the last byte must wait for the message to be inserted onto the 56-Kbps WAN facility prior to it being received at the frame relay access module. This is the insertion delay shown in the second row of Table 5.3. The last byte traverses the frame relay network in the last 53-byte cell, which experiences four insertion delays on the frame relay 1.5-Mbps trunks and five switching delays through the five frame relay switches in the reference connection. The last byte must then wait for the entire (original) message to be transmitted out onto the egress frame relay line to the other WAN attached router. The message is forwarded through the router and then inserted onto the other Ethernet LAN segment. Finally, we must account for the propagation delay due to the geographic separation between the two end systems. This succession of delay components is listed in the first set of rows in Table 5.3.

A similar reasoning applies to the output message. The difference here is due to the fact that this message is larger and is carried in two separate packets. Therefore pipelining considerations must be taken into account. This is the reason that the table shows one insertion delay of 1024 byte onto the frame relay access line and another 512 byte delay onto the frame relay egress line (referred to as a pipelining delay in the table).

This single transaction estimate relies on the assumption that the access lines were idle for a period of time sufficient to totally replenish the access control filter's token pools. Given this assumption, then the access control filter will not delay the messages. Therefore, our estimate above represents a best case estimate for this reference connection. Because the value of the CIR does not come into play in evaluating the delay components in Table 5.3, changing the CIR in this reference connection will not have any impact on the transaction delay. The main way to improve the delay performance for this transaction is to increase the speed of the frame relay access and/or egress facilities. Therefore, if this best case transaction delay is not good enough, the frame relay design should change by increasing either of these two frame relay access facilities. Increasing the CIR will have no effect on improving the transaction delay performance.

Finally, it is worth remembering that our analysis in this section depended on the token pool being set to a value greater than the message size of both the input and the output messages. If this were not true, then the access control filter would always buffer part of the message for a period of time before the last piece of the message gains access to the frame relay network. The amount of delay is dependent on the difference between the message size and the token pool and the value of the CIR. If the message is significantly larger than the token pool, then computation of the transaction delay is best accomplished by treating the message as if it were a file transfer and estimating the throughput achieved over the reference connection. Then from this throughput value, estimate the period of time required to send the entire transaction.

5.4.1.5 Bandwidth Sizing Summary

We have suggested a methodology to develop an initial frame relay network design. First, the ports and VCs are sized to carry—at a minimum—the necessary traffic as specified in a traffic matrix. This results in the minimum port and VC sizes to support this traffic. Then, we estimate the throughput achievable for a given transport protocol over this initial design. If the throughput estimates are too small, then the CIRs associated with the VCs will need to be increased if the CIR is determined to be the throughput bottleneck. Otherwise, the port speeds may have to be increased. Finally, the interactive delay performance for various transactions should be considered. If the interactive delays are too large, port speeds may be further increased in speed in order to satisfy delay targets. This level of analysis results in an initial frame relay design that has been fine-tuned based on a relatively simple set of throughput and interactive delay estimates. In the following sections, we discuss more intricate considerations in engineering frame relay networks.

5.5 Traffic Discrimination

5.5.1 Congestion Shift From Routers Into the Network

The speed mismatch between ports and PVCs is a unique characteristic of frame relay, especially in many-to-few connectivity scenarios. This has the consequence of sometimes "shifting congestion from the routers into the network buffers." Consider a many-to-one frame relay network with a T1 frame relay port at the central site and a 56-Kbps port at the remote sites. Typically, the PVC CIR speeds will be less than the remote port speed of 56 Kbps—say, 16 Kbps or 32 Kbps. See Figure 5.17. The central site router, router A, sends data into the network at T1 speeds. However, the maximum rate at which the network accepts data from the router is 56 Kbps.[5] For traffic moving from the central site to a remote location, it is useful to think of central site router

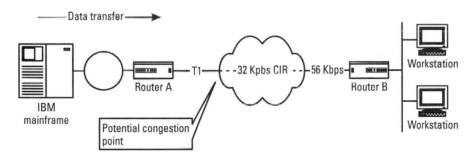

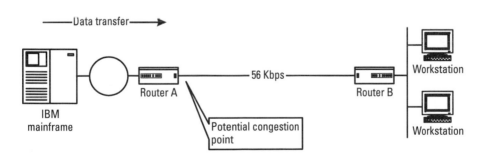

Figure 5.17 Congestion shift in frame relay networks.

---

5. In the open-loop implementation, the network accepts data at T1 speed, but data egresses the network at 56 Kbps. In both cases (open loop and closed loop) the maximum data rate cannot exceed 56 Kbps.

"filling" the ingress PVC buffer at T1 speeds. Correspondingly, the PVC will "drain" the PVC buffer at a rate between the minimum rate (the CIR) and maximum rate of 56 Kbps, the remote site's port speed. Again, in the open-loop implementation, the maximum "drain" rate is also 56 Kbps, except the drain occurs at the egress buffer.

The speed mismatch between fill rate at T1 speeds and drain rate at 56 Kbps may result in buffering of data at the ingress PVC buffers. (In an open-loop implementation, the buffering of data will occur at the egress buffer.) Thus, if the PVC is carrying latency-sensitive traffic (say, interactive SNA or TCP telnet) and file transfer traffic (say, TCP FTP or Web downloads), it is possible for latency-sensitive traffic to be "trapped" behind file transfer traffic at the network buffers. This will occur in spite of routers being able to giving interactive traffic higher priority. This is not the case in the corresponding private line connections where potential congestion points are only within the routers.

The following perspective will help clarify the issue further. Assume the frame relay connection is replaced by a private line connection between the routers at 56 Kbps, and assume the link is carrying interactive and file transfer traffic. The accumulation of file transfer traffic is likely to be at the router serial port. See the lower reference connection in Figure 5.17. This gives router priorities a better chance to be effective. Congestion can shift from the router to the network buffers, when a private line network is replaced by a frame relay network. Whether or not this congestion shift occurs depends on the nature of the background traffic (say, TCP FTP, Web downloads, and so on). It also depends on file transfer window sizes and number of simultaneously active file transfers.

5.5.2 Response to the Congestion Shift

Separating latency-sensitive traffic, or another traffic stream that needs higher priority, on its own PVC can help address the congestion shift just discussed. There are also a few related options listed next that are somewhat restricted in their applicability but which may be useful:

- Giving latency-sensitive traffic a separate PVC;
- Limiting application/protocol window size;
- Prioritizing PVCs in the frame relay network;
- FECN/BECN and DE bit marking support; and
- Traffic shaping and bandwidth management.

The separate PVC option and the other approaches are described in the following paragraphs.

5.5.2.1 Giving Latency-Sensitive Traffic a Separate PVC

Separating latency-sensitive and bandwidth-sensitive traffic streams on different PVCs is one way of addressing the congestion shift problem. It has the highest probability of success in a closed-loop implementation, where the switch buffers data at the ingress on a per-PVC basis. This places a burden on the routers to map different traffic classes on application-specific PVCs. Ordinarily, it is not a problem to have a router try to distinguish TCP/IP, Novell IPX, AppleTalk, or DECnet protocols. However, it is somewhat more complex to separate TCP/IP encapsulated SNA and other IP traffic on separate PVCs, and it is router vendor dependent. There are many occasions where telnet, HTTP, or FTP traffic, when mixed together on a speed-mismatched frame relay network, produces unacceptable echoplex response times for telnet.

The previous discussion raises an interesting issue. Under what conditions, if any, should latency-sensitive traffic be on its own PVC when mixing with bandwidth intensive traffic? There are some guiding factors to consider:

- Interactive traffic is typically light and requires consistent response times. Therefore, the interactive traffic is more vulnerable than file transfers to large variations in PVC loading.

- The applications and protocols that constitute noninteractive traffic typically present large volumes of data sent in the direction from the fast port to the slow port. Typically, these are file transfers or Web downloads. In some cases, these can be produced by periodic broadcasts, such as NetBEUI and Novell SAPs. As we have discussed, large volumes of data are determined by window size, not the size of the files. In this context, the window size is more important to consider than the actual file transfer size. Window size represents the simultaneous amount of data that a single session can send. Therefore, the window size is the maximum amount of data per session that can be resident in any frame relay network buffer at any given time. Correspondingly, the larger the window size of the file transfers, the more contention the interactive traffic is likely to experience. In general, whenever interactive traffic is mixed with FTP, HTTP file transfers, and Novell file transfers, caution must be exercised.

- The smaller the CIR and egress port speeds, the slower the potential drain rate from the VC buffers. This increases the possibility of congestion from file transfer traffic and may require a separate VC for interactive traffic. In other words, for higher CIRs, significant congestion can only occur when there are a large number of simultaneous file transfers. For example, if the port speed/CIR is 1.544 Mbps/256 Kbps, separate VCs may not be as necessary, compared with a 256 Kbps/ 16 Kbps port speed/CIR combination.

- Strict guidelines relative to the level of speed mismatch or central site frame relay port oversubscription that justify a separate VC for interactive traffic are nonexistent.

- Cost considerations surround the use of a second PVC. If PVCs are priced by CIRs, splitting a "large" PVC into two "smaller" PVCs should result in minimal additional cost. It is certainly not true that separating protocols on different PVCs "doubles the PVC cost," as has been suggested in some trade publications. However, pricing can always change. At the time of this writing, the cost penalty of multiple PVCs is minimal. The more important factor to consider—especially for large many-to-one frame relay networks—is the impact on router/FRAD capacity. Consider a 300-location many-to-one frame relay network carrying interactive and file transfer traffic. Assume for a moment 300 PVCs can terminate on a single router—although on several interfaces. The strategy of mapping interactive and file transfer traffic on different PVCs will double the number of PVCs that the router terminates. This may stretch router resources to a point where performance and capacity could be impacted.

5.5.2.2 Limiting Protocol Window Sizes

As discussed earlier, window size plays a key role in this issue. It is natural, then, to inquire whether limiting window sizes for file transfers will reduce potential congestion experienced by interactive traffic sharing the same PVC. Although the answer to this question is yes, this solution is hard to implement and, hence, not recommended. Nevertheless, this approach is worth elaborating on, because it leads to a more complete understanding of the underlying dynamics.

Consider, for example, a 300-node network with classical transaction-oriented applications supported on a centrally located mainframe. These are the lifeblood of the company's business. Good, consistent response times for these applications are critical. Now assume TCP/IP is supported on the mainframe to allow remote workstations to download files from the mainframe using FTP. Ordinarily, file transfers are scheduled during off-hours when there is very little interactive traffic activity. However, there are no guarantees that file transfers and interactive traffic will not mix. What should the network manager do to guarantee acceptable interactive response times at all times?

One option would be to limit the window size for the TCP/IP file transfers. Where should this window size reduction take place? On the mainframe or the remote clients? Note that the files are being transmitted from the mainframe to the clients. To reduce the outstanding data from TCP/IP file transfers, the receive window size advertised by the clients should be reduced. Realistically, in a large network, one is likely to find many different varieties of PCs, laptops, and workstations that need to be supported. Typically, TCP implementations on these devices are often not the same. Some TCP packages allow window size tuning, but many others do not. Even if the window size can be tuned, it may be hard to do. Besides, what should the window size be? Too small a window size will limit throughput. Too large a window size will affect other traffic on the network without improving file transfer performance. For these reasons, this approach has limited applicability and is not recommended.

5.5.2.3 Priority PVCs in the Frame Relay Network

Consider a frame relay network with three hub sites, as shown in Figure 5.18. Here, one hub site has a mainframe, one provides e-mail and intranet/Internet access, and the last provides external data feeds, such as financial market data. A remote location would typically need access to all three hub sites. Suppose the remote location port speed is 64 Kbps, and all three PVC speeds are 32 Kbps. The confluence of these PVCs at a remote location is likely to congest that frame relay port, especially considering the fact that remote frame relay port is oversubscribed by a factor of 3 to 2 ($3 \times 32 = 96$ Kbps > 64 Kbps). Hence, if there is enough traffic on all three PVCs, in the direction from the hub sites to the remote location, to exceed the overall drain rate of 64 Kbps, congestion at that frame relay port is inevitable.

One can try to minimize the possibility of egress port overload through an appropriate choice of port oversubscription. Relative to oversubscription, there are no hard and fast rules, although utilization levels, time-of-day, and time zone characteristics determine the oversubscription level to choose. However, do not be overly conservative in this approach, because there exists an economic advantage to considering a higher value of the overbooking factor. This represents a classic trade-off between performance and price.

No matter how conservative one is in designing the egress port oversubscription factor, periods of temporary traffic overload will occur at the egress port. If these periods are significant (as determined through actual measurements), then one should look to take advantage of tools within the frame relay network technology to address the traffic contention at the egress port. One possibility is to use priority queuing or bandwidth-sharing at the outbound transmit queue in the egress frame relay port. The most effective way to address this congestion issue is to designate some PVCs in the network as high priority

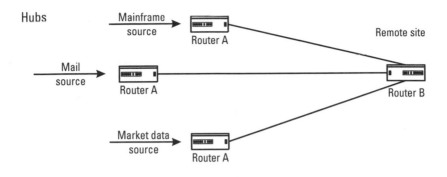

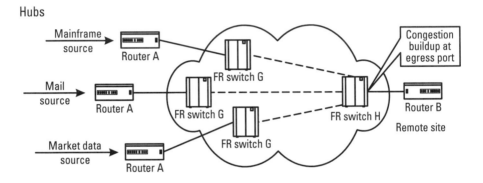

Figure 5.18 Traffic convergence at the frame relay egress port.

compared to the others. In the previous example, the mainframe PVC and the financial market data feed PVC need to be prioritized over the intranet/Internet/e-mail PVC. Most frame relay carriers support PVC priorities in the network. Some implementations offer minimum bandwidth guarantees for lower priority PVCs.

5.5.2.4 FECN/BECN and DE Bit Marking Support

FECN/BECN bits are set on a per-PVC basis by the frame relay carrier when the PVC experiences congestion in its path. In a frame relay network carrying transaction-oriented applications, FECN/BECN congestion notification is usually caused by nontransaction protocols carried on the same PVC. Routers can optionally tag a protocol *discard eligible* (DE) when encapsulating the respective protocols within a frame relay packet. For instance, all frames on a PVC belonging to FTP can be set DE.

It has been claimed that responding to FECN/BECN indications from the network will address the congestion issue when transaction-oriented traffic is mixed with file transfer traffic on the same PVC. Another suggestion that has been put forth is that the non-SNA traffic can be tagged as DE by the router. Both of these techniques can have limited effect. The effect depends on the frame relay carrier's implementation of FECN and BECN and DE thresholds. In reality, they tend to address frame discards rather than delays. In other words, responding to FECN/BECN or setting DE bits for file transfer traffic will likely protect against frame discards rather than guarantee that delay spikes—and perhaps time-outs—do not occur in the transaction-oriented protocol.

Most FRADs and many routers respond to congestion notification on a PVC from the frame relay service provider and take corrective action for that PVC. For example, one popular implementation is to send "receiver not ready" frames to SNA devices configured on the router/FRAD on receipt of a frame with the BECN bit set. The devices are notified about congestion in the forward directiThistects SNA frames from poy being discarded when the frame relay network becomes congested, and thereby prevents potential session loss.

The reason frame discards are protected—but not delayed—is that frame relay implementations typically send FECN/BECN information only when a significant amount of frames have been accumulated in network buffers. When such accumulation occurs, draining these buffers takes a relatively long time. For example, if an egress frame relay port is 56 Kbps, then draining 35 Kbytes of data accumulated at that port takes 5 sec.

Responding to FECN/BECN and setting DE bits are ineffective responses to this problem. The more important question that must be asked is the reason for an inordinate amount of data accumulating in network buffers. This could be due to several reasons—unfiltered router broadcasts, underconfigured CIRs, or excessively large transport layer windows.

5.5.2.5 Traffic Shaping and Bandwidth Management

Traffic shaping is a very broad term that is applicable in many contexts. This term is also used interchangeably with bandwidth management. There are some differences between the two approaches that we will attempt to distinguish.

There are essentially two approaches in the context of frame relay: router-based traffic shaping and bandwidth management using external devices. In router-based traffic shaping, more appropriately called *rate throttling,* the router can be configured not to transmit traffic on a PVC at greater than a specified rate, say, CIR. This eliminates or reduces the congestion in PVC buffers due to speed mismatches, and gives router priorities a better chance to be effective. Hence, traffic shaping is most applicable at hub sites rather than at remote sites. Bandwidth management, on the other hand, is accomplished through external devices, colocated with routers, that are capable of very granular schemes for allocating bandwidth and provide some schemes to support rate throttling.

Router-Based Traffic Shaping

In its simplest form, traffic shaping can be implemented on a per-PVC basis at the frame relay layer. Some routers have the capability of rate throttling to specified levels. Rate throttling can be accomplished by setting the excess burst size B_e (see Chapter 2) appropriately. For instance, setting

$$B_e = T_c \times [\min(\text{Ingress port speed}, \text{Egress port speed}) - \text{CIR}]$$

where the T_c is the measurement interval, would enable the router to take advantage of full bursting to port speed in the connection. The same approach can be used to limit the bursting to a fraction of the CIR. Setting $B_e = 0$ implies a complete rate throttle to CIR on a connection.

In some situations it may be necessary to throttle the rate on a PVC to a rate lower than the CIR. Imagine two separate frame relay ports of 128 Kbps with PVEs of 48 Kbps and 16 Kbps. Assume that the two PVCs terminate on a remote 64 Kbps frame relay port in a triangle configuration. Although the sum of the CIRs is equal to 64 Kbps, congestion will occur at the remote port because both PVCs will attempt to burst to 64 Kbps. If the 16 Kbps PVC is carrying time-sensitive traffic, then it makes sense to throttle the rate of the other PVC to a value less than 48 Kbps.

Router-based rate throttling should be used in conjunction with prioritization because rate throttling pushes the congestion back to the routers. The method of prioritization varies between router vendors.

Some routers allow for rate throttling to occur in response to receipt of FECNs and BECNs. While this approach might seem reasonable, it may not be very useful depending on how long the network buffers wait before signaling congestion. For time-sensitive traffic like TCP telnet, even the slightest congestion will be noticeable.

Note also that BECNs are, in a sense, more important than FECNs. Because BECNs notify congestion in the direction opposite to the direction of data transfer, rate throttling will be effective because it will reduce the data transmitted into the network. Rate throttling in response to receipt of FECNs will not be effective.

Bandwidth Management Techniques That Use External Devices

In this subsection, we present a brief overview of bandwidth management techniques as implemented in devices by Xedia, Packeteer, and others. The approach to bandwidth management and traffic shaping adopted in these devices appears to have promise in providing an overall QoS strategy for applications over a WAN.

Bandwidth management provides a flexible approach to first classifying user traffic and then assigning bandwidth characteristics to each traffic class. The classification can be quite granular in that applications can be sorted according to well-known TCP port numbers, UDP, source/destination addresses, and so on. Although similar bandwidth allocation techniques are available in routers today, the classification is usually not very granular. For instance, different classes of Internet or intranet traffic cannot be treated differently. More important, unlike router implementations, bandwidth management combines this granular prioritization with rate throttling, that is, these devices can be configured to limit bandwidth allocated to different classes. To illustrate this point, routers cannot be configured to allocate a maximum bandwidth of 4 Kbps for each Internet user at a given location; this is, however, feasible using external bandwidth devices.

Thus the distinction between bandwidth management and traffic shaping is that the former allows an explicit bandwidth allocation scheme at a granular level, whereas traffic shaping implies rate throttling. The two schemes should be combined to truly achieve differential services between classes of applications. Bandwidth management devices implement these features more efficiently than routers.

Bandwidth management devices are active devices in that packets from end systems flow through them before they are processed at the router. Conceptually, the easiest way to think of them is that they implement the bandwidth management and traffic shaping function outside the router.

Queuing Versus TCP Rate Control

A variety of external bandwidth management devices are available in the marketplace, and all of them have one primary purpose: to effectively allocate and manage bandwidth for each traffic class. However, they use vastly different approaches to achieve that objective. Broadly speaking, there are two approaches: queuing and TCP rate control.

Let us consider the case of a device that implements queuing. When a packet comes into the device, it is first classified. If the traffic class has not yet used all of its bandwidth, the packet flows immediately onto the outbound link. If a packet comes in and the class is attempting to use more than its committed rate, there are two possibilities: either the packet is placed in a queue and it is rate shaped or it is allowed to "borrow" from the currently idle bandwidth of any other traffic class (if the class is specified to be "borrowable").

In a TCP rate control scheme used by Packeteer's PacketShaper®, the granularity of bandwidth management is at the level of a specific TCP flow between a client and a server. For traffic exceeding the specified level, TCP rate control uses two methods to control TCP transmission rates: delaying acknowledgments and reducing advertised TCP window sizes. The PacketShaper "sits in the middle" of a TCP flow and intercepts the TCP packet and literally changes the header information that contains acknowledgment numbers, sequence numbers, and window advertisements. Thus, in a real sense, the delayed acknowledgments and reduced window advertisements provide a rate throttling mechanism by informing the sending TCP stations to slow down.

This issue of device placement in the network is important. Devices using the queuing approach are usually placed behind WAN routers at specific locations in the network, such as data centers, hub locations, or other potential points of congestion. For devices relying on TCP rate control, it does not matter where the device is placed—at the hub site or at the remote locations—because it manages traffic at a TCP level. However, in this case, it might make more sense to place the device at the remote locations because individual TCP sessions are better managed at the end user level rather than at the hub site.

Note also that it is likely that multiple devices would need to be placed at a single campus location, and that these devices typically work independently. For instance, consider two devices placed in the campus with one configured to manage Internet traffic and the other to manage client/server traffic. As far as the WAN is concerned, both traffic streams could share the PVC to the remote location. Because the two devices act independently of each other, one would necessarily have to bandwidth manage the Internet traffic stream to ensure good client/server performance. Over frame relay, this implies that the inherent burst capability of the network may not be fully utilized.

5.6 Global Versus Local DLCI

We have discussed how to design frame relay networks to ensure good delay and throughput performance. We have discussed traffic considerations, traffic discrimination issues, and considerations related to mapping IP subnets onto underlying PVC topologies. Here, we consider frame relay PVC addressing and CPE limiting design issues. We first introduce the concept of global versus local data link connection identifiers (DLCIs) and the impact on customer provisioning and maintenance processes. In the following section, we finish with a consideration of various scaling issues in designing large frame relay networks.

Typically, specific frame relay DLCIs only have local significance. That is, a given PVC on a port is randomly assigned a DLCI number at provisioning time. In this case, there is no relationship between the local DLCI number and the remote site. Therefore, each router on the frame relay network will have a unique table mapping the DLCIs to the IP address of the remote site. This complicates customer provisioning and troubleshooting. This is illustrated in Figure 5.19.

Given the random relationship between DLCI numbers and remote IP addresses, it becomes confusing to the network designers and engineers when laying out the design for new locations or when troubleshooting connectivity issues on existing locations.

To simplify the network design, frame relay providers often support global DLCIs within their service. When implementing global DLCIs, there is a strict relationship between the DLCI number and the far end site. Figure 5.20 shows the network from Figure 5.19, except that now global DLCIs are in use.

In Figure 5.20, a simple algorithm is used to determine the far-end IP address from the global DLCI number; that is, the DLCI number is equal to the last digit of the far-end site IP address, which is written in dot-decimal notation. The exact algorithm is not important, just the fact that there exists

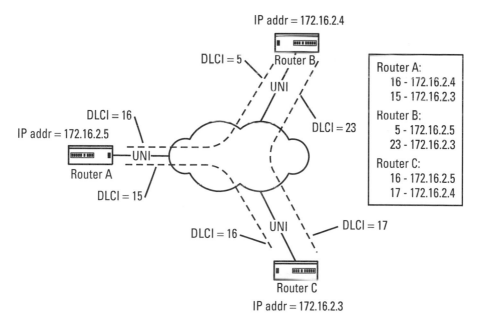

Figure 5.19 DLCI to remote IP address mapping for local DLCIs.

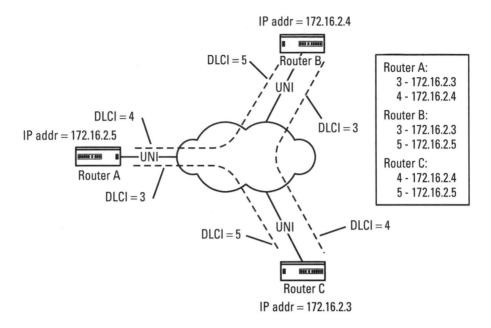

Figure 5.20 DLCI to remote IP address mapping table for global DLCIs.

a strict relationship. This simplifies the router configuration. Because there is essentially one master table, it is easier to build the router configurations. All table mappings will be the same for all routers, less their own entry. This is not an obvious advantage with a three-node network; however, when addressing 25 to 100 nodes, this scheme provides for simple DLCI-to-IP address configuration management.

5.7 Virtual Circuit Scaling Issues

Several issues arise when the size of the frame relay deployment exceeds several hundred sites. Although most implementations are much smaller than this, it is useful to understand various limitations to large-scale deployments. We discuss two such limitations, one related to the size of the DLCI field in the frame relay header and its impact on global DLCI use, and the other is due to the maximum number of PVCs that a typical router can support.

Figure 2.7 in Chapter 2 shows the header format for a frame relay packet. This shows that the field for the DLCI numbers is 10 bits in length. This means that there are at most $2^{10} = 1024$ DLCI numbers supported on a given frame relay interface. In fact, there are less. DLCI numbers 0 and 1023 are used by the LMI protocols, while DLCIs 1–15 and 1008–1022 are reserved for frame relay multicasting applications. This leaves $1024 - 32 = 992$ point-to-point DLCIs for a given frame relay port. For large frame relay deployments, this limit has implications on the use of global addressing schemes.

Most frame relay deployments are roughly star topologies, with one or two data centers to which the remaining remote locations connect. The data center locations must terminate one or two PVCs per remote location. The total number is dependent on potential connectivity, redundancy, and performance considerations. Therefore, straightforward applications of global DLCI schemes will scale to hundreds of remote locations. In the event that the number of remote locations is a thousand or more, the DLCI limit of 992 will be exceeded, and the global addressing scheme will fall apart. In these cases, it is necessary to divide the network into "logical" subnetworks, such that a global addressing scheme can be implemented within each of the logical subnetworks. To interconnect subnetworks, a smaller set of DLCIs is reserved for this purpose.

Another limit encountered when scaling to large frame relay deployments is due to the limit in routers to terminate large numbers of VCs. Routers resident within these corporate data centers are expected to terminate large numbers of frame relay circuits. Typically, routers view these individual circuits as separate point-to-point links. Routers were not originally developed to terminate hundreds of point-to-point links. High-end routers from vendors can typically support several hundred or less links per router and tens to a hundred links per interface card. In data center architectures supporting thousands of remote frame relay locations, this requires 10 or more frame relay ports coming into the data center and would require five or more routers due to these PVC limits. When dealing with networks of this size, other considerations come into play as well (e.g., throughput limits or routing protocol design issues). Both of these issues may force more routers within a given data center architecture. When planning router deployments of this size, it would be wise to engage the specific router vendor to aid in the design and planning phase of your implementation to ensure that proper consideration has been given to all router limitations.

5.8 Summary

Carrier pricing for public frame relay services is attractive. While distance, speed, and configuration variables make one-for-one cost comparisons between private lines and frame relay difficult, experience suggests that customers realize savings when migrating from leased lines to public frame relay services. This is particularly true when multiple networks are consolidated. This is not always the case in one-for-one mappings, especially when additional equipment and management costs are considered. However, for a large number of networks frame relay is an economically attractive technology.

There are several issues to remember when planning a migration from a private line based WAN to a frame relay based WAN. This chapter touched on many issues to consider when making this migration. We first discussed a straightforward comparison of frame relay and private line environments. The list of considerations discussed when comparing these two WAN designs were:

- Delay and delay variation differences;
- Open-loop versus closed-loop congestion control strategies;
- Speed mismatch between frame relay ports and VCs;
- Overbooking;
- Egress port buffer traffic convergence.

We next discussed several issues unique to global frame relay deployments. These issues included (1) additional delays due to the existence of NNI to interconnect the frame relay carrier networks, and (2) end-to-end management and monitoring difficulties resulting from the concatenation of multiple frame relay services.

Following this, we focused on network congestion, congestion points, and design considerations to mitigate congestion-induced delays in frame relay designs. These design considerations included:

- Separating delay-sensitive applications onto their own PVC;
- Limiting application/protocol window size;
- Prioritizing VCs in the frame relay network;
- FECN/BECN and DE bit marking schemes;
- Traffic shaping and bandwidth management at the router.

Finally, we addressed several PVC topology considerations within frame relay networks. These centered around global DLCIs versus local DLCIs and scaling issues with respect to DLCI limits on frame relay ports and on routers terminating the frame relay connections on customer premises.

References

[1] Stevens, W. R., *TCP/IP Illustrated, Volume 1: The Protocols*, Reading, MA: Addison-Wesley, 1994.

[2] Feller, W., *An Introduction to Probability Theory and Its Applications*, Volume 1, 3rd ed., New York: John Wiley & Sons, 1968.

6

Using Pings for Performance Analysis

6.1 Introduction

Network managers and technicians usually resort to pings to measure latency, and FTP to measure throughput on a network connection, in order to draw conclusions about network performance. The network managers may be troubleshooting a specific performance problem or performing a network validation test.

To be able to draw the right conclusions about network performance from pings, one should compare the measured performance against expected performance based on calculations and some reasonable assumptions. In this chapter, we discuss how to calculate expected ping delays and how these calculations can be used to estimate WAN latency. We will also demonstrate how large pings can be used for rudimentary measurements of bursting available for PVCs in a frame relay network. We will discuss leased lines and frame relay WANs.

The calculation of expected throughput for a network connection using FTP is discussed in Chapter 7.

This chapter is organized as follows. In Section 6.2 we describe the ping program and discuss important aspects such as protocol overhead. In Section 6.3 we discuss ping delay calculation for leased line and frame relay connections. In Section 6.4, we will demonstrate how small pings can be used to highlight network latency issues. Section 6.5 discusses general issues and caveats in using pings for estimating network latency. In Section 6.6 we show how large pings can be used to demonstrate whether or not PVCs can burst above their CIR for a frame relay connection, and some overall issues in the use of large pings for calculating throughput.

6.2 Pings

We start with the ping program. (For an excellent and detailed discussion of ping and FTP and, in fact, of TCP/IP networking in general, see Stevens [1].) The ping program (named in analogy to a sonar ping) sends a number of Internet Control Message Protocol (ICMP) *echo_request* packets to a given destination. The destination is required by the ICMP protocol to reply with an ICMP *echo_reply* message. Pings can be sent between any two hosts and are often used to test network reachability.

In addition to reachability, the ping program returns the round-trip time, which, along with information such as ping size, line speeds, and distance, can be used to measure performance of the connection. The round-trip time is measured by the ping program at the source that clocks the time between when *echo_request* is initiated to the time when the *echo_reply* is received.

Figure 6.1 shows the output from the execution of the ping program. The first line shows the command line options when executing the program. The following lines show the program output. The program, in this case, continues until the packet count, indicated with the "–c 10" command line option,[1] is

```
>ping 10.10.10.10 -c 10 -s 56 -i 2
PING enskog (10.10.10.10): 56 data bytes
64 bytes from 10.10.10.10: icmp_seq=0 ttl=32 time=0.9 ms
64 bytes from 10.10.10.10: icmp_seq=1 ttl=32 time=0.8 ms
64 bytes from 10.10.10.10: icmp_seq=2 ttl=32 time=0.7 ms
64 bytes from 10.10.10.10: icmp_seq=3 ttl=32 time=0.8 ms
64 bytes from 10.10.10.10: icmp_seq=4 ttl=32 time=0.8 ms
64 bytes from 10.10.10.10: icmp_seq=5 ttl=32 time=0.8 ms
64 bytes from 10.10.10.10: icmp_seq=6 ttl=32 time=0.8 ms
64 bytes from 10.10.10.10: icmp_seq=7 ttl=32 time=0.7 ms
64 bytes from 10.10.10.10: icmp_seq=8 ttl=32 time=0.8 ms
64 bytes from 10.10.10.10: icmp_seq=9 ttl=32 time=0.8 ms
--- enskog ping statistics ---
10 packets transmitted, 10 packets received, 0% packet loss
round-trip min/avg/max = 0.7/0.7/0.9 ms
>
```

Figure 6.1 Typical output from the ping program.

---

1. The example ping program discussed in this section is specific to the LINUX operating system. Other operating systems may have different command line options.

met. This is shown on the last line. The output lines list the destination address, whether the destination returns the ping, and, if it does return the ping, the measured value of the round-trip delay. The program provides the minimum/average/maximum round-trip delays of the successful pings at the conclusion of the program output.

The ping program is standard in all TCP/IP host implementations. One can use many options with the program. As an example, the ping program distributed with LINUX has command line options that allow the user to specify a number of variables when executing the program. These typically include:

- The number of *echo_request* messages sent (this is specified through the –c option in the command line);

- The interval between sending the ping messages (this is specified through the –i option in the command line and generally defaults to once a second);

- The size of the ICMP *echo_request* data field (this is specified through the –s option in the command line and generally defaults to 56 bytes of ICMP data); and

- Whether to allow for packet fragmentation or not (the default is usually not to allow for packet fragmentation).

These capabilities make the ping program a simple and extremely useful diagnostic tool.

One final piece of information regarding the ping program is required before we can put it to use in building a delay model of a network reference connection: message format and protocol overheads. This discussion is necessary in order to calculate insertion delays of the *echo_request* and *echo_reply* messages.

ICMP messages are carried directly inside IP packets. The ICMP echo message consists of an 8-byte header and is followed by the ICMP data. Within the ping program, the size of the ICMP data field is specified on the command line. Figure 6.2 shows an example of an ICMP echo message, which is fragmented. Link layer technologies, for example, Ethernet, token ring, and frame relay, have maximum frame sizes associated with them. The IP protocol addresses this fact through the implementation of IP datagram fragmentation. If a router or host wants to transmit a datagram onto a subnet having a maximum frame size smaller than the datagram, then the router or host may fragment the datagram into smaller multiple datagrams.

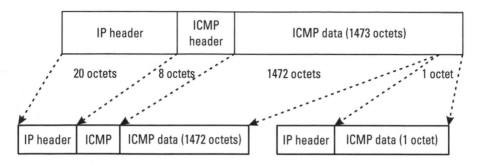

Figure 6.2 Message format for a fragmented ICMP echo message.

6.3 Calculating Ping Delays

We first consider a leased line connection and then a frame relay connection.

6.3.1 Leased Line Connection

Consider reference connection RC#1 shown in Figure 6.3. It shows a leased line connection between location A and B at 56 Kbps.

We will make the following assumptions for the ping calculation:

- The distance between the routers is roughly 1000 (airline) miles.
- Host A pings host B with a ping size of 100 bytes (i.e., ICMP echo packets are 100 bytes).
- The WAN connection is lightly loaded.
- LAN, router, and host delays are negligible.

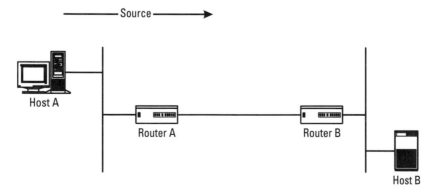

Figure 6.3 Reference connection RC#1 for calculating ping delay.

We now have enough information to calculate the expected ping delay. The following computations apply:

- Frame size = 100 + 20 (IP) + 8 (ICMP) + 8 (PPP) = 136 bytes.
- Insertion delay on the 56 Kbps link = 136 × 8/56,000 sec = 19 msec.
- One-way propagation delay (10 msec for every 1000 airline miles) = 10 msec.

Hence expected ping delay for this connection is 2 × (19 + 10) = 58 msec. The factor 2 accounts for the fact that the ICMP packets are echoed.

The total path traveled by the ping message, along with the associated delay estimates, is shown in Figure 6.4.

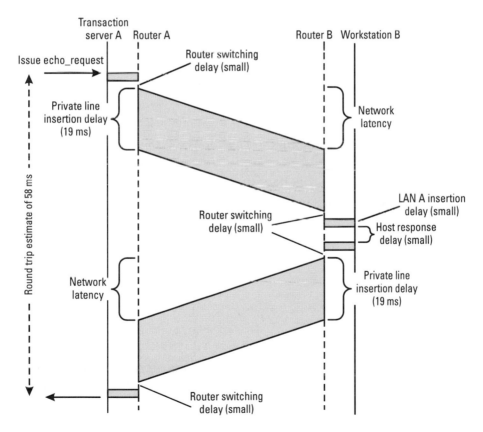

Figure 6.4 Calculation of the minimum ping delay for RC#1.

6.3.2 Frame Relay Connection

Consider a frame relay connection RC#2, as shown in Figure 6.5. Calculating ping delays for the frame relay case is a little more involved. One has to take into account issues such as pipelining, store-and-forward delays at the network edges, switch delays, and burst levels in the PVC.

For the sake of illustration, we will make the following assumptions:

- The frame relay connection is lightly loaded.
- Host A pings host B with a ping size of 100 bytes (excluding headers).
- The distance between the routers is roughly 1000 (airline) miles.
- The PVC is carried via four hops (five switches) and each switch adds 1 msec of latency.
- The carrier provides full bursting to the port speed without frame drops in both directions.
- LAN, router, and host delays are negligible.

The following computations apply for frame relay:

- Frame size = same as for leased line = 136 bytes.
- Insertion delay on 512 Kbps port = $136 \times 8/512,000$ sec = 2 msec.
- One-way network latency = 10 msec (1000 miles) + 5 msec (switch latency) = 15 msec.
- Insertion delay on 128 Kbps port = $136 \times 8/128,000$ sec = 8.5 msec.
- Expected ping delay = $(2 + 15 + 8.5) \times 2 = 51$ msec.

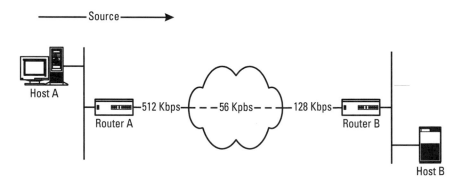

Figure 6.5 Reference connection RC#2 for calculating ping delay.

The calculation for the minimum ping for RC#2 is shown in Figure 6.6.

6.3.3 Observations

Note that propagation delay is a major contributor to the overall ping delay. For larger ping sizes, the relative contribution of the propagation delay will decrease.

The service provider (leased line or frame relay) should be able to provide the network latency (which includes propagation delay) number—in our case, 10 msec for leased line and 15 msec for frame relay. It is important to understand how the carrier measures latency—switch to switch or point of presence

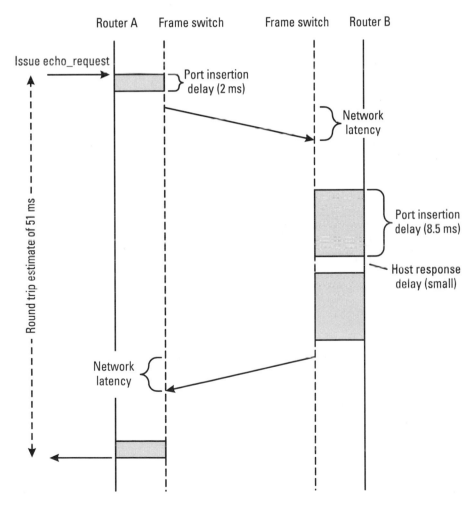

Figure 6.6 Calculation of the minimum ping delay for RC#2.

(POP) to POP. Most carriers advertise POP-to-POP delay objectives for frame relay, but cannot track these delays because not all POPs have frame relay switches. In other words, frame relay connections may have to be back-hauled to the nearest available POP.

Comparing the ping delay for the leased line and frame relay connections, we find that the leased line connection delay is actually longer. This is because we assumed full bursting to port speed. This offsets the extra delays over frame relay.

If, for some reason, the carrier limits the bursting to CIR in both directions, then the ping delay calculation needs to be changed slightly, as follows (assuming that frames are pipelined into the network using ATM cells):

- Insertion delay for a 136-byte frame in the A to B direction = $136 \times 8/56,000 + 0.015 + 136 \times 8/128,000 = 42.5$ msec;

- Insertion delay for a 136-byte frame in the B to A direction = $136 \times 8/56,000 + 0.015 + 136 \times 8/512,000 = 36$ msec; and

- Total ping delay = 80 msec (approximately).

If the underlying network is more complex (e.g., a hierarchical backbone of routers), then one must account for the fact that the forward and return paths may vary depending on what routes are chosen. In particular, the return path may not be the same as the forward path.

We have assumed that host and router delays are negligible, but they may not always be. One can use pings to estimate the additional latency due to these components. We show a simple method using the connection in Figure 6.3.

Suppose we send a ping from router A to host B. The ping delay components are (1) router processing delay at router A, (2) network delay, (3) router processing delay at router B, (4) LAN delay between router B and host B, and (5) host B processing delay.

To isolate the network delay component, one can first get independent measurements of router processing delay at router A and router B by sending pings between the local interfaces of the routers. These numbers can be subtracted from the ping delay between the two routers A and B. The remainder should be a reasonable estimate of the network delay. Next, if we were to subtract the router to router ping number from the router A to host B ping number, this gives an estimate of the host and LAN delays at the location B. Alternatively, one can ping router B from host B and subtract the processing delay on router B to get another estimate of the same metric.

One should perform these measurements a number of times to get an idea of the average and variability. Also note that the router processing times obtained thus are not representative of the overall processing delays in the router. For instance, an SNA packet requiring TCP/IP encapsulation may require more CPU cycles than pings. In addition, pings may be treated as lower priority packets for processing.

For the leased line case mentioned earlier, a simple way to account for background load would be to double the insertion delay. This implicitly assumes a 50% load on the WAN link in both directions and an M/M/1 queuing model [see (3.5) in Chapter 3]. If more accurate utilization numbers are available, then the following formula can be used to estimate the delay in any one direction:

Insertion + Queuing delay = Insertion delay on the 56-Kbps link / $(1 - U)$

where U is the utilization of the link in the direction in question.

For the reference connection in question, the assumption of 50% load in both directions will mean an increase in the ping delay estimate from 58 to 96 msec.

For the frame relay case, the calculation of ping delay under load is more complex and will not be discussed here.

6.4 Using Pings to Verify Network Latency

One useful application of the ping program is to ping between various location pairs within a data network and to compare the measured results to the expected delays. If the expected and measured delays differ significantly (say, more than 20%), then two possibilities arise—either the network is not performing as designed, or the assumptions underlying the calculations (for example, the link is assumed to have low load) are inaccurate. In both cases, investigations into the discrepancy should take place.

We first need to show how network latency can be estimated from the results of a ping.

In Section 6.3, we showed how ping delays can be calculated using some assumptions about distances and switch delays for an unloaded leased line or frame relay connection. Conversely, one can estimate the network latency given information about the ping delay, the ping size, and line speeds for the connection. Again, one would have to make some assumptions regarding LAN and router delays, and background load.

We will illustrate with two actual examples. These examples are taken from a real network with headquarters in Pittsburgh, Pennsylvania. Please also see Case Study 1 in Chapter 11.

Example 1: Pings Reveal That the Network Connection Is Healthy

The first example is a 512-Kbps leased line circuit between Pittsburgh, Pennsylvania, and Anaheim, California. The observed ping delay between the WAN routers at these locations for a 100-byte ping is 56/59/60 (min/avg/max). The connection is lightly loaded (peak utilization under 20%).

Let us use the minimum ping delay number, 56 msec, to calculate the best possible network latency. The average and maximum delays can be used to estimate representative numbers as well as a measure of delay variability.

The estimated network latency in this case is:

$$= \text{Observed ping delay} - 2 \times \text{Insertion delay}$$
$$= 56 - 2 \times (136 \times 8/512,000) = 52 \text{ msec (approximately), or about}$$
$$26\text{-msec one-way delay.}$$

Is this consistent with the fact that the connection is a leased line between Pittsburgh and Anaheim? The distance between Pittsburgh and Anaheim is about 2300 miles. Hence the propagation delay is expected to be about 23 msec one way. This is fairly close to the measured number and hence the conclusion is that there are no unusual latency issues in this connection, other than delays that could occur due to overutilized physical connections.

Example 2: Global Frame Relay Connection

The second example is an international frame relay connection between Pittsburgh and Australia. The frame relay ports at the two locations are 256 Kbps with a 128-Kbps PVC between them. As before, the connection is lightly loaded. For a 100-byte ping between the two WAN routers, the ping delay is measured as 384/390/404 msec (min/avg/max).

As in Example 1, the round-trip network latency between Pittsburgh and Australia can be estimated as

$$0.384 - (136 \times 8/256,000 + 136 \times 8/256,000) \times 2 \text{ sec} = 367 \text{ msec}$$

Is this reasonable? Rationalizing the latency using distances is very hard for global frame relay connections. For instance, even if we use a conservative distance estimate of 10,000 miles from Pittsburgh to Australia, the round-trip latency estimate is only 200 msec. The reason for the discrepancy is that global frame relay connections between the United States and other countries usually traverse three "clouds"—a U.S. domestic frame relay network, a global backbone, and a third frame relay network in that country. These clouds are connected via NNIs that typically clock at T1 or higher speeds. Hence there are several store-and-forward delays along with switch delays in the clouds.

Rather than attempting to validate the 367 msec, the best approach would be to repeat the ping tests multiple times and arrive at the network latency (which, for Australia, could well be in the vicinity of 350 msec).

6.5 General Comments Regarding the Use of Pings to Estimate Network Latency

While it is easy to use pings to estimate network latency, one should keep some key issues in mind.

Should We Use the Minimum, Average, or Maximum Ping Delay?

As we have seen, the min/avg/max delays are reported when a ping is success-fully executed. If the ping delay exceeds a time-out value, then it is discarded from the min/avg/max delay reporting. The question is which numbers should be used for estimating network latency?

The minimum round-trip delay is well defined. In fact, it is the sum of all the fixed delays in the connection for the packet (as shown, for example, in the timing diagram in Figure 6.4). It is the best delay possible on the connection during the period of observation. Thus to obtain an estimate of network latency (i.e., delay on an unloaded circuit), one can use the minimum ping delay. In addition, if a large number of ping measurements result in a signifi-cant number of the round-trip measurements close to the minimum delay, then one can be fairly certain about the accuracy of the measurement of the mini-mum delay.

The other fixed delays include the switching delays in the routers, which tend to be relatively small in general, and the host response delay. One way to estimate (or back out) some of these delays is to run pings on intermediate points along the reference connection, for example, ping the routers along the path.

Finally, the maximum ping delay measurement, while giving some indi-cation of the variation in the round-trip delays, is not very meaningful and will likely vary significantly with each separate run of ping measurements.

Effect of Background Load

The inference of network latency from ping delays are most accurate when the underlying connection is lightly loaded. By network latency, we mean switch delays (for frame relay) and propagation delay. Hence, as such, using ping to estimate network latency will assist in uncovering unusual problems in the connection (such as excessive backhaul for the carrier's POP to the nearest frame relay switch). Drawing inferences about network latency using ping delays under load is very difficult and subject to error.

Using Ping Delays to Represent Application Delays

This is quite obvious, but it must be stated: Ping delays cannot be automatically used to infer how applications will perform over the connection. One would have to understand application transactions, message sizes, frequency, and so on, to quantify response times. However, ping delays (as mentioned before) gives an estimate of network latency, which can then be used to characterize application performance.

Pings May Be Treated as Low-Priority Packets

Some hosts, including routers, may treat ICMP echoes as lower priority packets, thereby potentially misrepresenting the delays. Allocating lower priority on a slow-speed WAN interface on a router can impact the delays more than if ICMP echoes are treated as lower priority packets in the processor. Some routers allow for explicitly treating ICMP echoes as high-priority packets.

Using Pings to Estimate Delays for Global Connections

Pings are especially useful when trying to estimate latency for global frame relay connections. Please see discussion in Chapter 5, Section 5.3.

6.6   Calculating Delays for Large Pings

Default ping sizes on most systems are small (less than 100 bytes). However, pings can be quite large, sometimes as large as 65 Kbytes. Small and large pings may be generated in response to performance problems, while the network manager is attempting to validate delays in the underlying WAN. Large pings can also be used quite effectively over frame relay connections to verify whether or not the carrier is allowing PVCs to burst, as we will demonstrate in this section.

If the ping size is less than the serial line MTU (maximum transfer unit) on the routers, then the ping delay calculation proceeds exactly as shown in the last section. For instance, if the ping size is 1000 bytes, then one would use a frame size of 1036 bytes and perform the calculations, since the serial line MTU on most routers defaults to 1500 bytes. If the ping size exceeds 1500 bytes, then the router will have to fragment the IP packet. The calculation of ping delays then becomes a little more involved, especially for frame relay, as shown in the following examples.

6.6.1 Example 1: Leased Lines

Assume the reference connection RC#1, a leased line connection with 56 Kbps of WAN bandwidth between locations A and B. Suppose an 11,832-byte ping is sent from A to B. Figure 6.7 shows the timing diagram for this ping packet. Due to the fact that the ping message is now larger than the MTU size on the Ethernet LAN segment, that is, 1500 bytes, the message must be fragmented into eight IP packets. The maximum data field of an Ethernet frame is 1500 bytes. From this we must subtract 20 bytes for IP header information within each Ethernet frame. This yields 1480 bytes per frame for higher level data. Finally, given that the ping message is fragmented, only the first fragment carries the ICMP header of 8 bytes. Therefore, our ping message should fragment into eight separate frames, the first with 1472 bytes of ping data and the last seven with 1480 bytes of data (or a total of 11,832 bytes).

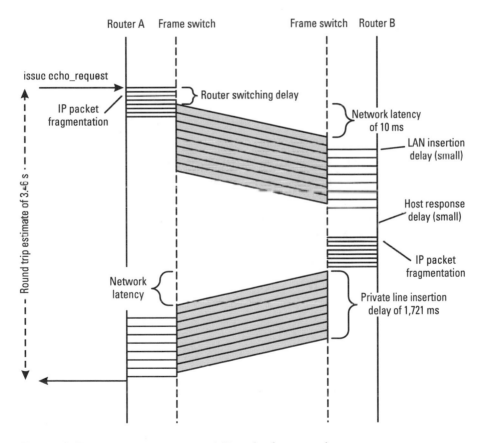

Figure 6.7 Timing diagram of a ping with IP packet fragmentation.

For large ping messages, the transmitting workstation performs the fragmentation necessary for the message to be transmitted over the local Ethernet segment. The eight separate IP packets pipeline their way across the reference connection. The destination host must accumulate all eight packets in order to reconstruct the entire ping message prior to issuing the ICMP *echo_reply* message. Because this reply is equal in size to the ICMP *echo_request* message, it must also be fragmented by the destination host. Again, these eight IP packets pipeline their way back to the original sender, are accumulated, and the message is fully reconstructed prior to stopping the ping round-trip timer.

Adding up all of these delays, we get a total round-trip delay of roughly 3.46 sec. Here are the details for the calculation:

$$\text{Insertion delay for large packet} = 8 \times [(1472 + 8 + 20 + 8) + 7 \times (1480 + 20 + 8)] / 56,000 \text{ sec} = 1.721 \text{ sec}$$

$$\text{Propagation delay} = 10 \text{ msec}$$

$$\text{Total delay} = 1.731 \text{ msec} \times 2 = 3.46 \text{ sec}$$

6.6.2 Example 2: Calculating Large Ping Delays Over Frame Relay

This example is the same as Example 2 in Section 6.4.2 of this chapter. The frame relay connection is between Pittsburgh, Pennsylvania, and Australia with a 256-Kbps port at both ends and a PVC with 128-Kbps CIR connecting them. The connection is lightly loaded. The frame relay connection goes through two NNI links at T1 speeds (three frame relay networks). An 18,024-byte ping was issued from the Pittsburgh router to the router in Australia. The observed delay was 1688/1810/2052 msec. How does it compare with the target (i.e. calculated) delay?

The 18,024-byte payload will be broken up into 13 fragments—the first being 1472 bytes, the next 11 being 1480 bytes, and the last being 272 bytes. Each will be sent as an IP datagram with 20 bytes of header and 8 bytes for frame relay, with the first packet having an extra 8 bytes of header for ICMP echo. Hence the individual frame sizes over the WAN are 12×1508 plus 300 bytes for a total of 18,396 bytes.

The next step is to assume a number for network latency. We can use the same numbers from Example 2 in the previous section—367 msec. However, this number slightly underestimates the latency because the 1508-byte frame has to traverse the NNI links in a store-and-forward manner. Thus the increase should be $4 \times 1508 \times 8/1,536,000$ sec (four hops through a T1 for a 1508-byte frame at T1 speeds), which is equal to 31 msec. Hence the total network latency will be about 400 msec.

To calculate the expected ping delay, let us first assume that the connection is bursting to the full port speed (best case). For this case,

Delay from Pittsburgh to Australia = 8 × 1508/256,000 + 0.2 (one-way latency) + 8 × 18,396/256,000 sec = 822 msec

Ping delay = 2 × 0.822 = 1.64 sec

Notice how close the calculated ping delay is to the minimum observed ping delay—1.64 sec versus 1.69 sec, a 3% error. This is remarkable considering the fact that a number of assumptions have been made regarding the network and the latency.

What conclusions can be drawn from this observation? It is clear that the PVC is bursting to the full port speed in both directions (at least during the time the ping test was done). If it were not, the observed ping delay should be much larger. For instance, if there was no bursting at all on the 128-Kbps PVC, then the calculated ping delay will be

Delay from Pittsburgh to Australia = 8 × 18,396/128,000 + 0.2 + 8 × 300/256,000 (last frame) sec = 1.36 sec

Ping delay = 2 × 1.36 = 2.72 sec

Clearly, the maximum observed ping delay is larger because of some intermittent background traffic skewing the average delay number. As stated before, the minimum delay number provides some insight into whether or not the PVC is bursting to port speed.

6.6.3   Some Comments Regarding the Use of Large Pings to Calculate Throughput

It is sometimes useful to perform a large ping test and use the results to make some low-level estimates of connection throughput. In the first example (Example 1) of this section, we were able to transfer 11,832 bytes in one direction in 1.73 sec (one-half of the ping delay). This assumes symmetry in the reference connection. (This is true in our analysis because we have assumed that there are no queuing delays along the path that would generate an asymmetry.)

Hence the calculated throughput is about 55 Kbps out of a maximum of 56 Kbps.

In Example 2 for the global frame relay connection, we found that a ping message with 18,024 bytes of data had a round-trip time of roughly 1.7 sec. Hence the connection transferred 18,024 bytes in one direction in roughly half the round-trip time, or 0.85 sec. Equivalently, the throughput is 18,024 bytes per 0.85 sec or 170 Kbps, out of a maximum of 256 Kbps.

Is this the best possible throughput for this connection? No. Throughput should really be measured by sending a continuous stream of packets in any one direction, as in a TCP bulk data transfer with a large enough window size. A large ping, no matter how large, is window limited (as in Novell burst mode), and therefore inappropriate for throughput calculations.

One other comment regarding determining burst levels in PVCs is in order. In Example 2, we were able to demonstrate that the PVC was indeed capable of bursting at port speed rate. We required that the connection be lightly loaded to facilitate the analysis. However, what is the guarantee that the service provider can sustain burst levels to port speeds when their backbone network is busy? This is an important question for most network managers. The only way to ascertain this is to continue to perform large ping tests at all representative periods of the day, and verify that the calculated ping delay is close to the observed minimum delay.

6.7 Summary

We discussed how pings work, how ping delays can be calculated, and how these pings can be used to estimate network latency and throughput. We showed that small pings can be used to estimate network latency and that large pings can be used to estimate throughput over a WAN connection, especially bursting over the CIR for frame relay. We also discussed some caveats for using pings to estimate network latency and throughput.

Reference ·

[1]   Stevens, W. R., *TCP/IP Illustrated, Volume 1: The Protocols,* Reading, MA: Addison-Wesley, 1994.

Part II
Specific Application/Protocol Suites

7

WAN Performance Analysis of TCP/IP Applications: FTP, HTTP, and Telnet

7.1 Introduction

Applications using TCP/IP services vary greatly in traffic characteristics—from bulk data transfers[1] like FTP, that predominantly send packets in one direction, to Telnet, that sends single-byte keystrokes across the WAN. Within that spectrum lie a plethora of TCP/IP applications that dominate data networking today—from SNA tn3270, Web transactions, to complex multitiered client/server applications.

Clearly, WAN performance issues depends on the traffic characteristic of the TCP application in question. For example, TCP Telnet and FTP present contrasting performance issues—Telnet performance is more sensitive to latency than bandwidth, whereas FTP performance is more often dictated by available bandwidth and not by latency. Some client/server applications exhibit a hybrid behavior, in which small packet exchanges between the client and the server may be interspersed with a one-way bulk data transfer from the server to the client. Web-based applications exhibit a bursty behavior with multiple image files being downloaded with multiple parallel TCP sessions.

In this chapter we study three representative TCP/IP applications—FTP, HyperText Transfer Protocol (HTTP), and Telnet—to demonstrate the performance analysis methodology. FTP represents a bulk data transfer

---

1. Bulk data transfers can be loosely thought of as file transfers, except that file transfers refer specifically to files being sent across a network. *Bulk data transfer* is a more general term that refers to a one-way transfer of user data, usually from the server to the client.

application, Telnet is an echoplex, and HTTP represents an application that exhibits an interactive nature as well bulk data transfer. Thus the three applications represent a continuum, from an extremely latency-sensitive application (Telnet) to an extremely bandwidth-sensitive application (FTP), with HTTP being somewhere in the middle. The focus is more on WAN and TCP/IP parameters that affect application performance,[2] rather than special characteristics of FTP, HTTP, and Telnet (such as, for example, the fact that FTP is a stateful protocol whereas HTTP is not).

There are two other important classes of TCP/IP applications not discussed in this chapter: client/server applications and SNA (using TCP/IP encapsulation or using gateways). Chapter 9 discusses the performance issues for client/server applications in more detail.

Chapter 10 discusses SNA performance issues.

We begin in Section 7.2 with a brief description of the essential aspects of TCP windowing and retransmission algorithms. We will use a sniffer trace to illustrate the description. The concepts presented here will be used extensively in subsequent chapters to discuss the impact of TCP parameters on performance.

In Section 7.3, we consider the performance of TCP bulk data transfers over private lines and frame relay. Specifically we address the question "What is the best possible throughput a TCP bulk data transfer can achieve on a given WAN connection?" In the process of developing formulas for calculating throughput and file transfer times, we discuss the impact of network parameters such as latency, bandwidth, protocol overhead, and background load, and TCP/IP parameters such as window size and segment size. Private line and frame relay are treated separately in this section. The concepts are illustrated with specific examples. This is then followed by general formulas that can be used for arbitrary WAN configurations and TCP/IP parameter settings. For simplicity of presentation, we assume unloaded WAN connections in this section.

In Section 7.4, we relax the assumption of unloaded serial lines and ports and PVCs. Calculating TCP throughput and delay under network load, and to capture the dependencies on all the variables involved, is a complex

---

2. Although we focus on TCP applications (FTP, HTTP, Telnet) in this chapter and on TCP-based client/server applications in Chapter 9, similar techniques can be employed to quantify UDP applications. Clearly, one has to account for the important differences between TCP and UDP, such as the "best effort" aspect of UDP. We mention TFTP briefly in a footnote and discuss the performance impact of SNMP (a UDP application) polling in Chapter 11.

mathematical exercise and beyond the scope of this book. We present a simple and intuitive approach that can be used for most practical situations.

In Section 7.5, we provide an approximate analysis for calculating performance for HTTP. This is an important area because applications using the Intranet model, where the end user accesses a set of applications using a Web browser, are becoming more prevalent in the industry. The objective in this section is to give the reader some information about the sequence of events that occurs when a Web page is accessed, and how one might calculate the performance impact of the WAN connection. Web performance, in general, is a complex topic, and the reader is referred to an excellent book [1].

In Section 7.6, we provide guidelines for calculating echoplex delays for TCP Telnet sessions over a WAN. We demonstrate that echoplex delays are reasonable over most terrestrial connections, and that the real issue for TCP Telnet performance arises when echoplex traffic mixes with bulk data transfers. This leads to a general discussion of traffic discrimination when supporting heterogeneous traffic streams over a WAN. This issue is discussed at some length in Chapters 4 and 5. In Section 7.7, we will summarize the issues and methods available for traffic discrimination in the context of TCP applications.

7.2 Some Essential Aspects of TCP Operation

We will start by summarizing how TCP window, time-out, and retransmission algorithms work, and how TCP segment sizes[3] are chosen. Please refer to Stevens [2] for an excellent discussion of various aspects of TCP.

TCP adds 20 bytes of protocol overhead (in addition to 20 bytes for IP). TCP hosts use a three-way handshake to establish a session between themselves. As part of this session establishment, window sizes are advertised (not really negotiated). Both hosts inform their partners about the maximum number of bytes that they can receive. This window advertisement can changed during the course of a transaction.

Figure 7.1(a) shows a sample sniffer trace of a TCP session establishment. In this trace, "140.x.x.x" is the client and the "139.x.x.x" is the server [see Figure 7.1(b)]. The server is on the U.S. West Coast and the client is in Western Europe.

The first three packets show the session establishment (SYN—SYN ACK—ACK). Note that the server is advertising an 8-Kbyte window size, and the client is advertising a 2880-byte window size. Although this trace does not

---

3. We will use the terminology *segment* size to refer to a transport layer protocol data unit (PDU), *packet* to a network layer PDU, and *frame* to a link layer PDU.

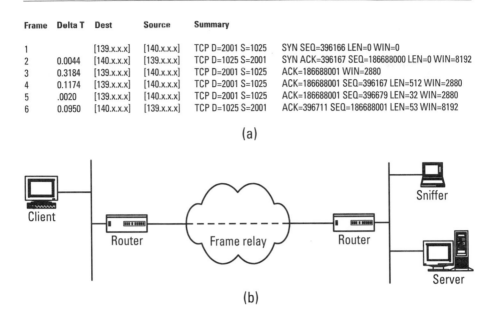

Frame	Delta T	Dest	Source	Summary	
1		[139.x.x.x]	[140.x.x.x]	TCP D=2001 S=1025	SYN SEQ=396166 LEN=0 WIN=0
2	0.0044	[140.x.x.x]	[139.x.x.x]	TCP D=1025 S=2001	SYN ACK=396167 SEQ=186688000 LEN=0 WIN=8192
3	0.3184	[139.x.x.x]	[140.x.x.x]	TCP D=2001 S=1025	ACK=186688001 WIN=2880
4	0.1174	[139.x.x.x]	[140.x.x.x]	TCP D=2001 S=1025	ACK=186688001 SEQ=396167 LEN=512 WIN=2880
5	.0020	[139.x.x.x]	[140.x.x.x]	TCP D=2001 S=1025	ACK=186688001 SEQ=396679 LEN=32 WIN=2880
6	0.0950	[140.x.x.x]	[139.x.x.x]	TCP D=1025 S=2001	ACK=396711 SEQ=186688001 LEN=53 WIN=8192

(a)

(b)

Figure 7.1 (a) Sample trace of a TCP session establishment. (b) Frame relay connection for the TCP session establishment trace.

show a change in these advertised window sizes, it can change during the course of the transaction.

If the client or server explicitly wished to use segment sizes larger than 512 bytes, this will be displayed in the three-way handshake. Its absence in this trace indicates that the maximum segment size used here is 512 bytes (see the LEN field in frames 4–6).

Notice the SEQ and ACK fields. These fields are comparable for packets going in opposite directions. They indicate bytes in a byte stream (starting at a number chosen, one for each direction, according to a preset pseudo-random algorithm when the session starts) in the send or receive direction. For instance, packet 4 is a request from the client to the server consisting of 512 bytes of payload starting at byte number 396,167. In this frame, the client also informs the server that it has successfully received from the server all bytes up to 186,688,000, and is expecting byte number 186,688,001.

Finally, the delta times (intervals of time between successive frames that the sniffer observed that was sent to and from the two stations for this TCP session) provide a rough estimate of WAN round-trip delays. The delta time for packet 2 is a mere 4 msec (because the sniffer is on the same LAN as the server), while the delta time for packet 3 is 318 msec. This is a good estimate of the round-trip latency through frame relay between the United States and Europe.

This is the time for a small packet to be sent out to the client from the server and another small packet to be received back at the server.

The maximum TCP window size is 65,535 bytes. The protocol allows larger window sizes via the use of specific TCP options.

TCP uses a sliding window algorithm. In a nutshell, sliding window means that each time an ack is received for a segment, the sending host can send one or more additional segments, ensuring that, at any time, the receiver-advertised window size is not exceeded. For instance, if four segments are sent out in a window and the sender receives an ack for segments 1 and 2, then two more segments can be sent. Thus the window of segments 1–4 *slides* to 3–6. Note that TCP does not explicitly send acks for segments, rather for the last byte in the byte stream successfully received; TCP is a byte-stream protocol. TCP also does not have to ack every segment. Some TCP implementations use *delayed ack*, whereby a receiving host waits for a period of time (typically 200 msec) before sending an explicit ack. If, however, user data is to be sent to the sending host, then an ack is piggybacked.

No discussion of the TCP windowing algorithm is complete without a reference to the slow start/congestion avoidance algorithm. In a sense, this algorithm makes the end-to-end TCP session "network aware." The receiver advertised window size is the maximum amount of data that a transmitter station can send to the receiver. Rather than sending the entire window, and risk packet drops and time-outs on the connection (especially, in the case of a WAN), the sending station uses a congestion window to regulate how many segments are sent at once. Initially, the congestion window is set to one (sometimes two, in some TCP implementations), and then increased exponentially when ACKs are received: 2, 4, 8, and so on. If time-outs occur, then the window size at which the congestion occurred is marked. The congestion window is again dropped to one, and slow start is employed until the congestion window increases to half of the point at which time-outs occurred (an exponential increase), at which point the increase in the window size is more linear. Please see Stevens [2] for a very detailed discussion of this topic.

Slow start/congestion avoidance is well suited for bulk data transfers across congested WAN links. For frequent and short-lived transactions like HTTP over a WAN connection, there is a penalty to be paid for using slow start in terms of network latency, as we will demonstrate in Section 7.5.

Figure 7.2 shows a portion of a trace of a bulk data transfer occurring from a server (port # 1527) to a client (port # 1040). Note that the server is advertising a 32-Kbyte window size, but since the server is transmitting the bulk data to the client, the operational window size is that advertised by the client, 8760 bytes. Because acks are being delivered in time, one can observe the window sliding one at a time, keeping the window movement smooth.

Frame	Rel Time	Bytes	Summary	
8	2.91647	1514	TCP D=1040 S=1527	ACK=1973292 SEQ=657541770 LEN=1460 WIN=32768
9	2.99540	642	TCP D=1040 S=1527	ACK=1973292 SEQ=657543230 LEN=588 WIN=32768
10	2.99589	60	TCP D=1527 S=1040	ACK=657543818 WIN=8760
11	3.18485	1514	TCP D=1040 S=1527	ACK=1973292 SEQ=657543818 LEN=1460 WIN=32768
12	3.37353	1514	TCP D=1040 S=1527	ACK=1973292 SEQ=657545278 LEN=1460 WIN=32768
13	3.37428	60	TCP D=1527 S=1040	ACK=657546738 WIN=8760
14	3.56219	1514	TCP D=1040 S=1527	ACK=1973292 SEQ=657546738 LEN=1460 WIN=32768
15	3.56304	60	TCP D=1527 S=1040	ACK=657548198 WIN=8760
16	3.77217	1514	TCP D=1040 S=1527	ACK=1973292 SEQ=657548198 LEN=1460 WIN=32768
17	3.88290	60	TCP D=1527 S=1040	ACK=657549658 WIN=8760
18	3.96095	1514	TCP D=1040 S=1527	ACK=1973292 SEQ=657549658 LEN=1460 WIN=32768
19	4.08356	60	TCP D=1527 S=1040	ACK=657551118 WIN=8760
20	4.14966	1514	TCP D=1040 S=1527	ACK=1973292 SEQ=657551118 LEN=1460 WIN=32768

Figure 7.2 Trace of a bulk data transfer. The columns, from left to right, refer to the frame number, incremental time since the last frame on this conversation (Delta T), the protocol decode (TCP), and source and destination port numbers. SYN packets denote the establishment of a TCP connection (three-way handshake). SEQ is the sequence number in the byte stream for the packet currently being transmitted. ACK is an indication of the last successfully received byte. Notice the relationship between SEQ and ACK numbers in the to and from directions. LEN is the packet size, not including any protocol overhead. WIN is the window size advertisement.

Note also that the segment size is larger than 512 bytes. Segment sizes of 512, 1024, and 1460 bytes are popular in many TCP implementations.

7.3 Calculating TCP Bulk Data Transfer Times and Throughput

In this section we explain how one can employ simple techniques to calculate TCP bulk data transfer times. TCP FTP is a classic example of a bulk data transfer occurring between a client and a server. Client/server applications, e-mail, and Web browsing are other examples of instances where bulk data transfers under TCP/IP occur. As we show in Chapter 9, some client/server applications have interactive traffic as well as bulk data transfer traffic embedded within a single transaction, such as requesting a screen update.

First, we consider a simple point-to-point WAN connection to illustrate how one can compute the best possible file transfer time, and hence the best throughput, for the connection. We then discuss the effect of TCP window size, segment size, WAN bandwidth, and latency on performance. This will help us derive a general formula for computing throughput for this type of a WAN connection.

Following that, we discuss the case of frame relay. Although many of the concepts and methods to calculate TCP performance carry over without significant change from private lines to frame relay, some important differences exist.

7.3.1 Variables Affecting TCP File Transfer Performance

Two classes of parameters have a direct bearing on file transfer performance: TCP parameters and WAN parameters.

Important TCP-related parameters are window size and segment size. One also needs to consider TCP time-out and retransmission algorithms. However, time-outs and retransmissions come into play when considering issues such as packet drops due to bit errors and network congestion. Bit error occurs relatively infrequently and therefore can be ignored. Packet drops due to network congestion (e.g., packets dropped in routers, WAN switches) can be accounted for in the calculation of TCP file transfer times. However, the analysis required is beyond the scope of this book. Therefore we will ignore this aspect as well. The interested reader is referred to [3] and [4] and references therein. We provide a heuristic argument for computing TCP throughput under background load in Section 7.4.

In what follows, we will not explicitly account for the effect of slow start/congestion avoidance described in the previous subsection. For a large file transfer (that is, many windows' worth of data) and relatively small probability of packet drops due to bit errors and/or congestion, the congestion window will reach the receiver advertised window size within a few round trip delays. Thus the initial effect of slow start will be negligible. Correspondingly, the analysis we present below can be construed as "the best possible throughput" that TCP can achieve for the WAN in question.

As far as the WAN is concerned, the important parameters are bandwidth, latency, and background load (or WAN link utilization).

7.3.2 Computing File Transfer Times for a Simple Point-to-Point Connection

Consider the two point-to-point connections between a remote branch office and a data center/hub site shown in Figures 7.3(a) and (b). Figure 7.3(a) shows a 56-Kbps private line connection and (b) shows a frame relay connection with a T1 port at the hub site and 56-Kbps frame relay port for the branch office with a 32-Kbps CIR (committed information rate) between them.

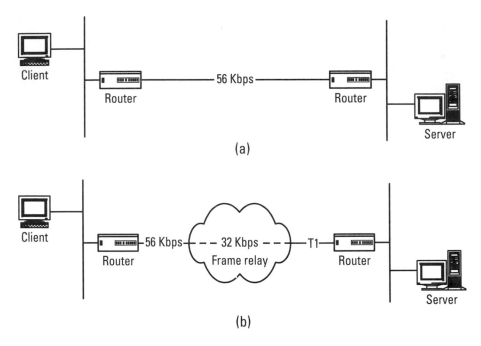

Figure 7.3   (a) Private line 56-Kbps connection. (b) Frame relay connection with T1/56-Kbps ports and 32-Kbps CIR.

Suppose that a client at the remote office requests a 1-Mbyte file from the file server at the data center. How long will the file take to get to the client over the private line connection and over the frame relay connection? To simplify the discussion and to illustrate the concepts involved, let's assume that the WAN connection is unloaded. We will deal with the case of loaded connections in Section 7.4.

We also need to make an assumption regarding network latency from router to router. Suppose that the one-way network latency is 30 msec.[4] This represents a 3000-mile leased line connection across the United States. For the frame relay connection, assume that the network latency is about 40 msec.

---

4. How does one arrive at a specific number for network latency? Here's a simple approximation—estimate the distance in miles between the two locations. Multiply this by 0.01 (msec per mile) to estimate the propagation delay between the locations. If the underlying WAN is packet switched, like frame relay, then some additional delay is added due to packet switches (about 1–2 msec per switch). For a 3000-mile connection over frame relay, the PVC may traverse four or five switches. Hence about 10-msec extra delay might need to be added. Please see Chapter 3.

Another important assumption is that the router delays, LAN delays, and client/server delays are negligible. This is true in most cases, although a busy server may add significant delay in processing packets.

Assume that the receiver (client) advertises a window size of 4096 bytes and assume that the TCP segment size is 512 bytes. This means that the server will send the client eight segments of 512 bytes each. Also assume that the receiving TCP host sends an ack packet that does not contain actual data (this assumption simplifies the round-trip delay calculation).

To sum up, here are the major assumptions so far:

- WAN connections (private line, ports/PVCs) are unloaded;
- End-to-end network latency is 30 msec for private line, 40 msec for frame relay;
- Router, LAN, and client/server delays are negligible;
- TCP segment size is 512 bytes, TCP receive window size is 4096 bytes; and
- TCP acks (46 bytes) are sent separately—not piggybacked with actual data traffic.

Now we compute the WAN protocol overhead factor. For the network connection in question, three types of protocol overhead are carried over the WAN—TCP (20 bytes), IP (20 bytes), and link layer (8 bytes).[5] The link layer is either PPP (point-to-point protocol) or some HDLC variant in the private line case, or frame relay. In the latter case, RFC 1490 might be employed to encapsulate IP datagrams with frame relay headers. The total overhead per packet is 48 bytes. Hence each full TCP data packet of 512 bytes will be transmitted on the WAN link as a $512 + 48 = 560$-byte frame.

Hence protocol overhead factor is $560/512 = 1.09$, or about 9%.

A general formula for calculating protocol overhead factor is given here:

Let S be the PDU size under consideration.

Let O be the overall protocol overhead.

Protocol overhead factor F is $1 + O/S$.

---

5. Eight bytes for link layer is an approximation. It is not usually necessary to be very precise about this aspect (e.g., should you include starting and ending flags?) because a difference of a few extra bytes can be ignored in calculating response times.

As another example, consider the case of character-mode TCP telnet where every keystroke from the client is echoed from the remote telnet server. In this case $S = 1$ byte and $O = 48$ bytes. The protocol overhead factor is

$$1 + 48 / 1 = 49, \text{ or } 4900 \% \text{ !!!!}$$

Protocol overhead factors depend on what protocol is being carried over the WAN. In Chapter 10, where we discuss SNA performance issues, we will encounter protocol overhead factors in a different setting.

7.3.3 Private Line Analysis

We will analyze the 56-Kbps private line case first. The timing diagram in Figure 7.4 illustrates the data transfer sequence. From the timing diagram,[6] it is clear that the time to send the entire window is

$$= [(8 \text{ frames} \times (560 \text{ bytes/frame}) \times 8 \text{ bits/byte})] / 56,000 \text{ sec} = 640 \text{ msec}$$

and that the round-trip time for a TCP segment to be sent and ack to be received, including network latency, is

$$= (8 \times 560/56,000 + 0.030) + (8 \times 48/56,000 + 0.030) \text{ sec} = 147 \text{ msec}$$

Hence it is clear that the window will not close, that is, the window is large enough so that the sender always has packets to send to keep the 56-Kbps WAN link busy 100% of the time. Since any resource in the network that is busy 100% of the time determines the throughput (that is, the bottleneck resource), the maximum data throughput for this connection is

$$(56,000 \text{ bps}) / \text{Protocol overhead factor} = 56,000/1.09 = 51,376 \text{ bps} = 6422 \text{ bytes/sec}$$

Hence the best possible file transfer time is

$$\text{File size} / \text{Maximum throughput} = 1,024,000 \text{ bytes} / 6422 \text{ bytes/sec} = 159.5 \text{ sec} = 2 \text{ min } 40 \text{ sec (approximately)}$$

---

6. Henceforth, we will not consider LAN delays in timing diagrams.

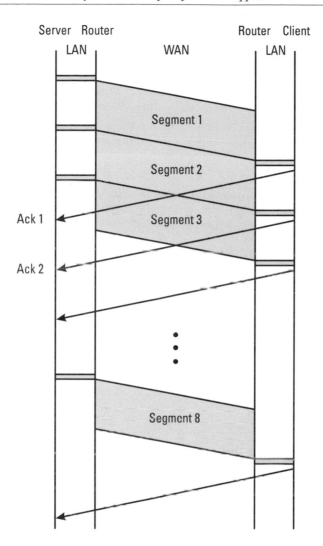

Figure 7.4 Timing diagram for TCP file transfer for 56-Kbps private line.

7.3.3.1 Impact of Window Size and Optimal Window Sizes

One critical assumption we made in the previous analysis was that the window is large enough to "fill the pipe." To assess the impact of changing window sizes on file transfer performance, first consider the following. If the window size is too small (say, 1 TCP segment of 512 bytes), then we would be wasting WAN resources because there is not enough data to send across. On the other hand, if the window size is too large, then no improvement in throughput will result because the slowest resource (that is the WAN link) would already be saturated.

Thus, making the TCP window size very large might result in excess buffering in network buffers—a "hurry up and wait" scenario. Indeed, this buffering may sometimes cause TCP time-outs and retransmits for those TCP implementations that use fixed timers.[7] Thus, there is an *optimal window size* for maximizing the throughput while at the same time avoiding an oversaturation of the slowest WAN link. However, before we actually compute this optimal window size, let us calculate the throughput as a function of the window size.

Window Size = 1[8]

Assume that the window size is fixed at one TCP segment of 512 bytes. Using the timing diagram in Figure 7.5, it is easy to see that the throughput in bits per second is given by

$$\text{Throughput} = 512 \times 8/R$$

where R, the round-trip delay, is given by

$$R = [(512 + 48) \times 8/56{,}000 + 0.03] + (48 \times 8/56{,}000 + 0.03) \text{ sec} = 147 \text{ msec}$$

Hence, the best possible throughput is

$$T = 512 \times 8 \text{ bits } / 0.147 \text{ sec} = 28{,}055 \text{ bps or 28 Kbps (approximately)}$$

Correspondingly, the best possible file transfer time is

$$\text{File size } / \text{ Throughput} = 1{,}024{,}000 \times 8/(28{,}055 \text{ bps}) = 292 \text{ sec} = 4 \text{ min } 52 \text{ sec}$$

Window Size = 2

What would happen if we increased the window size to 2 (or two TCP segments of 512 bytes)?

---

7. The authors experienced this issue while troubleshooting a client/server application performance problem. The client was running Novell LAN Workplace for DOS®, and the TCP timer was fixed at 1 sec.

8. This discussion can be used almost verbatim (except for protocol overhead) for computing transfer times for Trivial File Transfer Protocol (TFTP). TFTP uses UDP services instead of TCP, with an overhead of 8 bytes for UDP, all other overhead characters remaining the same. TFTP restricts segment size to 512 bytes.

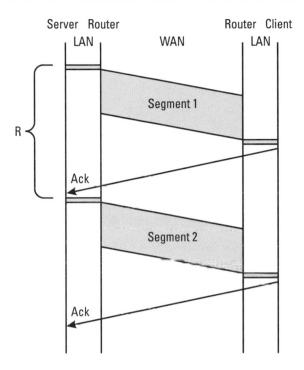

Figure 7.5 Timing diagram for window size = 1 data transfer.

Time to send the window = $(2 \times 560 \times 8/56{,}000) = 160$ msec

R = Round-trip delay = 147 msec (same as for the window = 1 case)

Therefore an ack will be received by the time the TCP window closes. Thus it is clear that a window size of 2 is sufficient to fill the pipe, but a window size of 1 is not sufficient. The throughput with window size of 2 will be exactly the same as that obtained earlier with a window size of 8, that is, 51.4 Kbps, approximately.

Thus it appears that the optimal window is two TCP segments or 1024 bytes. However, as mentioned earlier, TCP window sizes are adjusted in bytes, not in segments. A more refined number for the optimal window can be derived if we consider window size from the "number of bytes" point of view. The formula for the optimal window size is

$$W^* = (\text{Slowest link speed in the connection}) \times (\text{Round-trip delay for a segment})$$

This is called the *bandwidth-delay product.* For the 56-Kbps private line network under consideration,

$$W^* = 7000 \text{ bytes/sec} \times 0.146 \text{ sec} = 1022 \text{ bytes}$$

This is consistent with using a TCP window size of two segments of 512 bytes each.

Note that the optimal window size is always greater than the TCP segment size. A window of one TCP segment can never be optimal.

$$W^* > \text{TCP segment size}$$

Impact of Segment Size

The impact of segment sizes is easier to analyze. Basically, increasing segment size results in lower protocol overhead and hence better throughput. In the previous example, suppose we change the segment size from 512 to 1024 bytes. The protocol overhead factor becomes

$$(1024 + 48) \ / \ 1024 = 1.047 \text{ or } 4.7\%$$

Hence the best throughput (for a large enough window size) is $56,000/1.047 = 53486$ Kbps.

What is the impact on the optimal window size? The bandwidth delay product changes somewhat:

$$W^* = \text{Round-trip delay} \times \text{Link speed} = [(1024 + 48) \times 8/56,000 + 0.03\ ) + (48 \times 8/56,000 + 0.03)] \text{ sec} \times 7000 \text{ (bytes/sec)} = 0.220 \text{ sec} \times 7000 \text{ bytes/sec} = 1540 \text{ bytes}$$

Again, the optimal window size turns out to be two TCP segments.

7.3.3.2 A Comment About Window Size Adjustments

The previous discussion clearly shows that TCP window sizes have an important bearing on the maximum achievable throughput for the connection. Indeed, in certain situations, window size tuning (upwards or downwards) may be the only recourse to address WAN performance issues. At the same time, it must be noted that it is not always easy, and sometimes impossible, to adjust

TCP window sizes on clients and servers. This is because TCP implementations are many and it may not be possible to configure TCP windows. However, certain popular platforms such as Windows NT® (default window size of 8 Kbytes) allow the window size advertisements to be changed. The same issue holds for Novell and SNA protocols as well.

7.3.3.3 Impact of Network Latency on TCP Throughput

Many companies have a need to send or receive files across global connections. Occasionally, access to corporate data networks may be via satellite connections, especially in countries where high-speed and high-quality digital terrestrial links are not available. In these situations, understanding the impact of longer network latencies on file transfer throughput is very important. Equally important is the realization that TCP window sizes need to be increased to get the maximum possible throughput.

It is fairly easy to perform the calculations using the simple formulas established in the previous sections of this chapter.

Assume the same private line 56-Kbps connection as in Figure 7.3(a), except that the client and server are separated by a 10,000-mile connection (e.g., from the United States to the Pacific Rim). Let us also assume that the window size is 1024 bytes (2 × 512-byte segment size). To calculate the throughput, compare the window insertion time and the round-trip acknowledgment delay.

Without even performing a detailed analysis, it is easy to show that the window size is too small. The window insertion time is $2 \times 560 \times 8/56,000 = 0.16$ sec. However, the round-trip propagation delay alone is likely to be 200 msec (20,000 miles for the round-trip; 10 msec for every 1000 miles). Therefore, the window will definitely close, that is, the window size will be too small.

The optimal window size is easily calculated, as discussed earlier:

$$W^* = \text{Round-trip delay} \times \text{Slowest WAN link speed} = (560 \times 8/56,000 + 0.1 + 48 \times 8/56,000 + 0.1) \text{ sec} \times 56,000 \text{ bps} = 0.286 \text{ sec} \times 56,000 \text{ bps} = 2000 \text{ bytes (approximately)}$$

Hence, the recommended window size is four TCP segments of 512 bytes or two TCP segments of 1024 bytes.

It is clear from the preceding discussion that the lower bound for the optimal window size is

$$W^* > \text{Round-trip propagation delay} \times \text{Slowest WAN link speed}$$

For the WAN connection under question, this lower bound is 56,000 bps ×
0.2 sec = 1400 bytes.

For WAN connections over long distances (global or satellite), this lower
bound provides a quick estimate for the minimum window size. Indeed, the
optimal window can then be obtained by adding the segment size and acknowledgment packet size to this minimum window size.

Satellite Links

Assume that the 56 Kbps connection between the two routers in Figure 7.3(a)
is a satellite link (see Figure 7.6). Satellites are situated approximately
26,400 miles above the Earth in a geostationary orbit. However, it is incorrect
to calculate propagation delay for satellite links at the rate of 10 msec for every
1000 msec. The correct number is 5.35 msec for every 1000 msec resulting in
approximately 280 msec of propagation delay (see Chapter 3 for details). Note
that this is the one-way delay from the end user point of view—the signal needs
to propagate on the satellite uplink and then down to the earth station. The
round-trip network latency (for a segment to be sent and an ack to be received)
is 560 msec.

A quick estimate of the minimum window size for a 56-Kbps satellite
link is

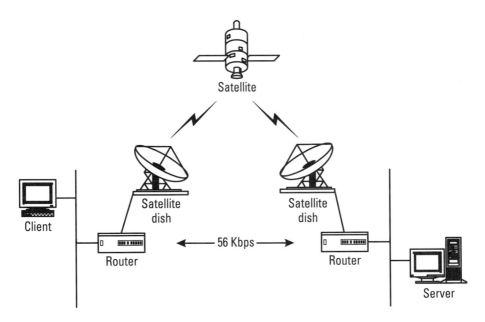

Figure 7.6 56-Kbps satellite connection.

$$0.56 \text{ sec} \times 56,000 \text{ bps} = 3920 \text{ bytes}$$

The optimal window size for this case is

$$3920 \text{ bytes} + \text{Segment size} + \text{Ack packet size} = 3920 + 604 = 4524 \text{ bytes}$$

Hence the recommended window size for a 56-Kbps satellite link is nine TCP segments of 512 bytes or five TCP segments of 1024 bytes each.

7.3.4 Frame Relay Analysis

In the last section, we discussed the various aspects of calculating TCP file transfer throughput for an unloaded private line point-to-point connection. In this section, we demonstrate how these techniques can be applied to point-to-point frame relay connections, and also point out some special aspects of frame relay that need to be taken into account.

Consider the point-to-point frame relay connection of Figure 7.3(b). The server is situated in the data center where the frame relay port speed is T1 (1536 Kbps). The client is at a remote branch office where the frame relay port speed is 56 Kbps. The two locations are connected via a 32-Kbps CIR.

There are some important characteristics of frame relay that differentiate it from private line as related to TCP throughput.

- *Bandwidth entities.* There are three WAN bandwidth entities for frame relay (two port speeds and a CIR) versus a single WAN bandwidth entity for private line. As far as throughput calculations are concerned, the slowest link in the connection must be the focus of attention. For frame relay, this can be either the CIR or minimum of the two port speeds, depending on burst capabilities.

- *Bursting.* Perhaps the biggest difference is the fact that frame relay allows bursting above the CIR. No such concept exists in private line. The CIR is the minimum guaranteed bandwidth[9] for the connection and bandwidth in excess of the CIR allocated to the PVC is "best

---

9. Frame relay service providers can oversubscribe bandwidth in the network. For example, a 2-to-1 oversubscription means that all PVCs are provisioned with half of their CIRs. Because not all PVCs are likely to be active at the same time at full throttle, it is reasonable to expect that a specific PVC at any given time will get the full CIR bandwidth, and more. We will assume that the minimum CIR is always available to the PVC.

effort." Indeed, there are "zero CIR" implementations in which delivery of packets is totally best effort.

- *Best and worst case conditions.* We will account for this best effort delivery by considering the *best* and *worst* case conditions. The best case is "full bursting" when the PVC is consistently bursting to port speed. The worst case is "no bursting" when the PVC bandwidth is limited to its provisioned CIR and no more.[10] It is almost impossible to consider anything intermediate between best case and worst case, because frame relay service providers typically do not engineer their backbone network for a specific bursting level on PVCs (say, bursting at twice the CIR during off-hours). In addition, it is theoretically possible that the frame relay PVC is bursting to port speed in one direction, but not in the reverse direction. We will ignore this possibility. In a sense, we are considering the absolute best case and the absolute worst case.

  Note that the worst case analysis is really only applicable to closed-loop implementations. For open-loop implementations of frame relay, the worst case analysis is hard to do because one is required to compute TCP throughput assuming a random number of packet discards.

- *Extra insertion delay.* There is an extra serial line insertion delay for frame relay. For instance, in the direction of server to client, a TCP segment has to be inserted (serialized) twice, once at the T1 port and next at the 56-Kbps remote port. This adds to network latency. Clearly, the T1 port insertion delay is small, but there are many frame relay networks with much lower speed ports. In addition, network switches in the backbone (e.g., ATM switches) add some latency, perhaps a millisecond or two. Hence, overall network latency may increase somewhat over frame relay. However, we will demonstrate that this has no impact on TCP throughput as long as the window sizes are set appropriately.

We now discuss TCP file transfer throughput for frame relay, optimal window size issues, and impact of network latency on throughput. It will be helpful to think of the frame relay connection in Figure 7.3(b) as a replacement to the private line connection in Figure 7.3(a).

Assume the following:

---

10. Note that if the PVC is unloaded, it does not necessarily mean that the PVC can burst to port speed. No bursting above CIR really means that the service provider is metering traffic into network at CIR rate. This has nothing to do with whether or not the PVC is loaded.

- Server sends a 1-Mbyte file to the client.
- TCP segment size is 512 bytes and window size is $2 \times 512 = 1024$ bytes (recall that this was found to be optimal for 56-Kbps private line).
- Frame relay network latency is 40 msec—a typical frame relay connection across the United States (compare with 30 msec for private line).
- The connection is unloaded, that is, there is no background traffic.

The protocol overhead is the same as for private line.[11] For example, a 512-byte TCP segment has overhead given by

$$512 + 20 \; (TCP) + 20 \; (IP) + 8 \; (FR) = 560 \text{ bytes}$$

As mentioned earlier, frame relay allows bursting above the CIR. We will consider the best case (consistent bursting to port speed) and the worst case (no bursting above the CIR).

Best Case Analysis

The timing diagram[12] in Figure 7.7(a) illustrates the sequence of events.
The round-trip time to send a segment and receive an ack is[13]

$$(560 \times 8/56,000 + 0.04 + 560 \times 8/56,000) + (48 \times 8/56,000 + 0.04 + 48 \times 8/1,536,000) = 0.247 \text{ sec}$$

The time to insert the window on the slowest link (56 Kbps in this case) is

$$2 \times 560 \times 8/56,000 = 0.16 \text{ sec}$$

---

11. It is not necessary to be very accurate with the overhead calculation if the difference is a matter of a couple of bytes. The insertion delays and protocol overhead will not change significantly.

12. The timing diagrams are not drawn to scale. So, for instance, the insertion delay on a T1 line is more than 24 times faster than on a 56 Kbps. This difference is not shown in the timing diagrams.

13. An implicit assumption in this calculation is that the maximum data rate out of the ingress switch buffer is 56 Kbps, not T1. Closed-loop frame relay implementations are built in this way.

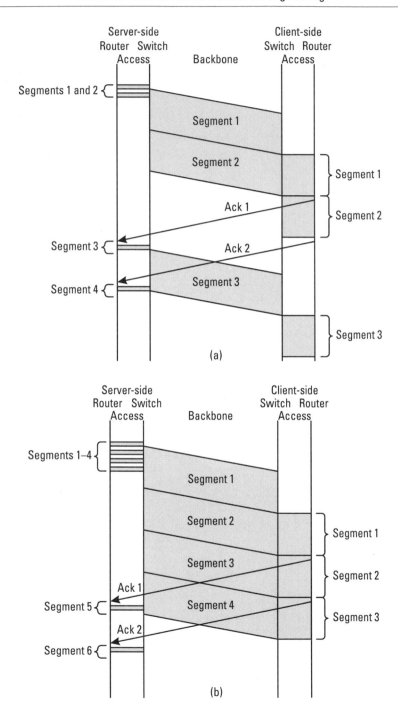

Figure 7.7 Timing diagram for frame relay with full bursting and (a) window size of 2 or (b) window size of 4.

Thus, it is clear that the window size of 2 segments of 512 bytes each is not large enough. The throughput for this case is

$$2 \times 512 \times 8 \text{ bits}/0.12 \text{ sec} = 48.2 \text{ Kbps}$$

Note that we can improve on this throughput by increasing the window size. If the window size is 4×512 bytes = 2048 bytes, then the time to insert the window on the slowest link is

$$4 \times 560 \times 8/56,000 = 0.32 \text{ sec}$$

See the timing diagram in Figure 7.7(b). Hence the 56-Kbps link will limit the throughput. That is,

$$\text{Throughput} = 56,000/\text{Protocol overhead} = 56,000/1.09 = 51.4 \text{ Kbps}$$

This is consistent with the following optimal window size calculation:

$$\text{Optimal window size} = \text{Round-trip delay} \times \text{Slowest WAN link} = 0.247 \text{ sec} \times 56,000 \text{ bps} = 1729 \text{ bytes}$$

or rounded off to four TCP segments of 512 bytes. Equivalently, the window can be two TCP segments of 1024 bytes.

Observe that frame relay needed a slightly larger TCP window size compared to private line. This is consistent with the longer network latency that we assumed for frame relay.

Worst Case Analysis

If the assumption is that the network limits the PVC bandwidth to the CIR (32 Kbps, in this case), then one needs to make some changes to the throughput and window size calculations. The trick is to consider the PVC as a private line with link speed equal to CIR. The additional factor that needs to be taken into account is pipelining. In the direction from the server to client, frames containing TCP segments are sent at T1 speed into the network, but the receiving frame relay switch does not store-and-forward the frame on the PVC—it pipelines the frame by sending partial packets of data on the PVC. For example, if the underlying transport network is ATM, then the receiving frame relay switch will pipeline on ATM cells.

To start off, consider the timing diagram shown in Figure 7.8. Assume that the window size is 4×512 byte segments = 2048 bytes. It is easy to see that the round-trip delay for sending a TCP segment and receiving an ack is

$$(560 \times 8/32,000 + 0.04 + 560 \times 8/56,000) + (48 \times 8/32,000 + 0.04 + 48 \times 8/1,536,000) = 0.312 \text{ sec}$$

Note that in the forward direction (server to client) the T1 link speed is ignored because of pipelining. Similarly the 56-Kbps port is ignored in the reverse direction. Actually, one needs to account for the delay until a pipelining packet (such as an ATM cell) has been accumulated. This delay is negligible (53 bytes / 7 = 7.5 msec on a 56-Kbps link). For higher speed links, this delay will be even smaller.

The time to insert the window on the slowest link (32 Kbps in this case) is

$$4 \times 560 \times 8/32,000 = 0.56 \text{ sec}$$

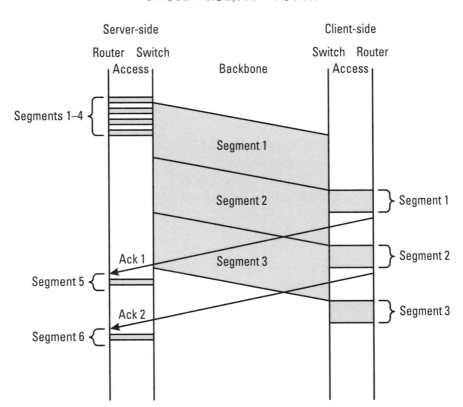

Figure 7.8 Timing diagram for frame relay with no bursting above CIR.

Hence the window will not close. The throughput is therefore determined by the slowest link, that is, 32 Kbps:

$$\text{Throughput} = 32{,}000 \text{ bps} / 1.09 = 29.4 \text{ Kbps}$$

The optimal window size is given by

$$0.312 \text{ sec} \times 32 \text{ Kbps} = 1248 \text{ bytes}$$

or three TCP segments of 512 bytes each or two TCP segments of 1024 bytes each.

Satellite Access to Frame Relay

Satellite access to frame relay is a likely option for companies with far-flung branches (e.g., manufacturing sites in Eastern Europe or Latin America) that need to connect via frame relay to a data center (see Figure 7.9).

The impact of latency on application performance is usually severe. However, for bulk data transfer applications, that is, applications that are more bandwidth sensitive than latency sensitive, this issue can be addressed by having the receiver advertise a larger window size.

Let us address TCP throughput issues here. The formulas developed in this section for best case and worst case situations can be applied to this case by simply adding in the extra delay due to satellite propagation. Essentially, the access delay can be transferred to the network delay and the calculations proceed from there.

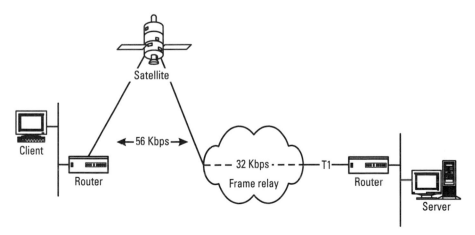

Figure 7.9 Satellite access to frame relay.

We now perform the calculations for the best case scenario. If, as in Figure 7.6, we assume that the 56-Kbps access link is via a satellite connection, and the underlying frame relay backbone latency is 40 msec, then the round-trip time to send a segment and receive an ack is

$$(560 \times 8/1,536,000 + 0.28 + 0.04 + 560 \times 8/56,000) + (48 \times 8/56,000 + 0.28 + 0.04 + 48 \times 8/1,536,000) = 0.73 \text{ sec}$$

Hence the optimal window size is

$$0.73 \text{ sec} \times 56 \text{ Kbps} = 5110 \text{ bytes}$$

or 10 TCP segments of 512 bytes or 5 TCP segments of 1024 bytes each. With at least this window size, the throughput will be the same as for cases discussed earlier, that is,

$$\text{Throughput} = 56,000 \text{ bps} / 1.09 = 51.4 \text{ Kbps}$$

7.3.5 General Formulas for Calculating TCP Throughput for Private Lines and Frame Relay

In the previous sections, we showed how to calculate TCP throughput and optimal window sizes for a specific point-to-point connection under private line (56-Kbps) and frame relay (T1/56-Kbps port speeds, 32-Kbps CIR).

Clearly, these techniques can be applied in general for arbitrary link speeds and CIRs. They can also be applied to a wide variety of WANs besides private line and frame relay, including ATM, X.25, ISDN, and low-speed dial-up connections. We next present some general formulas for TCP through-put computations for private line and frame relay. The reference connections we will use are shown in Figures 7.10(a) and (b).

We need to establish some notation and state assumptions. For the basic input parameters, let

S = TCP segment size in bytes

A = TCP-level ack = 48 bytes (no data)

O = protocol overhead bytes (TCP + IP + Link layer) = 48 bytes

W = window size (Advertised by the receiver)

D_p = one-way network latency for private line in seconds

D_f = one-way network latency for frame relay in seconds

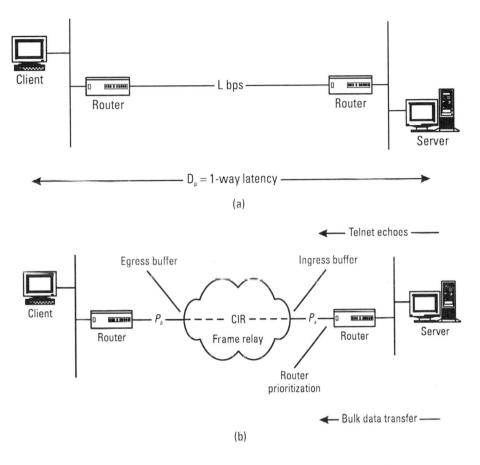

Figure 7.10 Reference connection for (a) a private line and (b) frame relay.

L = link speed for private line in bytes/sec

P_a = port speed at frame relay location A in bytes/sec

P_b = port speed at frame relay location B in bytes/sec

CIR = PVC rate between locations A and B in bytes/sec

F = file size

For the computed performance metrics, let

T_w = time to insert the window W on the slowest WAN link

R_d = Round-trip delay to send a TCP segment and receive an ack

W^* = optimal window size

O_f = overhead factor

X = throughput for the connection

We make the following assumptions in decreasing order of importance:

- *WAN connections (private line, ports/PVCs) are unloaded.* This is usually not the case, but we will discuss how to deal with loaded connections in the next section of this chapter.
- *Router, LAN, client/server delays are negligible.* Router and LAN delays are usually negligible. Sometimes server delays can be significant—100 to 200 msec processing delays for servers are not unusual. One can get a rough estimate of these delays using pings or by examining protocol traces.
- *TCP acks (48 bytes) are sent separately—not piggybacked with actual data traffic.* This is not a very serious assumption. If acks are piggybacked, then the round-trip delay will be slightly longer.
- *Assume that $P_a > P_b$ > CIR.* That is that port speed at location A is larger than at location B.
- *Assume that the window size $W = N_W \times S$, an integer multiple of the segment size S.* This is just for convenience. The formulas become cumbersome to write if, for example, the window size is 1000 bytes and the segment size is 512 bytes.

With these notations and assumptions, here are some general formulas:

Time to insert window W on a link of speed $L = N_W \times (S + O) / L$ sec

The optimal window size W^* is that value of W that just saturates the slowest WAN link. If $W < W^*$ then increasing W will increase throughput. If $W > W^*$ then any increase in W will not result in increase in throughput.

Optimal window size $W^* = R_d \times$ Slowest WAN link $/ O_f$

Protocol overhead factor $O_f = 1 + O/S$

Throughput: $X =$ Slowest WAN link$/ O_f$, if W is greater than or equal to W^*

Throughput: $X = W / R_d$, if W is less than or equal to W^*

File transfer time $= F / X$

Table 7.1 separates the specific formulas for private line and frame relay.

7.4 Calculating TCP Throughput for Loaded WAN Links

The calculation of TCP file transfer throughput in Section 7.3 assumed an unloaded connection for the private line and frame relay cases. This assumption is clearly unrealistic. In reality, WAN serial links (private line and frame relay ports) have moderate utilization levels, say, in the range of 20% to 40%. In some instances, utilization levels can be as high as 70% to 80%. Utilization beyond 80% is relatively uncommon, especially in private networks carrying mission-critical traffic. Very often, when the utilization levels exceed 80%, the links are upgraded. In the case of frame relay, however, CIR utilization can be in excess of 150%, say, in the range of 200% to 300%. This indicates bursting on the PVCs and the level of bursting depends on the carrier's admittance policies and also on frame relay port speeds. For instance, in a case where the minimum frame relay port speed is 56 Kbps and the CIR is 32 Kbps, the maximum bursting one can achieve on the PVC is 56/32 = 175%.

It is a relatively complex mathematical exercise to provide exact formulas for computing TCP throughput under load. There are many dependencies—link utilization in the inbound and outbound direction (two variables for private line and six variables for frame relay), segment size, window size, propagation delay, and so on. Instead of attempting to capture the complexities, we will make some reasonable assumptions and provide formulas based on intuitive reasoning. For most practical cases where the utilization levels are moderate (20% to 70%), the formulas in this section can be used as a good starting point.

Table 7.1
Specific Frame Relay and Private Line Formulas

	Private Line	Frame Relay	
		Best Case	Worst Case
Time to insert window	$T_w = N_W \times (S + O)/L$	$T_w = N_W \times (S + O)/P_b$	$T_w = N_W \times (S + O)/CIR$
Round-trip delay for a segment + ack	$R_d = (S + O)/L + 2\ D_p + A/L$	$R_d = (S + O)/P_a + 2\ D_f + (S + O)/P_b + A/P_b + A/P_a$	$R_d = (S + O)/CIR + 2\ D_f + (S + O)/P_b + A/CIR + A/P_a$
Optimal window size	$W^* = R_d \times L/O_f$	$W^* = R_d \times P_b/O_f$	$W^* = R_d \times CIR/O_f$

The most important assumption we will make is that the window size is set at the optimal level or larger for the connection in question. This is the window size that can "fill the slowest pipe" in the connection. The rationale here is to exclude from consideration cases where the window size is too small to even fill the available bandwidth in a loaded connection. The following example illustrates this concept more clearly.

Consider the private line and frame relay connections in Figures 7.3(a) and (b).

7.4.1 Private Line Case

Let us first consider the case of the private line connection. Assume that the file transfer occurs between the server and remote client over the 56-Kbps WAN link, which has 40% utilization in the direction of server to client[14] as shown in Figure 7.11(a). Assuming that the network latency is 30 msec, we calculated the optimal window size as 1022 bytes (see Section 2.3.1), or two TCP segments of 512 bytes or 1024 bytes.

The available bandwidth on a 56-Kbps link that is 40% utilized is
Available bandwidth = $56 \times 0.6 = 33.6$ Kbps

If the window size is sufficient to fill the 56-Kbps link when it is unloaded, it will be sufficient to fill the available bandwidth.

Hence the throughput for the TCP file transfer for a loaded connection will be

Available bandwidth on the slowest link / Overhead factor =
Slowest link $\times (1 - U)$ / Overhead factor

Thus for the specific example being discussed, the throughput and file transfer time for a 1-Mbyte file are

33.6 Kbps / 1.09 = 30.8 Kbps (1,024,000 bytes $\times$ 8 bits/byte) / 30,800 = 266 sec

It is not surprising that

---

14. This can be average or peak utilization. Please refer to Chapter 4, Section 4.2, for a discussion about the use of average and peak bandwidth utilization for performance and capacity management.

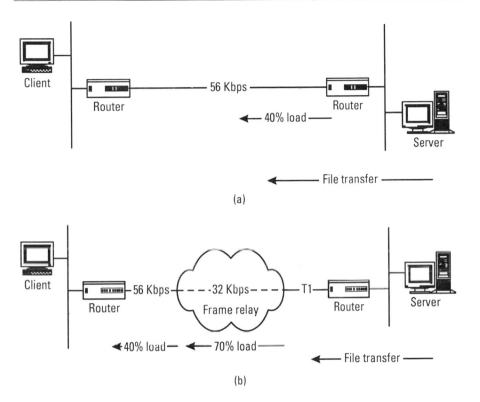

Figure 7.11 TCP throughput for (a) a loaded private line connection and (b) a loaded frame relay connection.

$$266 \text{ sec} = 159.5 \text{ sec} / (1 - 0.4)$$

because 159.5 sec is the file transfer time for an unloaded 56-Kbps connection as derived in Section 7.3.3. Hence, the file transfer time for a loaded connection with utilization U is

(File transfer time without load) $/ (1 - U)$

7.4.2 Frame Relay Case

Let us now consider the case of frame relay. This is somewhat more complex because of three different links that make up the WAN connection. Each of the three links may have different utilization levels.

However, the overall guideline remains the same, that is, we need to isolate the slowest link in the connection. In addition we have to account for the fact that the PVC can burst above the CIR. Let us consider a specific case first and then attempt to derive some general formulas.

Consider the frame relay connection of Figure 7.11(b). Assume that the file transfer occurs in the direction of server to client, and that the 56-Kbps remote frame relay port has 40% utilization (in that direction). Assume also that the 32-Kbps CIR has 70% utilization[15,16] in the same direction.

Assuming that the PVC can burst fully up to port speed, the slowest link is 56 Kbps, of which 60% is available for the TCP file transfer. If we assume that the PVC cannot burst up to the port speed, then the slowest link speed is 32 Kbps, of which 30% is available. Hence the best case throughput is a function of

$$\max(0.3 \times 32 \text{ Kbps}, 0.6 \times 56 \text{ Kbps}) = 0.6 \times 56 \text{ Kbps} = 33.6 \text{ Kbps}$$

The worst case throughput is a function of

$$\min(0.3 \times 32 \text{ Kbps}, 0.6 \times 56 \text{ Kbps}) = 0.3 \times 32 \text{ Kbps} = 9.6 \text{ Kbps}$$

In most practical situations, the worst case is likely to be extreme. In the preceding example, 70% utilization of the PVC corresponds to 22.4 Kbps of traffic offered on the PVC. If the average burst rate on the PVC is 150%, then the available bandwidth is

$$\text{CIR} \times \text{Burst rate} - 0.7 \times \text{CIR} = 25.6 \text{ Kbps}$$

Therefore if the burst rate were known, it would be possible to narrow the range of the available bandwidth. However, burst rates are hard to measure. It is also not advisable to use loosely stated burst rates when designing frame relay networks to meet specific performance criteria.

In the preceding example, then, the throughput and file transfer time can be calculated as

---

15. If there is a single PVC between the client and server locations, then the bandwidth consumption at the PVC should match that at the remote port. If the remote port has PVCs to other locations, then clearly this is not true.

16. Just because the PVC utilization does not exceed 100% does not mean that PVC is not bursting above the CIR. It is possible that not enough traffic is being sent on the connection.

Best throughput = 33.6 Kbps/Protocol overhead = 30.8 Kbps

Best file transfer time = File size / Throughput = 1,024,000 × 8/30,800 = 266 sec

and

Worst case throughput = 9.6 Kbps / Protocol overhead = 8.8 Kbps

Worst case file transfer time = File size / Throughput = 1,024,000 × 8/8800 = 931 sec

The foregoing ideas and calculations can be captured in succinct formulas in the case of frame relay. Assume a fixed direction for the file transfer (say, central/hub site to client site). Let

P_a = frame relay port speed at client site

P_b = frame relay port speed at hub site

C = CIR

U_a, U_b = utilization of the frame relay ports at the client and server side, respectively

U = utilization in the given direction for the PVC

O_f = protocol overhead factor

F = file size being transferred

X_{max} = maximum throughput achievable on this connection

X_{min} = minimum throughput achievable on this connection

T_{max} = maximum file transfer time in the given direction

T_{min} = minimum file transfer time in the given direction

Then one has, for throughput and file transfer times, the following equations:

$$X_{max} = \min([1 - U_a] \times P_a, [1 - U_b] \times P_b] / O_f$$

$$X_{min} = (1 - U) \times C / O_f$$

$$T_{min} = \text{File size / Maximum throughput} = F/X_{max}$$

$$T_{max} = \text{File size / Minimum throughput} = F/X_{min}$$

7.5 WAN Performance Issues for HTTP

The explosion of Internet technologies in recent years has made HTTP an important member of the family of applications supported by TCP/IP. Just as FTP is used to transfer files and telnet is used for remote terminal access, HTTP is used for exchanging hypertext documents between clients and servers. This section briefly discusses HTTP performance over WANs. We also provide some approximations for calculating how long it would take to download a HyperText Markup Language (HTML) document with multiple embedded images/documents (henceforth, we will use the term *embedded images*) from a Web site (such as a home page) over a private line or frame relay network.

Our focus is strictly the wide-area network. The issue of HTTP performance is complex and has been the subject of extensive study, analyzing HTTP performance from various points of view: client/server implementations, caching, proxy server issues, WAN characteristics, Internet performance, and so on. Please see [1].

Some background information about HTTP is relevant to the discussion here. Please see [1] for more details. Unlike, telnet and FTP,[17] whose performance characteristics are discussed in this chapter, the data transfer characteristics of HTTP are somewhat more complex. Two versions of HTTP are in use today: HTTP/1.0 and HTTP/1.1. The original version, HTTP/1.0, is a fairly straightforward protocol. When a client requests an HTML document from a server, a TCP connection is first established. The client then issues a HTTP Get Request to retrieve the HTML document, and then issues separate TCP connections for each image embedded in the HTML document.

HTTP/1.0 suffers from two primary WAN-related performance problems. First, the use of a separate TCP connection for retrieving the embedded images is clearly inefficient, especially considering that HTML documents can contain multiple small embedded images—HTML documents with 20 or 30 images, of average size of less than 5 Kbytes, are not uncommon. Second, a

---

17. Although the WAN characteristics of an FTP file transfer are relatively straightforward, some complexities are not immediately apparent. For instance, since FTP is stateful protocol, two or three TCP sessions need to be opened before an FTP file transfer takes place (see [2]). For a high-speed link and/or a relatively small file, the network latency will be a major contributor to FTP performance.

penalty is paid in the use of TCP slow start, even when the images are relatively small (a 5-Kbyte image will contain about 9 or 10 segments of size 512 bytes). This is because slow start is most efficient for relatively long data transfers. For short data transfers the penalty is paid in terms of round-trip latency.

To address these concerns, three approaches have emerged. The first two are enhancements to HTTP/1.0 and the third is a full extension to HTTP/1.1.

- *Use multiple TCP connections (three or four) between the client and the server.* This approach is adopted by some common browsers. Although this solves the issue of having to retrieve images sequentially by opening and closing several TCP sessions, the trade-off is bursty traffic on the WAN (the effective window size is now increased three- or four-fold per HTTP session), not to mention the severe penalties on server resources.

- *Use a persistent TCP connection.* Rather than open a separate TCP connection to retrieve each embedded image in an HTML document, the client (browser) sends a Connection Keep-Alive along with the HTTP request, requesting the server to "keep the TCP connection alive" for a specified amount of time.

- *Use HTTP/1.1.* HTTP/1.1 provides other performance benefits in addition to persistent TCP connections. It supports *pipelining;* that is, multiple HTTP requests can be submitted simultaneously, unlike HTTP/1.0, which requires that the requests be submitted sequentially. HTTP/1.1 also supports byte range downloads, that is, the entire image does not have to be downloaded. This also helps in resuming data transfers following failures. HTTP/1.1 needs to be supported by both the client and the server to be effective.

A note of caution regarding HTTP/1.1. The protocol implies that all data transfers take place on a single TCP session. In reality, the implementation of HTTP/1.1 in clients and servers may be different from the standard. For starters, at the time of this writing, many servers and browsers still support HTTP/1.0 with multiple TCP connections. Microsoft IE4.0® appears to support HTTP/1.1. However, HTTP/1.1 support on servers appears to be not very wide spread. This observation is based on a small study of packets flowing between an IE4.0 browser and www.whitehouse.gov, www.cisco.com, www.sun.com, and www.bloomberg.com. For the one site (www.apache.com) that appears to support HTTP/1.1, IE 4.0 uses two parallel TCP sessions to retrieve the HTML document and for the embedded images. For HTTP/1.1, it may actually be beneficial to use two TCP sessions rather than one if there is sufficient bandwidth in the connection and the window size is less than the optimal size for that connection.

7.5.1 Summary of HTTP WAN Performance Issues

The preceding discussion can be summarized as follows:

- Multiple images can be embedded in the HTML document, each of which needs to be retrieved.
- Each data transfer could require a separate TCP session.
- Bursty traffic can result from the use of multiple parallel TCP sessions.
- TCP slow start results in extra delay in the context of HTTP.
- HTTP/1.1 uses persistent TCP connections and pipelining HTTP requests to improve performance.

7.5.2 A Sample Trace of an HTTP Transaction

Figure 7.12 shows a sample protocol trace of an HTTP transaction.[18] The action captured was to load an HTML document on the client PC. The ACK, SEQ, and WIN numbers are omitted after a few packets. The memory and the disk cache in the client PC were cleared prior to the protocol trace. Subsequent to the download of the Web page, the cache revealed about 20 images (.gif files) with an average size of about 5 Kbps. The HTML document itself was about 12 Kbps. This was discovered by saving the HTML page onto a disk. The sniffer was at the client side.

We make the following observations:

- The entire process, performed at an off-peak hour, consists of 153 packets.
- The HTTP requests appear to be about 400 bytes long (obtained from the TCP-level information of the HTTP requests—the size field includes overhead).
- The maximum segment size appears to be 1460 bytes and the advertised window about 8 Kbytes.

---

18. The protocol trace was collected from the authors' client PC using Netscape Navigator® 4.5 to the URL www.att.com/data. The server appears to be Netscape as well, although the version is unknown.

#	Source	Dest	Layer	Size	Delta Time	Summary
1	Client	Server	TCP	64	0 ms	D=80 S=1214 SYN SEQ=1657881 LEN=0 WIN=8192
2	Server	Client	TCP	64	59 ms	D=1214 S=80 SYN ACK=1657882 SEQ=179941713 LEN=0 WIN=8760
3	Client	Server	TCP	64	0	D=80 S=1214 ACK
4	Client	Server	HTTP	430	3	C Port=1214 GET /data/ HTTP/1.0
5	Server	Client	TCP	64	87	D=1214 S=80 ACK
6	Server	Client	HTTP	1518	276	R Port=1214 HTML Data
7	Client	Server	TCP	64	09	D=80 S=1214 ACK
8	Server	Client	HTTP	1518	97	R Port=1214 HTML Data
:						
14	Server	Client	HTTP	1518	8	R Port=1214 HTML Data
15	Client	Server	TCP	64	41	D=80 S=1215 SYN SEQ=1658680 LEN=0 WIN=8192
16	Client	Server	TCP	64	48	D=80 S=1216 SYN SEQ=1658729 LEN=0 WIN=8192
17	Server	Client	HTTP	1280	2	R Port=1214 HTML Data
:						
21	Client	Server	TCP	64	19	D=80 S=1217 SYN SEQ=1658680 LEN=0 WIN=8192
22	Client	Server	HTTP	478	6	C Port=1215 GET /images/attlogo_n9050.gif HTTP/1.0
:						
25	Client	Server	HTTP	479	3	C Port=1216 GET /data/images/title_dns.gif HTTP/1.0
:						
32	Client	Server	TCP	64	390	D=80 S=1214 FIN
:						
45	Client	Server	TCP	64	54	D=80 S=1215 FIN
:						
60	Client	Server	TCP	64	21	D=80 S=1216 FIN
:						
153	Last packet					

Figure 7.12 A sniffer trace of an HTTP download.

- Client and server processing delays appear to be significant in some instances. This is to be expected because the HTTP clients are "fatter" than HTTP servers—they are responsible for parsing the HTML information returned from the server and presentation to the end user.

- Notice how one TCP session is used for retrieving the HTML document (port 1214) and, prior to a complete receipt of this document, three other TCP sessions are opened on ports 1215, 1216, and 1217. These are used to retrieve specific .gif files and are closed immediately after they are retrieved. The total number of TCP ports open at any given time appears to be four.

- It appears that (not shown in the traces given) the server does not support Connection Keep-Alive. The client requested a Connection Keep-Alive in the first HTTP Get request. The server response does not contain a Connection Keep-Alive in the HTTP response. Thus the client proceeded to open three additional TCP sessions.

7.5.3   Estimating HTTP Performance Over a WAN Connection

In this section, we provide some approximate guidelines for estimating how long it would take to retrieve an HTML document from a Web server, under some assumptions. An exhaustive treatment of this topic is beyond the scope of this book.

The reference connection for the calculations is always useful. Assume that a client browser is connected to an intranet Web server via a frame relay connection with speeds as shown in Figure 7.13, and that the client is requesting an HTML document from the Web server.

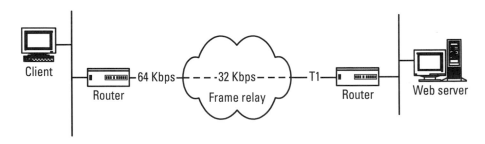

Figure 7.13 Reference connection for an intranet between a client and a Web server.

It is impossible to provide quantitative estimates of WAN performance without making assumptions. The assumptions are discussed next.

First, an intranet model is assumed. We do not model the Web download process through a corporate network and a proxy server to the Internet, although this is an important aspect of end user use of Web technologies.

Unlike the calculation of file transfer times using FTP where the size of the file is easily available, the information regarding HTTP transactions is harder to obtain. Three sets of variables are needed to characterize an HTTP transaction: size of the HTTP request, size of the HTML document, and the number and size of the embedded images. The size of the HTTP request can only be obtained via a protocol trace, as shown earlier. If unavailable, an assumption of a few hundred bytes should suffice. The size of the HTML document can be obtained by saving the document to disk. Typical HTML documents are less than 10 Kbytes, although larger documents (fancier) can exist. As for the number and size of the embedded images, one way to obtain this information is to make sure the cache is cleared prior to loading the document and observing the cache contents after the download.

For the calculations that follow, we assume that the HTTP requests are approximately 400 bytes (without headers), that the HTML document is about six TCP segments of 1460 bytes each (or about 8 Kbytes), and that there are 20 images with an average size of about three TCP segments of 1460 bytes each (or about 4 Kbytes).

We also made some network-related assumptions. Assume that the round-trip latency is 60 msec and that there is minimal background load (always the first assumption in baselining performance). Assume also that all other network components (routers, LANs, DNS) do not add significant delays.

As shown in the trace in Figure 7.12, the client and server contributions to overall delays can be significant. However, we ignore this delay in this calculation and focus on the WAN contribution.

Our TCP-related assumptions include that the maximum frame size for a frame carrying a TCP packet is 1500 bytes (roughly equivalent to assuming that the TCP maximum segment size is 1460 bytes), the ACK, SYN, and FIN frames are 48 bytes (at the frame relay layer), and the window size advertised by clients and servers is 8 Kbytes.

In addition, the effect of slow start needs to be factored in. For this, we assume that slow start begins with one TCP segment, then two, four, eight segments, and so on, after each ACK. Some implementations of slow start begin with two segments, but we ignore this.

Here is our "cheat sheet" for delay calculations:

Insertion delay for a 1500-byte frame on a T1 link = 1500 × 8/1,536,000 = 8 msec;

Insertion delay for a 1500-byte frame on a 64-Kbps link = 1500 × 8/64,000 = 187 msec;

Insertion delay for a 48-byte frame on a T1 link = 0.25 msec;

Insertion delay for a 48-byte frame on a 64-Kbps link = 6 msec;

Insertion delay for a 448-byte HTTP request on a T1 link = 2.3 msec;

Insertion delay for a 448-byte HTTP request on a 64-Kbps link = 56 msec; and

One-way latency = 30 msec.

We now estimate the performance of the Web download for the three cases mentioned above: HTTP/1.0, HTTP/1.0 with multiple TCP connections, and HTTP/1.1.

7.5.3.1 Estimating HTTP/1.0 Performance

Although this case may not be encountered in practice, it is nevertheless useful for comparison purposes. The sequence of events that occurs in the process of retrieving an HTML document and the embedded images is as follows (refer to Figure 7.14):

- Get the HTML document (all times and bit rates are expressed in msec, for the sake of clarity)
- Establish a TCP session with the server

Delay = SYN + SYN ACK = (6 + 30 + 0.25) + (0.25 + 30 + 6) = 72.5 msec

- HTTP Get Request

Delay = ACK from client (from the 3-way handshake) + 448 bytes request from client = [6 + 30 + (448 + 48) × 8/64] = 98 msec (Note that the ack and request are pipelined.)

- Get HTML Document

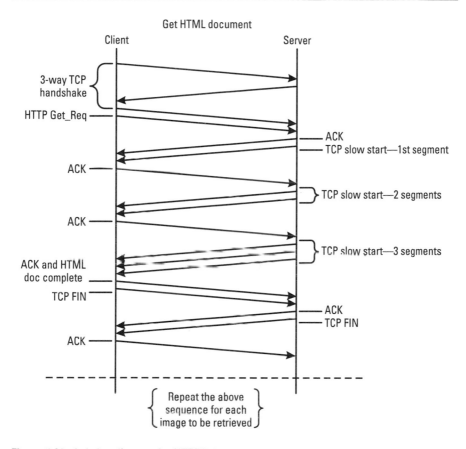

Figure 7.14 A timing diagram for HTTP/1.0.

Delay = ACK from server + 6 TCP segments of 1500 bytes each from server (taking into account slow start) = 0.25 + 30 + [{(1500 + 48) × 8/64} + ACK(from client) + (8 + 30 + 2 × 187) + ACK(from client) + (8 + 30 + 3 × 187)] = 1307.25 msec

- Close the TCP connection

Delay = FIN (from client) + [ACK(from server) + FIN(from server)] + ACK(from client) = (6 + 30 + 0.25) + (0.25 + 30 + 12) + (6 + 30 + 0.25) = 114.75 msec

Total delay = 1.59 seconds

- Get the embedded images (see previous discussion for calculations) Establish a TCP session with the server

$$Delay = 72.5 \text{ msec}$$

- HTTP Get Request

$$Delay = 98 \text{ msec}$$

- Get image from server

$$Delay = \text{ACK from server (to HTTP Get request)} + 3 \text{ TCP segments of}$$
1500 bytes each from server using slow start $= [0.25 + 30 + (1500 + 48) \times$
$8/64] + \text{ACK(from client)} + (8 + 30 + 2 \times 187) + \text{ACK(from client)} =$
708.25 msec

- Close the TCP connection

$$Delay = 114.75$$

Total delay for 1 image = 993 msec
Total delay for 20 images = $0.993 \times 20 = 19.86$ sec

- Total WAN contribution to the delay = $1.59 + 19.86 = 21.5$ sec (approximately)

7.5.3.2 Estimating HTTP/1.0 Performance with Multiple TCP Sessions

Clearly, using multiple TCP sessions will improve performance compared to the previous case. The interactions between the multiple TCP sessions and slow start mechanisms superimposed on a single WAN link are hard to capture in a simple formula. Therefore, we provide some intuitive estimates under the following additional assumptions. A more accurate model would require a more careful analysis, empirical observations, and a simulation model.

- A TCP session is set up to retrieve the HTML document. The image transfers are initiated only after the HTML document is completely received (this may not be true in some cases, where a TCP session for retrieving an image may be set up earlier).

- Four parallel TCP sessions are then set up for image retrieval. Each session issues a separate HTTP request for an image. After the images have been received, additional TCP sessions are set up to retrieve the next batch of the images, and so on until all the images have been received.

- The effect of slow start for a single image retrieval may be significant, but when parallel TCP sessions are established, we will assume that the server has enough data to send to the client to saturate the WAN link.

Under these assumptions, here is a method to quickly estimate the Web page download time for the example we are studying. First, estimate the time to get the HTML document. Second, estimate the time to download the first batch of four images using four TCP sessions. Third, estimate the total time to download the Web page by multiplying this time by the number of iterations needed to retrieve all images. Refer to Figure 7.15.

- Get the HTML document using a single TCP session: Delay = 1.59 sec (from previous example);
- Time for four images: Delay = Time to establish the first TCP session + time to send HTTP request + time to transfer the four images from the server = $72.5 + 98 + (1500 \times 8/1536 + 30 + 4 \times 3 \times 1500 \ 8/64) = 2.46$ sec;
- Time to get all the 20 images: Delay = $5 \times 2.46 = 12.3$ sec; and
- Time for Web download: $1.59 + 12.3 = 14$ sec (approximately).

7.5.3.3 Estimating HTTP/1.1 Performance

To estimate the performance of the Web page download, we proceed in a method similar to the case of HTTP/1.0 with multiple TCP sessions. First, the HTML needs to be downloaded and then the HTTP requests have to be sent to the server in a pipelined fashion. When the server receives the first HTTP request, it starts to download the first image, and from then on, the HTTP responses are streamed one after another, filling the WAN connection. Hence (refer to Figure 7.16):

- Time for the HTML document = 1.59 sec;
- Time for the first HTTP request = $448 \times 8/64 + 30 = 86$ msec;

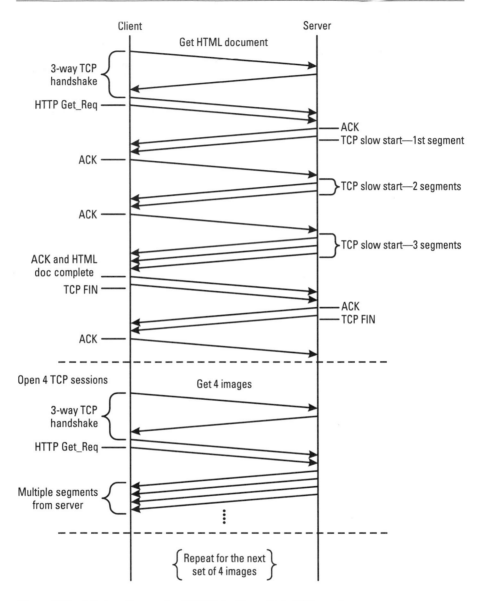

Figure 7.15 A timing diagram for HTTP/1.0 with multiple TCP sessions.

- Time to download all the images on a single TCP session = 1500 × 8/1536 + 30 + 20 × 3 × 1500 × 8/64 = 11.29 sec; and

- Total time for the Web page download = 13 sec (approximately).

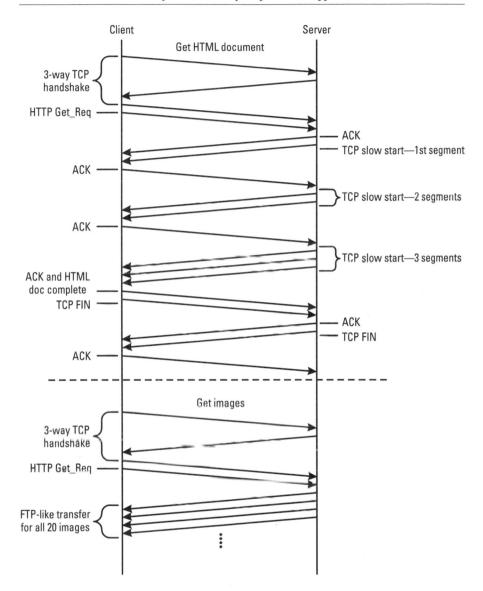

Figure 7.16 A timing diagram for HTTP/1.1.

7.5.3.4 Some Comments on the Performance Estimates

Clearly, the performance estimates for HTTP transfers are harder to compute than for FTP. The complexity arises as a result of many factors—multiple images within an HTML document, multiple TCP sessions, slow start, and so on. Of these, the case of the multiple TCP sessions appears to be most complex.

Using HTTP/1.0 with a single TCP session is clearly the worst performer. However, the fact is that HTTP/1.1 appears to give only slightly better performance than HTTP/1.0 with multiple TCP sessions. In reality, it is likely that both methods have comparable performance on the WAN, from a purely data transfer perspective. The key is the fact that both methods do a reasonable job of keeping the slowest link busy, which then determines the transfer time.

HTTP/1.0 with multiple TCP sessions is nonoptimal from a broader perspective of WAN performance, especially when the window size advertised by the client is large enough so that a single TCP session can saturate the WAN connection (as is the case in our example). In this case, using multiple TCP sessions will, in fact, be detrimental to other application traffic sharing the same WAN connection.

HTTP/1.1 with a single TCP session may be nonoptimal for the converse reason. If the WAN connection is high speed (think of a T1 connection between the client and the server), and if the window size advertised by the client is less than the optimal (as would be the case in our example), then it would make sense to use two or three parallel TCP sessions to "fill the pipe."

7.6 TCP Telnet Performance Issues

Telnet (and Rlogin for Unix) uses TCP/IP services to allow remote hosts to log in to servers as though they are locally attached. Mission-critical applications based on telnet are not very common, but they do exist. The performance issues for telnet discussed later are also applicable in the context of remote presentation techniques, such as Citrix Winframe/Metaframe, which is a commonly used method to improve response times for many client/server applications. Client/server applications are discussed in Chapter 9.

The single most important characteristic of telnet is the fact that it is "echoplex" in nature, that is, every character typed at the remote host is sent to the server, which echoes the character back to the host.[19] Thus if delays on the WAN are long, either due to long latency or due to load, the end user is likely to type "ahead." In other words, there will be a marked delay between the time the character is typed at the keyboard to the time that character appears on the screen. Needless to say, this can be a frustrating experience for the end user.

---

19. Telnet can also operate in line mode, but this is not as prevalent as character mode (or echoplex).

In this section we discuss how to calculate echoplex delays for private line and frame relay networks and demonstrate how telnet performance can be adversely affected by background traffic. Consider Figures 7.3(a) and (b), which show private line and frame relay point-to-point connections. In both cases, assume that the remote PC is a telnet client attached to a telnet server at the central site. Assume also that the network is unloaded and let us calculate the echoplex delay. Because each character (1 byte) is sent across the WAN as a TCP/IP packet, the frame size for this packet is

$$1 + 20 \text{ (TCP)} + 20 \text{ (IP)} + 8 \text{ (PL or FR)} = 49 \text{ bytes}$$

Hence to compute the echoplex delay all we need to do is calculate the round-trip delay to send and receive a 49-byte frame. This is no different from calculating ping delays.

For private line (using the same assumptions about network latency as in Sections 7.3), the estimated echoplex delay is

$$49 \times 8/56,000 + 2 \times 0.03 + 49 \times 8/56,000 = 74 \text{ msec}$$

ignoring, of course, additional delays introduced by routers, LANs, and clients/servers.

For frame relay, the corresponding number is

$$2 \times (49 \times 8/56,000 + 0.04 + 49 \times 8/1,536,000) = 94 \text{ msec}$$

Note that these delays are more a function of network latency than bandwidth. Increasing bandwidth will have little effect on end user telnet performance.

Are these numbers—74 sec for private line and 94 msec for frame relay—tolerable for character-mode telnet users? To answer this question, consider a typical telnet user. Users are likely to type at the average rate of two to three characters per second, with a peak rate of four to five characters per second. Thus the interval between successive character entries is 300 to 500 msec in the average case, and 200 to 250 msec at the peak rate. Hence, for telnet performance to be acceptable, echoplex delay should be in the 200- to 500-msec range allowing the user to type at a rate of two to five characters per second. Clearly both the private line and frame relay numbers are acceptable.

How about the case of a global or satellite connection in which a remote telnet client has logged on to a telnet host across continents? For a 10,000-mile terrestrial connection, the echoplex delay will be at least 200 msec (2×10 msec for every 1000 miles), perhaps in the 200- to 250-msec range for unloaded connections. As demonstrated, this is acceptable. However, for an unloaded satellite connection, the echoplex delay will be at least 560 msec (2×280 msec for a signal to propagate to the satellite and back). This is clearly unacceptable even without considering the impact of load and other delays on the connection.

We now turn to a discussion of the impact of background traffic on telnet response times. While telnet response time over terrestrial WAN connections may be satisfactory, it is severely impacted by background load as demonstrated next, using the example of TCP FTP and telnet sharing the same PVC over a frame relay network.

Figure 7.17 shows two clients at a remote site both connected via TCP/IP sessions over frame relay to a server at a central site. The port speeds and CIR for the connection are as indicated. Assume that client A has a telnet connection to the server, and that client B has an FTP session with the same server. Let us try to calculate echoplex delays when there is background TCP file transfer in the direction of server to client.

Assume that the receiver advertised window size is 4 Kbytes with a segment size of 512 bytes (i.e., window size of eight segments of 512 bytes each). A 4-Kbyte window size is more than sufficient for achieving maximum throughput. Assume also that a single file transfer is in progress and that telnet has the same priority as FTP on this connection. What is the maximum additional delay that the telnet user will experience?

Look at the timing diagram in Figure 7.18. Assume that a telnet echo from the server to the client is sent at a random instant of time while an FTP file transfer is in progress. Because of the speed mismatch between the T1 line and the PVC CIR and remote port speed (32 and 56 Kbps, respectively),

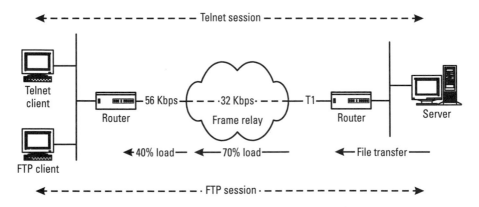

Figure 7.17 Mixing telnet and FTP over a frame relay connection.

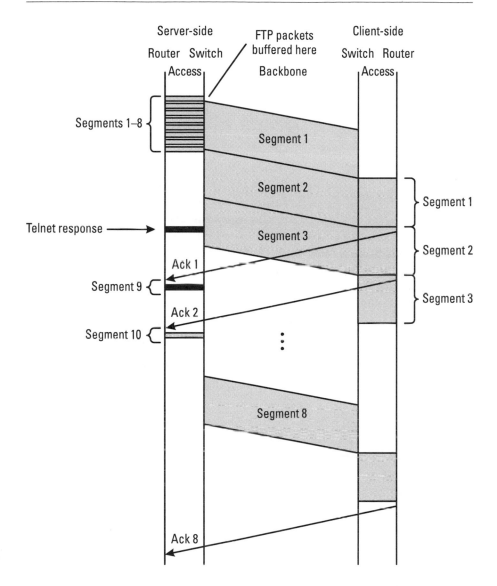

Figure 7.18 Timing diagram showing telnet echoes mixing with FTP.

buffering of the FTP data will occur at the ingress buffers and/or egress buffers at the network edges.

We need to calculate how much FTP data can be resident in network switch buffers at the ingress or egress at any given time, under the assumption of a single file transfer with a window size of 4 Kbytes. Note that the maximum amount of FTP data in the network switch buffers at any given time is the window size (plus overhead), that is, $8 \times (512 + 48) = 4480$ bytes. Hence, the maximum amount of time that the telnet echo will be delayed is equal to the time it takes to empty a window's worth of data on the slowest link. This is equal to

$$4480 \times 8/56{,}000 = 640 \text{ msec}$$

Actually, we can refine this analysis further. Note that the time between successive acks received is equal to the time required to insert a frame at the slowest link. Hence, the sender is allowed to send one segment every time an ack is received; the *net* input rate to network buffers is 0. Because the sender starts with a window of 8 Kbytes,[20] and the switch can empty $W^* = $ Round-trip delay $\times$ Slowest link (optimal window) while waiting for an acknowledgment, and one segment is injected for every acknowledgment, the switch buffer will have

$$\text{Advertised window size (plus overhead)} - W^*$$

For this example, the optimal window size is approximately two segments of 512 bytes. Hence, the switch buffer contains approximately six segments of 512 bytes each. The time to drain these data from the buffers at the slowest link speed is

$$(512 + 48) \times 6 \times 8/56{,}000 \text{ sec} = 482 \text{ msec}$$

This is an approximation for the additional delay that a telnet echo response will experience.

Clearly, these are unacceptable response times for the telnet end user. Note that this is the impact due to single TCP FTP in the background. When there are multiple parallel bulk data transfers in the background, the impact will be far greater.

This analysis also shows the impact of using a window size larger than the optimal window size for TCP applications. If the window size is set at or near

---

20. This is not strictly true—the sender will not send the entire window size advertised by the receiver. Rather, it will implement *slow start*, whereby the sending station will send one segment, receive an ack, send two more, four more, and so on. With a relatively small advertised window size, it will take only a few round-trip delays for the sender to send the full window of 8 Kbytes advertised by the receiver.

its optimal value W^*, then telnet echo will experience no additional delay, and yet TCP FTP will receive the best throughput possible. That is the beauty of the optimal window size setting!

To summarize, for bulk data transfers, using a window size $W > W^*$, the optimal window size, results in no additional gain in throughput. On the other hand, it increases the possibility of negatively impacting interactive application response time!

7.7   Review of Methods to Provide Traffic Discrimination for TCP/IP Applications

The preceding discussion shows the need to provide preferential treatment for certain applications over a WAN via traffic discrimination. The problem is most apparent when mixing response-time-sensitive TCP interactive applications (like telnet, Citrix Winframe/Metaframe, client/server applications, TCP encapsulated SNA traffic) and bulk data transfer applications (like TCP FTP and HTTP) on the same connection. In Chapter 5, Section 5.5, we discussed this issue in the context of frame relay networks and the congestion shift from routers to the frame relay cloud. In that section we also discussed several methods to ensure good performance for higher priority traffic. In this section, we review some of those methods, using TCP telnet and TCP bulk data traffic over frame relay as the example. Clearly, the same logic can be applied to other applications as well.

We review the following methods:

- Prioritize telnet over bulk data transfer traffic at the router;
- Separate telnet on its own PVC—priority PVCs;
- Traffic shape at the router; and
- Deploy more general bandwidth management techniques.

7.7.1   Prioritize Telnet Over Bulk Data Transfers at the Router

One can prioritize telnet over other traffic streams at the router WAN serial interface on the basis of TCP port numbers. Thus, for instance, telnet (port 23) can be prioritized over FTP (port 21), or telnet can be prioritized over Web (HTTP) traffic on port 80.

What is the maximum gain in delay for telnet echoplex under such a prioritization scheme? Given the direction of telnet and bulk data transfer traffic, prioritization needs to take place at the WAN router closest to the server (see

Figure 7.19). Again, due to the speed mismatch between the T1 frame relay port and the CIR/remote port speed (32/56 Kbps), there can be at most one TCP segment at the T1 port on the router that the telnet echo gets priority over. This is due to our assumption of a single background TCP FTP. Hence, the gain in delay is likely not to exceed the insertion delay of a TCP segment on the T1 line, that is,

$$(512 + 48) \times 8/1{,}536{,}000 = 3 \text{ msec}$$

Note that the specific implementation of router prioritization, whether strict priorities or bandwidth allocation with minimum guarantees for lower priority traffic, is irrelevant. Even weighted fair queuing (WFQ) will be ineffective because WFQ works well to allocate bandwidth for time-sensitive traffic only under congestion. Bandwidth allocation schemes appear to be attractive candidates to resolve the prioritization issue. For example, bulk data transfer can be allocated limited bandwidth at the router (say, 90% for telnet and 10% for FTP). Even this scheme will fail because routers will not limit the bandwidth for lower priority traffic when higher priority traffic is *absent*. Thus bulk data transfers will receive 100% of the available bandwidth during times when telnet is not present. Limiting the bandwidth allocated to a particular traffic stream is the function of traffic shaping, not router prioritization.

Therefore it is clear that router prioritization schemes, no matter how sophisticated, will not address the problem as stated above. This is because prioritization needs to take place at congestion points—the T1 port is hardly the congestion point. The congestion is more likely to take place at the network edges.

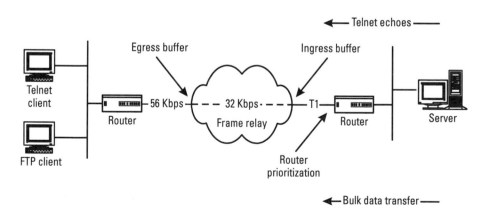

Figure 7.19 Prioritizing telnet over bulk data traffic.

This, however, does not mean that router prioritization should never be used. In reality, the WAN serial port may support several PVCs, each of which may be supporting bursty TCP traffic (like Web downloads). In such cases, the router should definitely be configured to prioritize telnet traffic. The point of the preceding discussion is that router prioritization alone is insufficient to guarantee good performance for telnet.

7.7.2 Separate Telnet on Its Own PVC

If telnet were to be separated on its own PVCs (some routers, if not all, can be configured to do this), then the "packet train" effect will be minimized. To see this, consider a closed-loop frame relay network (see Chapter 2, Section 2.4, for a discussion on closed- and open-loop implementations of frame relay). In such an implementation, the ingress buffer (see Figure 7.19) will carry bulk data and telnet packets. The effect of allocating a separate PVC for telnet will be to form two separate queues, served in round-robin format, for telnet and other traffic. Hence a telnet packet will not have to wait behind a packet train of non-telnet packets. However, if there is enough background traffic, a queue will build up at the egress buffer (see Figure 7.19). Hence, in addition to allocating a separate PVC for telnet, this PVC must also be prioritized in the network at the egress. Typical implementations of priority PVCs at egress frame relay ports use a single queue and assign a higher transmission priority to specific PVCs with bandwidth guarantees for lower priority PVCs.

In the open-loop implementation, the queue buildup will occur at the egress buffers. To be able to distinguish between telnet packets and non-telnet packets, telnet will have to be on a separate PVC and will have to be prioritized. In other words, the solution is exactly the same for both open- and closed-loop systems.

Although the separate PVC approach provides a reasonable solution, it can only be recommended for resolving specific situations in a small network (that is, a "band-aid" approach). It is not a universal solution for providing traffic discrimination over frame relay. It is limited by the capability of the router to separate traffic classes into different PVCs, requires PVC prioritization in the network, and increases the administrative overhead in the network, and is therefore not a scalable solution.

7.7.3 Traffic Shaping at the Router

In the previous example, if the router can be configured to throttle the transmit rate on the PVC carrying telnet and bulk data transfer traffic to 56 Kbps, then the speed mismatch between the frame relay port speeds (T1 versus 56 Kbps) will be eliminated. Thus the queue buildup will more likely happen at the router buffers rather than the network buffers. Because routers can distinguish between telnet and non-telnet traffic, router prioritization will become effective and telnet response times will improve. At the same time, bulk data transfer throughput will not be affected because the maximum throughput is capped at 56 Kbps.

Of course, depending the bursting capabilities of the PVC, there is still a chance for network buffers to accumulate packets. In this case, the router can be configured to rate throttle the PVC to the CIR (assuming, of course, that the carrier guarantees the CIR). The downside of this approach is that the PVC will never be allowed to burst.

In general, however, router-based traffic shaping also cannot be recommended as a universal solution for traffic discrimination. This approach is only applicable for frame relay, and traffic shaping is not commonly implemented in all routers. It is also not scalable because of its impact on router resources.

7.7.4 More General Bandwidth Management Techniques

Suppose in addition to telnet, the network has to support Web-based intranet applications, client/server applications, e-mail, and Internet access. Combine this with the fact that different users in the different parts of the organization require different service levels for the same applications. For instance, Internet access may be more important for R&D users than people in accounting. Conversely, SAP R3® is likely to be more important to accounting users compared to R&D.

Clearly, none of the techniques already mentioned—router prioritization, separate PVCs, and router-based traffic shaping—will be adequate. These are relatively straightforward layer 2 and layer 3 approaches. One needs a more sophisticated bandwidth allocation and traffic shaping scheme. At the time of this writing, a few vendors offer products that address this need. We mention three—Packeteer, Xedia Corporation, and CheckPoint Software Technologies Ltd. Packeteer uses a TCP rate control approach where individual TCP sessions are monitored and management occurs from the bandwidth point of view. Xedia and CheckPoint, on the other hand, use queuing techniques—class-based queuing and weighted fair queuing—to achieve the same goals. These products are placed at strategic locations in the network, typically at congestion points, and allow bandwidth allocation at a very granular level.

7.8 Summary

This chapter provides ways to quantify TCP/IP application performance over a wide-area network. Consistent with the rest of the book, the WAN technologies analyzed are leased lines and frame relay. The focus is on the techniques used to arrive at performance metrics, rather than on the underlying WAN technology itself, although the two are sometimes related. These techniques can be used in other contexts as well, such as dial-up connections, VPNs, and ATM WANs.

We first discuss TCP windowing characteristics and demonstrate how throughput for a TCP bulk data transfer can be calculated, for unloaded WAN connections, as a function of window size, network latency, and other parameters. The importance of optimal window size for a WAN connection is then discussed. We provide general formulas for throughput computation for private line and frame relay networks. We then discuss the impact of background load on TCP throughput. This is followed by a discussion of performance issues for HTTP and telnet applications. Finally, we briefly discuss the issue of traffic discrimination and the various approaches one can take to ensure satisfactory performance for a class of critical applications in the presence of less critical applications.

References

[1]   Killelea, P., *Web Performance Tuning,* Cambridge, MA: O'Reilly & Associates, 1998.

[2]   Stevens, W. R., *TCP/IP Illustrated, Volume 1: The Protocols,* Reading, MA: Addison-Wesley, 1994.

[3]   Padhye, J., et al., "Modeling TCP Throughput: A Simple Model and Its Empirical Validation," *Proc. SIGCOMM '98,* Vancouver, British Columbia, August 1998.

[4]   Madhavi, J., and S. Floyd, *TCP-Friendly Unicast Rate-Based Flow Control,* http://www.psc.edu/networking/papers/tcp_friendly.html, January 1997.

8

WAN Performance Considerations for Novell NetWare Networks

8.1 Introduction

Novell networking was the dominant LAN networking protocol suite in the early 1990s. With the deployment of LAN interconnections over wide-area networks, it became one of the dominant protocol suites to run over enterprise networks. The Novell protocol suite has its roots in the development of the Internet protocols, but is primarily a Novell proprietary networking protocol. With the emergence of the Internet and its protocols as the networking standard, Novell is in the process of embracing the TCP/IP protocols where appropriate.

In this chapter, we discuss the aspects of the Novell networking protocols that are relevant to performance engineering of WANs. In the next section, we present a brief overview of Novell NetWare protocol suite. We follow this with a presentation of packet formats and protocol specifics that affect the amount of bandwidth necessary to support NetWare. In the following section, we describe the evolution of the NetWare transport protocols through the various releases of NetWare from release 3.11 to 4.x. The final section of this chapter contains some simple formula for throughput and delay of NetWare applications in private line and frame relay environments, in the spirit of the formulas of the previous chapter on TCP/IP performance.

At the time of the writing of this book, NetWare 5.0 was just being released by Novell. With this release, NetWare can be configured to run over the TCP/IP transport protocols. In this situation, the formulas we presented in

Chapter 7 on TCP/IP are more appropriate. In the following material, we concentrate only on NetWare releases 3.11 to 4.x and on those release 5.0 implementations that rely on the older NCP transport capabilities, for example, non-burst-mode and burst-mode transport protocols.

8.2 Overview

The Novell NetWare protocols draw heavily on the Xerox Networking System (XNS). Figure 8.1 shows the dominant layers and their corresponding protocols, which comprise the NetWare protocol suite. (For a thorough discussion of Novell's NetWare protocols, see [1] and [2].) This figure shows the data link, packet, transport, and file handling layers of NetWare. NetWare supports numerous data link protocols, including Ethernet, token ring, and X.25. The packet layer is the internetwork packet exchange (IPX) protocol. It is a datagram protocol like IP and has its origins in the XNS internetwork datagram protocol (IDP). NetWare protocols, which rely on the services of IPX, include the routing information protocol (RIP), the service advertisement protocol (SAP), the sequenced packet exchange (SPX) protocol, and the network core protocol (NCP).

The NetWare protocol defines the interaction between Novell clients and servers. On the server side, the server runs the NetWare operating system and the networking protocols, which extend the operating system services to the clients over the network. The NetWare operating system consists of a kernel providing core services. Additionally, capabilities can be added to the server through additional software referred to as network loadable modules (NLMs) in NetWare 3.x or virtual loadable modules (VLMs) in NetWare 4.x. These loadable modules are software programs that are dynamically loaded when necessary and act as part of the operating system kernel. We will discuss some specific NLMs and VLMs that enhance the NCP transport protocols.

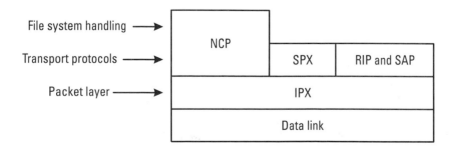

Figure 8.1 Novell NetWare release 3.x and 4.x protocol stack.

On the client side, the NetWare software extends the client operating system onto the network. This software is referred to as the NetWare shell. This is illustrated in Figure 8.2.

The protocol stack, as shown in Figure 8.1, is indicative of the NetWare release 3.x and 4.x versions. With the next release of NetWare, release 5.X, Novell more fully embraces the now-dominant TCP/IP protocol stack (see Figure 8.3). In this release, the Novell user can replace IPX with IP at the packet layer and rely on TCP as the primary transport protocol. Release 5.0 was due out near the end of 1998. With this release, much of our earlier discussions on TCP/IP performance will apply to NetWare. For more information on NetWare integration onto TCP/IP, refer to [3]. However, it will take some time for extensive deployment of 5.0 to occur. In the meantime, we expect earlier releases of NetWare to be commonplace in many enterprise networks.

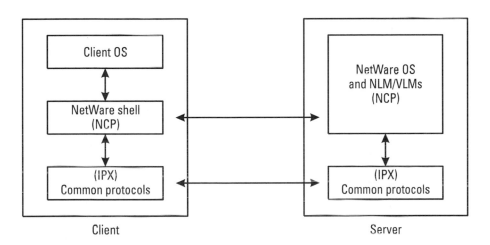

Figure 8.2 NetWare operating system client/server architecture.

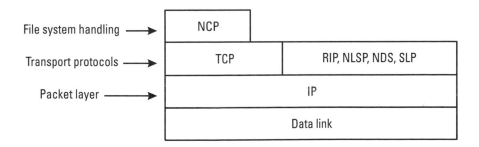

Figure 8.3 NetWare support for the TCP/IP protocol stack in release 5.X.

In this chapter we discuss some of the performance-affecting characteristics of the earlier releases, primarily NetWare releases 3.x and 4.x. We first discuss these protocols, packet formats, and bandwidth considerations on WANs in the next section of this chapter. We then discuss the evolution of the NetWare transport and windowing strategies and discuss the performance motivations for this evolution. We end this chapter with a brief discussion of some formulas to estimate the performance of these transport protocols over representative private line and frame relay environments.

Throughout this chapter, we will focus on NCP as the dominant NetWare transport protocol and RIP and SAP as the routing and discovery protocols. By doing so, we implicitly claim that the majority of the WAN performance and engineering issues with NetWare 3.x and 4.x are related to these three protocols.

8.3 Overhead and Bandwidth Considerations

In this section we present several aspects of the NetWare protocols and packet formats that affect bandwidth considerations over WANs. We concentrate on the NetWare NCP transport and RIP and SAP routing and service discovery protocols for NetWare releases 3.x and 4.x.

The network core protocol (NCP) provides basic communication, file access, print access, and queue management services within the NetWare networking environment. For example, the NCP handles read and write requests issued over the LAN/WAN. The NCP relies on the IPX protocol for its datagram-based networking. The IPX protocol is similar to IP in the services it supplies to the upper protocol layers. Figure 8.4 shows the format of the IPX packet header. The IPX header is 30 bytes in length and contains the source and destination network and node addresses. The NCP includes the transport protocol capabilities within it. These transport capabilities have evolved much

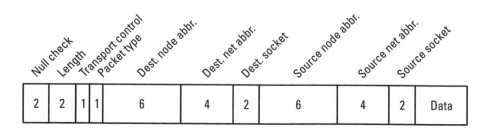

Figure 8.4 IPX packet header fields and lengths.

from NetWare release 3.x to 4.x and 5.x. We discuss this evolution and performance in the following section.

Most implementations of NetWare rely on the routing information protocol (RIP) as their routing discovery protocol. The RIP also runs over the IPX protocol. The RIP is a distance vector protocol (like the RIP within the TCP/IP protocol suite). Systems maintaining RIP routing databases will periodically (typically once each 60 sec) broadcast their routing table entries to all of their nearest neighbors. For each reachable network address, a routing table entry consists of the network address, the hop count to that network, and the time estimate to the network.

The RIP message format is shown in Figure 8.5. Here we see that the RIP message header is 4 bytes in length. This is followed by up to 50 network addresses, hop counts, and time estimate entries. Each of these entries is 8 bytes in length: 4 bytes for network address, 2 bytes for hop count and 2 bytes for time estimate. When these messages are transmitted over a typical WAN interface, they are encapsulated in an IPX packet, which is further encapsulated within the data link packet.

Assuming that the typical data link packet adds 8 bytes (that is, 6 bytes for link headers and trailers and 2 bytes for higher level protocol identifiers) to the packet and that we know that the IPX header is 30 bytes, we get for the best case scenario the following estimate for bytes transmitted per network advertised:

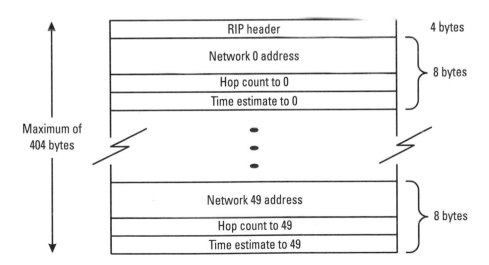

Figure 8.5 The NetWare RIP message format.

$$50 \text{ networks} = (8 \text{ bytes for data link}) + (30 \text{ bytes for IPX}) +$$
$$(50 \times 8 + 4 \text{ bytes}) = 442 \text{ bytes}$$

or

$$E(\text{RIP}) = 9 \text{ bytes} / \text{NetWare RIP network entry}$$

Let us now estimate the bandwidth requirements to support NetWare's RIP. The bandwidth requirement will reflect the amount of traffic to be transmitted per network entry (just derived) and frequency that this information is to be transmitted over the WAN facilities, that is,

$$\text{RIP bandwidth requirement} = N(\text{RIP}) \times F(\text{RIP}) \times E(\text{RIP}) \times 8 \text{ bits/byte}$$

$$\text{RIP bandwidth requirement} = 72 \times N \times F \text{ (in bps)}$$

where $N(\text{RIP})$ is the number of routing entries to be advertised, $F(\text{RIP})$ is the frequency of the routing advertisements, and $E(\text{RIP})$ is the number of bytes to be transmitted per routing table entry. We have used the just derived value of $E(\text{RIP}) = 9$ bytes/entry to obtain the second expression.

As an example, for a routing table of 1000 entries that is broadcast once every 60 sec, we get

$$72 \times 1000 \times 1/60 = 1.2 \text{ Kbps}$$

In many situations this does not appear to be an excessive bandwidth requirement. Most serial lines are 56 Kbps or higher, so this represents a fairly small percentage of these links. However, the instantaneous demand for bandwidth caused by the broadcast of RIP packets each 60 sec may be problematic. In cases where a single frame relay port supports multiple PVCs with a high overbooking factor, the RIP packets are replicated onto each of the PVCs on the frame relay port. For a port supporting 20 PVCs, this would represent an instantaneous demand for bandwidth due to the transmission of $9000 \times 20 = 180$ Kbytes. This may be enough to cause packet drops, FECN/BECN, and so on. In situations like this, it is probably wise to either increase the port speed of the frame relay interface if it is necessary to run dynamic updates over these VCs, or use static and default routes if not. For a discussion of routing design and its relationship to network topology, see [4].

The next NetWare protocol we discuss is the service advertisement protocol (SAP). The SAP is the NetWare protocol that advertises the various NetWare services and servers. Any server or service can be advertised, but common service advertisements include routers, gateways, and file and print servers. The SAP allows programs to register their services and clients to request the location of those services. Requests are broadcast to all nodes on the network.

Servers on a LAN segment typically send a SAP broadcast every 60 sec on that LAN. Routers use these broadcasts to populate SAP information tables (SIT). Routers then exchange SIT information over the WAN each 60 sec. The SAP packets containing these router exchanges are shown in Figure 8.6 and can contain multiple (up to seven) service advertisements.

Here we see that the message header is 2 bytes in length. This is followed by up to seven service advertisement entries. Each of these entries is 64 bytes in length. When these messages are transmitted over a typical WAN interface, they are encapsulated in an IPX packet, which is further encapsulated within the data link packet. As before, assuming that the typical data link packet adds 8 bytes to the packet and that we know that the IPX header is 30 bytes, we get for the best case scenario the following estimate for bytes transmitted per service advertised as:

$$7 \text{ SAP entries} = (8 \text{ bytes for data link}) + (30 \text{ bytes for IPX}) +$$
$$(450 \text{ bytes for SAP}) = 488 \text{ bytes}$$

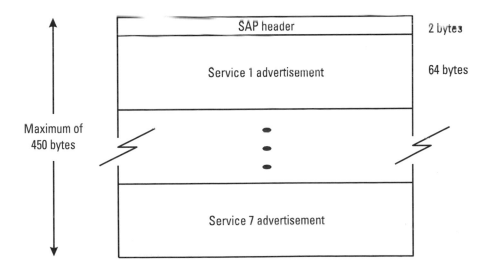

Figure 8.6 NetWare SAP message formats.

or

$$E(SAP) = 70 \text{ bytes / NetWare SAP entry}$$

Given that it is not uncommon to find 500 SAP entries in a SIT on a large NetWare network, updating a SIT can impose a high traffic load on WAN links. In this example, the background traffic is roughly

$$\text{SAP bandwidth requirement} = N(SAP) \times F(SAP) \times E(SAP) \times 8 \text{ bits/byte}$$

$$\text{SAP bandwidth requirement} = 560 \times N(SAP) \times F(SAP) \text{ (in bps)}$$

where $N(SAP)$ is the number of SAP entries to be advertised, $F(SAP)$ is the frequency of the SIT updates, and $E(SAP)$ is the number of bytes to be transmitted per SIT table entry. We have used the just derived value of $E(SAP) = 70$ bytes/entry to obtain the second expression.

As an example, for a SAP information table of 500 entries that is broadcast once every 60 sec, we get

$$560 \times 500 \times 1/60 = 4.7 \text{ Kbps}$$

Like the case of estimating the average load introduced by RIP broadcasts on the network, the average load cause by SAP broadcasts in most situations is not large either. However, in some situations the instantaneous transmission of (4.7 Kbps × 60 sec / 8 bits per byte =) 35 Kbytes per link can be problematic.

To lessen the burden of RIP and SAP broadcasts over WAN facilities, Novell introduced the NetWare link services protocol (NLSP). NLSP is akin to a link state routing protocol in that it only transmits changes in routing or services information tables. This is in contrast to RIP and SAP, which periodically broadcast their entire information tables. When implemented, NLSP replaces both the RIP and SAP protocols. Therefore, NLSP implementations do not suffer from the periodic and instantaneous transmission of large volumes of routing and service advertisement data.

8.4 Novell Windowing Schemes

In this section we discuss the evolution of the NetWare transport windowing capabilities. We begin with a discussion of NetWare 3.0, and then discuss the modifications Novell implemented in releases 3.11, 3.12, and 4.0. The majority of the NetWare services are handled through NCP. The NCP sits directly on top of the IPX protocol. NCP incorporates the services associated with the transport protocol, including flow control and error recovery on an end-to-end basis. As such, we focus on the transport capabilities implemented in the various releases of NCP.

8.4.1 NetWare Pre-Release 3.11

Prior to release 3.11, the NCP imposed a simple request/response dialog between a client and a NetWare server. When the client issued a read request for a file of 1024 Kbytes, NCP would divide the read request into individual NCP requests/responses based on the maximum packet size allowed in the NetWare router, for example, typically 512 bytes of data in a packet of 576 bytes total (see Figure 8.7). In this example then, the single client/server service transaction would require 1024 Kbytes / 512 bytes = 2000 separate NCP request/response exchanges. The performance implications of this ping-pong effect imposed by the pre-release 3.11 NCP are obvious. We discuss the performance impact in Section 8.5.

The NCP packet format is shown in Figure 8.7. This shows that the NCP overhead is 34 bytes and that the data portion of the packet supports up to 512 bytes of data, assuming a negotiated packet size of 576 bytes. Of course, larger packet sizes can be negotiated, resulting in larger data fields.

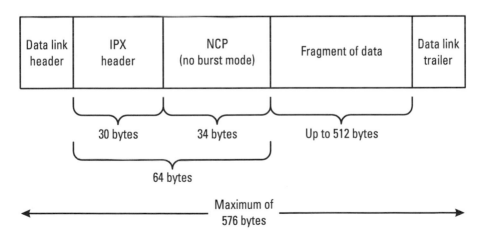

Figure 8.7 NCP packet format for pre-release 3.11 of NetWare.

8.4.2   NetWare Release 3.11

With NetWare release 3.11, Novell introduced its *burst-mode technology,* sometimes referred to as *packet burst.* This first introduction of a new NCP transport capability required the client to install the burst-mode version of the NetWare shell (BNET.EXE) and required the server to install the packet burst NLM (PBURST.NLM). This first burst-mode implementation added the following transport capabilities to NCP: an adjustable windowing protocol and a rate-based packet metering scheme.

Burst-mode technology introduced a nonsliding selective repeat windowing scheme to NCP. A theoretical maximum window size of 64 Kbytes (or 128 × 512-byte packets) was available. The initial, or minimum, window size was determined by the product of the BNETX shell's buffers (as determined in the NET.CFG) and the maximum data link packet size. The actual window size was bound from below by the minimum and from above by the window maximum. Based on the performance of the previous burst, the current window was adjusted either up or down for the next burst. The algorithm was:

- A successful burst transaction which causes the window to increase by 100 bytes. (A successful burst transaction is one with *no* dropped packets.)
- A burst with one or more dropped packets which causes the window to decrease to 7/8 of its current size.

The windowing scheme is not sliding. That is, the entire window (or burst) is transmitted and then the transmitter waits until the ack of the entire burst is received prior to adjusting the current window and then transmitting the next burst. In the event that the current burst experienced some packet loss, the response from the receiver would indicate the packets that were lost within the missing fragment list (see burst-mode packet formats in Figure 8.8). The transmitter will then selectively retransmit only those packets that were lost. This continues until all packets are successfully received, completing the burst. The network time-out value is dynamically adjusted based on a moving average and variance of the network delays.

In addition to the dynamic window scheme, NetWare 3.11 implemented an interpacket gap (IPG) to rate control the flow of packets. The IPG is the time between the end of one packet transmission to the time beginning the transmission of the next packet within the same packet burst. The client requests that a server use a given IPG to keep fast servers from overrunning slower clients. This IPG remains constant once it is established by the client. The initial IPG is determined by the median time between packet arrivals.

Figure 8.8 shows the NCP packet overhead for the burst-mode implementation of release 3.11 and later. The figure shows overhead factors similar to those shown in Figure 8.7 for the non-burst-mode implementations. However, the note in Figure 8.8 shows that there are additional overheads in the event that the packet is associated with read/write requests/replies. However, for simplicity in the following calculations, we will ignore these additional overhead items. This is because a single read_reply may generate hundreds of IPX packets, only the first of which carries the read_reply overhead. Figure 8.9 shows the relationship between the read/write request/reply transaction, termed a *service transaction*, and the underlying burst and packet structures. A service

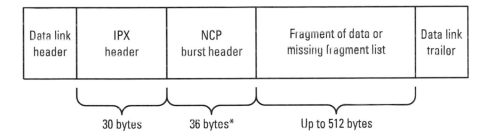

* Additional bytes are included if burst header is associated with
   1) read_req: 24 bytes,
   2) read_reply: 8 bytes,
   3) write_req: 24 bytes, or
   4) write_reply: from 4 to 24 additional bytes.

Figure 8.8 NCP packet format for release 3.11 of NetWare burst mode implementation.

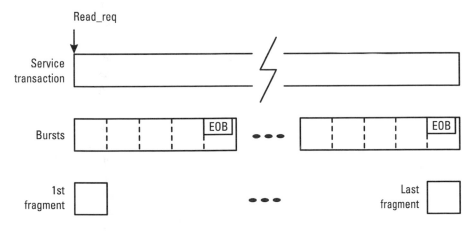

Figure 8.9 Relationship of the read/write request/reply transaction to the burst structure.

transaction would consist, for example, of a read_request and read_reply. The read_reply could consist of many megabytes of data being transmitted down to the client that initiated the read_request. The read_reply is segmented up into multiple bursts. The end of each burst is delimited by an end-of-burst (EOB) indicator. The individual bursts must be further segmented into smaller fragments based on the negotiated maximum packet size.

For more information on burst-mode packet formats refer to [5] and [6].

In summary, release 3.11 of NetWare implemented the following new features:

- A fixed IPG;
- A dynamic, nonsliding window with selective repeats for error recovery;
- Current window size adjustment for each burst based on a specific success criteria; and
- Initial window size determined by the product of the maximum data link packet size and the number of BNETX shell buffers.

8.4.3 NetWare Releases 3.12 and 4.0

NetWare releases 3.12 and 4.0 have integrated the burst-mode technology into the operating system of the servers. Burst mode is automatic, enabled on the client through the file input/output (FIO) virtual load module. With these releases, Novell also modified and enhanced the burst-mode capabilities. These enhancements included modifications to workstation buffering and windowing and IPG adjustments.

Unlike the BNETX shell (of release 3.11), the VLM dynamically configures buffering at the clients. This implies that there are no window size limitations as there were with the BNETX shell.

The first implementations of VLM (1.02) fixed the window size at 16 packets for reads and 10 packets for writes. Instead of adjusting the window size, VLMs earlier than 1.02 adjusted the IPG; the client sends a series of ping messages to the destination and back. Then the maximum IPG is set to one-half the fastest round-trip time. The initial IPG is set to 0. If a threshold of packet drops occurs, for example, two of six packets, the IPG is increased until either no more packets are dropped or until the maximum IPG is reached.

VLM 1.02 and greater changed this algorithm. Now both the IPG and the window size are dynamically changed during transmission. Here, the IPG is first adjusted eventually to its maximum value. Only then does the window adjustment occur. The initial IPG is set to one-half the maximum IPG. The algorithm then essentially halves or doubles the IPG if the transmission succeeded or failed, respectively. In the event that the IPG reaches the maximum, then further modifications would cause the window size to be adjusted. In this case, the maximum window for reads and writes is specified by the client in the NET.CFG file. The default is 16 for reads and 10 for writes. These values can be overridden by the NET.CFG file. The maximum window for reads and writes, in this case, can range from a low of 3 to a high of 255 packets. In the event that the IPG is maximized and the window is being adjusted, the following algorithm is followed:

If the window is decreasing: $W_{i+1} = W_i - (W_i - 3 \times \text{Max\_Pack\_Size}) / 4$

If the window is increasing: $W_{i+1} = W_i + 100$ bytes

Finally, let us briefly mention the sequenced packet exchange (SPX) transport protocol and its successor, SPX II. The SPX protocol sees limited use by Novell applications; it is used primarily in Novell printer applications. SPX sits on top of the IPX protocol and provides a guaranteed packet delivery service. SPX permits only a single outstanding packet to be in transit at a time. The maximum SPX packet size is 576 bytes. In this sense, the SPX transport protocol behaves similar to the NCP non-burst-mode transport protocol implementations. An enhanced version of SPX, SPX II, was introduced in NetWare 4.x releases. The primary performance enhancements with SPX II are (1) support for a larger maximum packet size and packet size negotiations, and (2) support for multiple outstanding packets in a single SPX II window. In this sense, SPX II behaves similar to NCP burst-mode implementations.

8.5 Private Line and Frame Relay Formulas

In this section we present some simple formulas to characterize performance of Novell file transfer applications. Our methodology draws heavily on the material in Chapter 7 on computing similar metrics for TCP/IP applications. We address here only those aspects unique to Novell applications and protocols.

In developing the formulas in this section, we make several assumptions in line with the assumptions made in Chapter 7. We list these assumptions up front so that this is clear:

- *Negligible packet loss.* This has been a common assumption throughout this book. If this assumption is not valid, then the root cause of the high packet loss should be addresses through capacity upgrades or other means.

- *End-to-end latency.* We assume, for simplicity, the private line latency to be 30 msec and the frame relay latency to be 40 msec. These are typical of U.S. domestic networks. For international networks, much higher rates are assumed.

- *Router, LAN, and client/server delays.* We assume these delays to be negligible.

- *IPG is negligible.* This assumption is a consequence of the "no packet loss" assumption. For VLM implementation of versions 1.02 and higher, the IPG is adjusted based on packet loss. In the absence of packet loss, the IPG will eventually be adjusted down to zero.

- *Burst-mode VLM version 1.02 and later.* We present formulas for non-burst-mode and for burst-mode VLM version 1.02 and later.

- *Private line reference connection.* The private line formulas are specific to the private line reference connection shown in Figure 8.10.

- *Frame relay reference connection.* The frame relay formulas are specific to the frame relay connection shown in Figure 8.11.

- *Non-burst-mode and burst-mode performance differ only in window size.* Given that the IPG is assumed to be zero in the burst-mode implementation, then burst-mode transfers consist of transmitting a window's worth of data and then waiting for an acknowledgment. This is

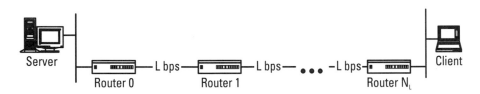

Figure 8.10 Private line reference connection for simple formula development.

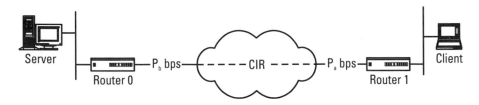

Figure 8.11 Frame relay reference connection for simple formula development.

the same for the non-burst-mode implementation except that the window is restricted to one packet.

Finally, before proceeding to our discussion of formula development, let us spend a few moments talking about the notion of optimal window size in relation to the windowing scheme implemented in NetWare pre-release 5.0. We spent some time in Chapter 7 discussing the optimal window size for TCP/IP and on how to estimate the optimal window size. Unlike NetWare NCP implementations, TCP/IP employs a sliding window scheme. NetWare NCP employs a nonsliding, simplex windowing protocol where the size of the window varies from one for pre-release 3.11 and greater than one for release 3.11 and later. However, because the window is nonsliding, the concept of an optimal window does not apply. For simplex windowing protocols, the larger the window the higher the throughput, up to the point that buffers begin to overflow and packet loss becomes an issue. For sliding window protocols, for example, TCP/IP, we argued that an optimal window size exists and is equal to the minimum window that maintains 100% utilization of the slowest link in the path. For simplex windowing schemes, no matter how large the window is, there will exist idle periods on the slowest link in the path. In this sense, simplex windowing protocols are always window limited. Therefore, we will not attempt to identify an optimal window size for NetWare NCP windowing schemes.

We next derive our formulas for private line environments. We follow this with a discussion of frame relay environments. We conclude this section by furthering our understanding of cross-application effects by analyzing the case of FTP TCP/IP applications competing for bandwidth with a Novell non-burst-mode file transfer.

8.5.1 Private Line Formulas

The formulas for the performance of NetWare file transfers are derived in conjunction with the reference connection shown in Figure 8.10 and the associated timing diagram shown in Figure 8.12.

This timing diagram assumes that the number of private line links, N_L, is three and that the window size is N_W. To the left-hand side of each of the timing diagrams in Figure 8.12 we indicate the size of the various quantities of interest, for example, R_d, T_W, and T_S. Referring to this figure, we can write the round-trip delay, R_d, as

$$R_d = T_W + (N_L - 1) \times T_S + N_L \times T_A + 2 \times D_p$$

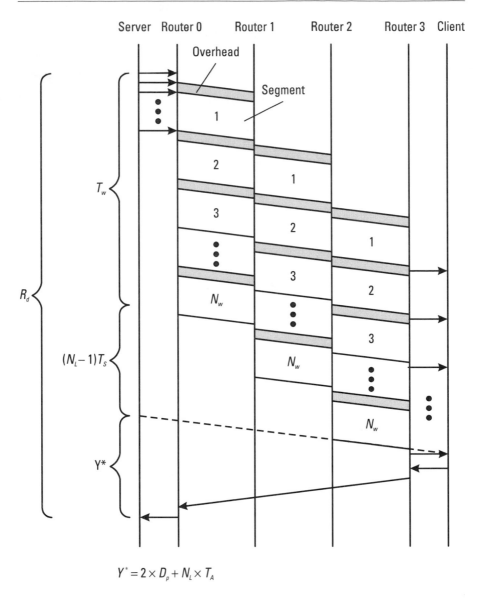

$$Y^* = 2 \times D_p + N_L \times T_A$$

Figure 8.12 Timing diagram for the reference connection shown in Figure 8.10.

where $2 \times D_p$ covers the latency for the data traveling to the client and $N_L \times T_A$ covers for the time to transmit the ack returning to the server. We see that T_W is essentially

$$T_W = N_W \times T_S$$

We now need expressions for T_A and T_S to complete our formulation. As in Chapter 7, we rely on an available bandwidth argument to derive these expressions and to account in a crude way for the background link utilization. Therefore, we write

$$T_S = S / \text{Available bandwidth} = S \times O_f / L \times (1 - U)$$

$$T_A = A / \text{Available bandwidth} = A \times O_f / L \times (1 - U)$$

where we have written the available bandwidth as $L\,(1 - U)\,/O_f$. So, we get

$$R_d = [(N_L - 1) + N_W] \times S \times O_f / L \times (1 - U) + (N_L \times A \times O_f)\,/L(1 - U) + 2 \times D_p$$

and

$$X = W / R_d$$

$$X = N_W \times S / [\{(N_L - 1) + N_W\} \times S \times O_f / L(1 - U) + (N_L \times A \times O_f)\,/L(1 - U) + 2 \times D_p]$$

where for non-burst-mode we set $N_W = 1$ and for burst-mode we set N_W to the appropriate window size. Remember, also, that there may be maximum packet size differences between non-burst-mode and burst-mode situations. This is reflected in the value of S in this expression.

As an example of the use of this formula, let us assume burst-mode with the following parameters:

$S = 1024$ bytes

$O = (36 + 30 + 8) = 74$

$O_f = 1 + O/S = 1 + 74 / 1024 = 1.07$

$A = 74$ bytes

$N_L = 1$

$N_W = 16$

$F = 1.0$ Mbytes $\times$ 8 bits/byte $= 8.0$ Mbits

$L = 56$ Kbps

$U = 0.40$

$D_p = 0.030$ sec

Inserting these values into our formula for throughput, we get

$$R_d = 4.3 \text{ sec}$$

$$X = 30 \text{ Kbps}$$

$$T = F/X = 4.4 \text{ min}$$

Thus, we see that it will take approximately 4.4 min to transfer the entire file over the reference connection found in Figure 8.10.

What if we assume that a similar file transfer occurs in a non-burst-mode scenario? We can assume the same parameters as given earlier, except that the window size is one and we will set the maximum negotiated packet size to 512. Therefore, we get for our parameter set:

$S = 512$ bytes

$O = (34 + 30 + 8) = 72$

$O_f = 1 + O/S = 1 + 72 / 512 = 1.14$

$A = 72$ bytes

$N_L = 1$

$N_W = 1$

$F = 1.0$ Mbytes $\times$ 8 bits/byte $= 8.0$ Mbits

$L = 56$ Kbps

$U = 0.40$

$D_p = 0.030$ sec

Inserting these values into our formula for throughput, we get

$$R_d = 0.22 \text{ sec}$$

$$X = W/R_d = 512 \times 8 \text{ bits} / 0.22 \text{ sec} = 19 \text{ Kbps}$$

$$T = F/X = 7.0 \text{ min}$$

Thus, we see that it will take approximately 7.0 min to transfer the entire file over the reference connection found in Figure 8.10. Defining the improvement in performance as

$$\% \text{ improvement} = 100 \times (T_{\text{NBM}} - T_{\text{BM}}) / T_{\text{BM}}$$

and comparing the two results for burst-mode versus non-burst-mode implementations, we see an approximately 60% improvement in file transfer performance. Of course, this estimate is a result of our specific assumptions regarding file size, packet and window size, and reference connection.

8.5.2 Frame Relay Formulas

We now concentrate on estimates for file transfer performance in frame relay environments. This section and associated assumptions and arguments rely heavily on the previous discussion found in Section 7.3.4. Again we will rely on the assumptions discussed at the beginning of this section. Further, our formulas will be specific to the reference connection shown in Figure 8.11. In Figures 8.13 and 8.14 we give two timing diagrams for two different realizations of the reference connection shown in Figure 8.11. The diagram in Figure 8.13 is for a reference connection where the VC CIR rate is equal to the speed of the frame relay port on the client side of the connection. The diagram in Figure 8.14 is for a connection where the VC CIR is the slowest link in the data path, that is, the frame relay ports are faster than the CIR of the VC.

To the left-hand side of each of the timing diagrams in Figure 8.13 we indicate the size of the various quantities of interest, for example, R_d, T_W, and T_S. Assuming that the CIR of the VC is always less than or equal to the frame relay port speeds on the client and sever side of the connections, we can write the expression for the round-trip delay as

$$R_d = T_W(\text{CIR}) + T_S(P_a) + 2 \times D_f + T_A(P_a) + T_A(P_b)$$

where $T_W(\text{CIR})$ is the time to transmit the entire window over the VC of rate CIR, $T_S(P_a)$ is the time to transmit a single segment over the link of rate P_a, and $T_A(P)$ is the time to transmit the ack over the frame relay port. We can write $T_W(\text{CIR})$, $T_S(X)$, and $T_A(x)$ as

$$T_W(\text{CIR}) = N_W \times T_S(\text{CIR})$$

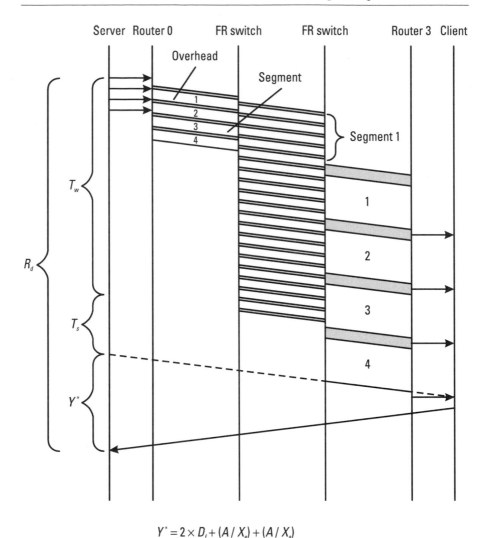

$$Y^{*} = 2 \times D_{f} + (A / X_{e}) + (A / X_{a})$$

Figure 8.13 Round-trip delay timing diagram for a frame relay connection with a CIR equal to the client-side frame relay port speed.

$$T_{S}(\mathrm{CIR}) = S / \text{ Available bandwidth} = S \times O_{f} / \mathrm{CIR} \times (1 - U_{\mathrm{CIR}})$$

$$T_{S}(P_{a}) = S \times O_{f} / P_{a} \times (1 - U_{a})$$

$$T_{A}(P_{a}) = A \times O_{f} / P_{a} \times (1 - U_{a})$$

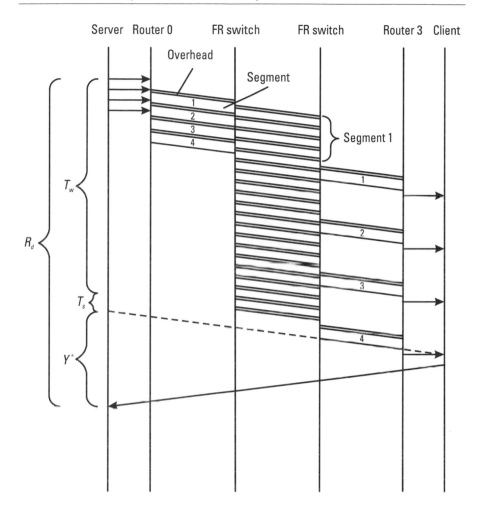

$$Y^* = 2 \times D_t + (A / X_e) + (A / X_a)$$

Figure 8.14 Round-trip delay timing diagram for a frame relay connection where the CIR is the slowest link on the data path.

Finally, inserting the above expressions into the formula for throughput, that is, $X = W / R_d$, we get

$$X = W / R_d = (N_W \times S) / [N_W \times S \times O_f / \text{CIR} \times (1 - U_{\text{CIR}}) + S \times O_f / P_a \times (1 - U_a) + 2 \ D_f + A \times O_f / P_a \times (1 - U_a) + A \times O_f / P_b \times (1 - U_b)]$$

As in the case of the formulas for the private line scenarios given earlier, for the non-burst-mode case, set $N_W = 1$, and for the burst-mode case, set N_W to the appropriate window size based on the discussion of burst-mode implementations in Section 8.4.

Let us run through an example computation to illustrate the use of our throughput estimate for frame relay reference connections. We will follow the example from Section 8.5.2 and compare the burst-mode case to non-burst-mode. For the burst-mode case, we assume the following parameters:

$$S = 1024 \text{ bytes}$$

$$O = (36 + 30 + 8) = 74$$

$$O_f = 1 + O/S = 1 + 74/1024 = 1.07$$

$$A = 74 \text{ bytes}$$

$$N_W = 16$$

$$F = 1.0 \text{ Mbytes} \times 8 \text{ bits/byte} = 8.0 \text{ Mbits}$$

$$P_a = 56 \text{ Kbps}$$

$$P_b = 1.536 \text{ Mbps}$$

$$CIR = 32 \text{ Kbps}$$

$$U_a = 0.40$$

$$U_{CIR} = 0.40$$

$$D_F = 0.040 \text{ sec}$$

Inserting these values into our formula, we get

$$R_d = 7.7 \text{ sec}$$

$$X_{MIN} = 17 \text{ Kbps}$$

$$T_{MAX} = F/X_{MIN} = 7.8 \text{ min}$$

where we have used the label X_{MIN} for the throughput because we have assumed that the throughput of the VC is strictly limited to 32 Kbps. However, as discussed in Section 7.3.4, it is common to find that the throughput of a VC will exceed the CIR due to the bursting capabilities afforded by the frame relay service provider. Therefore, a best case analysis is provided by assuming that the rate of the VC is equal to the slowest frame relay port speed on the ends of the VC, for example, 56 Kbps in our example. If we assume a VC rate of 56 Kbps and make an adjustment for the relative change in the VC utilization to 0.23 (instead of 0.4) in order to account for the higher rate, we get the following best case estimates for file transfer throughput and time:

$$R_d = 3.6 \text{ sec}$$

$$X_{MAX} = 36 \text{ Kbps}$$

$$T_{MIN} = F / X_{MAX} = 3.7 \text{ min}$$

Due to the variability in the VC rate caused by the bursting capabilities provided by the various frame relay providers, we have to bound the estimates for transfer times between 3.7 and 7.8 min.

For the non-burst-mode case, we assume the following parameters (following our example for the private line case from earlier):

$S = 512$ bytes
$O = (36 + 30 + 8) = 74$
$O_f = 1 + O / S = 1 + 74 / 512 = 1.14$
$A = 74$ bytes
$N_w = 16$
$F = 1.0$ Mbytes $\times$ 8 bits/byte $= 8.0$ Mbits
$P_a = 56$ Kbps
$P_b = 1.536$ Mbps
CIR $= 32$ Kbps
$U_a = 0.40$
$U_{CIR} = 0.40$
$D_F = 0.040$ sec

Inserting these values into our formula, we get

$$R_d = 0.48 \text{ sec}$$

$$X_{MIN} = 9 \text{ Kbps}$$

$$T_{MAX} = F / X_{MIN} = 15 \text{ min}$$

for the worst case analysis. For the best case analysis, we get

$$R_d = 0.35 \text{ sec}$$

$$X_{\text{MAX}} = 12 \text{ Kbps}$$

$$T_{\text{MIN}} = F/X_{\text{MAX}} = 11 \text{ min}$$

For the preceding examples of file transfer times for burst-mode and non-burst-mode implementations over frame relay networks, we see an improvement of roughly 90% in the transfer times within the worst case analysis. Again, this estimate of performance improvement is highly dependent on the reference connections, parameter settings, and our assumption set listed at the beginning of Section 8.5.1. For a more detailed estimate of performance improvements in various situations refer to [5].

8.5.3   Cross-Application Effects: Mixing Novell and TCP/IP

Section 7.6 discussed the impact of an FTP file transfer on a telnet session sharing the same data path on the reference connection shown in Figure 7.12. There, we argued that the additional delay caused by the file transfer traffic on the telnet session was roughly equal to the time to drain the additional file transfer data from the queue on the slowest link in the path. And the additional file transfer traffic in queue is given by

$$\text{Advertised window size (plus overhead)} - W^* = \text{Burden}$$

where W^* is the optimal window size for the file transfer over the given reference connection. We will call this the *burden* that the FTP file transfer places on the buffer at the slowest link. Then the burden divided by the rate of the slowest link is the additional delay caused by the FTP traffic.

A similar effect will occur when mixing Novell and FTP traffic over common links. This is especially a concern for non-burst-mode implementations, where the NCP simplex window is one segment. In this case, the FTP traffic will increase the round-trip delay experienced by the Novell traffic, causing a diminished throughput for the Novell traffic. Following our earlier discussion, we can estimate the impact that the additional FTP traffic will have on the

Novell file transfer by incorporating the additional burden placed on the round-trip delay experienced by the Novell packets, that is,

$$R_d(\text{FTP}) = R_d(\text{no FTP}) + \text{Burden} / L$$

where $R_d(\text{no FTP})$ is the round-trip delay for the Novell traffic when not competing with an FTP file transfer and $R_d(\text{FTP})$ is the round-trip delay for the Novell traffic when it is competing with the FTP traffic. We then obtain a revised estimate for the expected throughput:

$$X = W / [R_d(\text{no FTP}) + \text{Burden} / L]$$

We can now use this formula to estimate the impact on the Novell file transfer performance. Assuming the same non-burst-mode parameters and private line reference connections as in Section 8.5.1 and using the FTP assumptions from Section 7.6 for a window size of eight segments of 512 bytes and optimal window size of two segments of 512 bytes, we get

$$X = 512 \times 8 / [0.22 + 6 \times (512 + 48) \times 8 / 56 \text{ Kbps}] = 5.9 \text{ Kbps}$$

This is opposed to 19 Kbps throughput without the competition from the FTP file transfer. As one can see, Novell non-burst-mode traffic does not fare well in competition with sliding window protocols like TCP.

8.6 Summary

Novell NetWare is a networking protocol that saw a large deployment in the 1990s, first in LAN environments and then extending into WAN environments. NetWare is a proprietary networking protocol that has been evolving and incorporating open standards over the years. We overviewed the protocols that make up the NetWare suite. We discussed the specifics of the NetWare protocols that have an impact on the bandwidth utilization in WAN implementations. We then outlined the evolution of the NetWare NCP transport protocol. We finished this chapter with an analysis of the throughput performance of the NetWare NCP transport protocols, specifically the burst-mode implementations' improvements over the non-burst-mode implementation for private line and frame relay environments.

References

[1]   Chappell, L., and D. Hakes, *Novell's Guide to NetWare LAN Analysis*, 2nd ed., San Jose, CA: Novell Press, 1994.

[2]   Malamud, C., *Analyzing Novell Networks*, New York: Van Nostrand Reinhold, 1992.

[3]   Kimball, K., R. Holm, and E. Liebing, "NetWare Over TCP/IP: Integrating NetWare Services Into the TCP/IP Environment," *Novell AppNotes,* Vol. 8, No. 3, March 1997.

[4]   Ballew, S., *Managing IP Networks with Cisco Routers*, Cambridge, MA: O'Reilly & Associates, 1997.

[5]   Stevenson, D., and S. Duncan, "An Introduction to Novell's Burst-Mode Protocol," *NetWare Application Notes,* March 1992, p. 45.

[6]   Mosbarger, M., and D. Dixon, "Packet Burst Update: BNETX versus VLM Implementations," *Novell Research Reports,* November 1994.

9

WAN Performance Issues for Client/Server Applications

9.1 Introduction

Client/server applications constitute the most important class of TCP/IP applications to impact the corporate WAN in recent years. Several companies have implemented or are in the process of implementing enterprise resource planning (ERP) applications from well-known companies such as SAP AG, PeopleSoft, and Oracle. Other companies have chosen to implement applications from others such as Baan, J. D. Edwards, Hyperion, Lawson Software, QAD, and Seibel Systems. In addition, companies usually deploy "home-grown" client/server applications developed to address their specific needs. Some companies have completely replaced mainframe-based applications with ERP applications, the rationale being cost reduction and year 2000 issues. Millions of dollars have been spent on customizing these applications, redundant hardware, and on end user platform support and training. Focusing on business process automation, these applications have been deemed mission-critical by the companies adopting them.

However, for all the importance of these applications to the enterprise and the associated deployment costs, not enough attention has been paid to how these applications actually perform over a WAN, and setting end user expectations. This is particularly troublesome to the network manager, because the network is usually blamed when end users complain of dropped sessions or slow screen response times. Therefore, it is important that a WAN traffic study be undertaken first before full application deployment. This traffic study should consist of controlled testing and analysis of application behavior in a laboratory followed by a pilot test in the production network. Only recently have tools begun to be available that address the main concerns of the network manager supporting client/server applications: bandwidth per user, expected end user response times, and so on. However, tools are not a substitute for a basic understanding of the essential traffic and performance characteristics of client/server applications. As we will demonstrate in this chapter, it is possible to perform "back-of-the-envelope" calculations to estimate overall bandwidth requirements and expected response times, based on transaction-level information about the application.

We begin in Section 9.2 with an overview of client/server applications and tiered architectures. This section will clarify some terminology that is frequently used in the context of client/server WAN performance issues. In Section 9.3 we discuss traffic characteristics of client/server applications. This will be done using protocol traces of some client/server applications from SAP R3, PeopleSoft, Oracle, and a few others. Section 9.4 contains a discussion on data collection for client/server applications that is necessary for performance analysis. In Section 9.5 we provide some guidelines for computing bandwidth requirements for client/server applications. This is particularly useful for network managers facing the challenge of planning for network upgrades (and budgeting) before the applications are fully deployed. Estimating bandwidth requirements is relatively easy compared to estimating end user response times. However, response time is an important issue, and we provide some guidance in Section 9.5.

No treatment of client/server application performance is complete without a discussion of thin client technology. Thin clients have been receiving enormous attention lately because the cost of supporting full function PCs and workstations — loaded with word processing, spreadsheets, e-mail, terminal emulation, and client/server software — in an enterprise is expensive. Thin client architecture promises to remedy this situation by reducing support costs by as much as 80%, according to some estimates.

There are two popular thin client approaches: *remote presentation*, as exemplified by Citrix Winframe/Metaframe and variants, and *web-based network computing (NC)*, as implemented in Oracle's Network Computing Architecture® (NCA). The thin client approach introduces yet another aspect of concern regarding WAN performance for client/server applications. Remote presentation can add traffic to the network while making end user performance more vulnerable to network delay variations. Network computing, based on the

Java® computing paradigm, relies on HTTP, which has performance issues over the WAN, as we discussed in Chapter 7.

Thin client WAN performance issues are discussed in Section 9.6.

9.2 Client/Server Overview

Client/server computing refers to a division of tasks between clients and servers [1]. When one refers to a client/server application, the usual frame of reference is that of one or more PCs and workstations connected across a LAN or WAN to a central site database server, typically via TCP/IP (see Figure 9.1).

The application logic resides in the client and the server is the repository for the data (usually a relational database management system, RDBMS, like Oracle) from which the client periodically requests data to process the application request from the end user. This approach is in contrast to the mainframe-centric computing approach in the old IBM/SNA world, where the mainframes were responsible for all aspects of computing: data services, application processing, and data presentation. This gave rise to the term "dumb" terminal, as in the classic 3270 terminal.

The three tasks in client/server computing mentioned earlier are split between clients and servers. Data services are performed by database servers. The other two functions, namely, application processing and data presentation, are performed by the client. The term "fat client" is self explanatory in this context, because the primary aspects of the application—logic and presentation—now reside in the client.

Tiered architecture is a term that is often used in discussing client/server performance. A *two-tier* client/server model is one in which a client requests and receives information from a database server using SQL (Structured Query

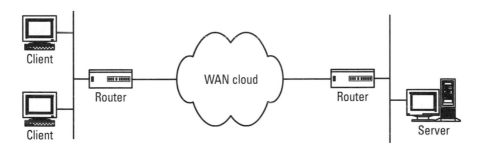

Figure 9.1 Reference connection for a client/server application.

Language) calls, and then processes and presents the information locally on a PC or workstation to an end user. For completeness, one may think of a *one-tier* model as the 3270 terminal-to-mainframe model, where all intelligence lies in the server. Another, perhaps more relevant model for client/server applications is one where a remote client accesses a server via a telnet connection. Applications from some vendors are built on this architecture.

There are many ways to describe a *three-tier* client/server architecture model. We use this term to refer to a client/server architecture in which the application processing is performed by an intermediate process called the *application server* (see Figure 9.2).

The client is relieved of the responsibility for the application logic, but it performs the data presentation and user interface functions. Indeed, one can think of extending the tiered concept to include more than three tiers by visualizing multiple application servers, each responsible for a different set of application processing tasks. Note that the application server and the database server functions can be performed in a single "box." A (usually) proprietary GUI (graphical user interface) runs on the client side.

The advantage of a two-tier architecture is that applications can be written to work with many different database engines such as Oracle, Sybase, and Informix, because the server does not perform application processing. However, a severe penalty is paid in storing application logic in the client: End user support becomes tedious, expensive, and complex. Furthermore, as we will demonstrate, these applications are not WAN friendly because of the chatty SQL calls that are exchanged between the database server and the client for each transaction. Three-tier architectures address this issue by centralizing the application processing function. Because no SQL calls traverse the WAN, three-tier applications are very WAN friendly.

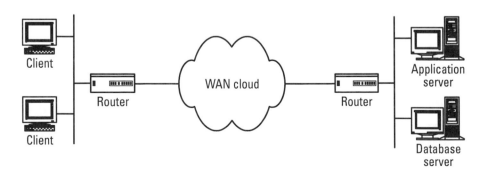

Figure 9.2 Three-tier architecture.

From the networking perspective, a three-tier approach is preferable. From an end user support perspective, a thin client approach (Web or remote presentation) is a major advantage because it implies minimal end user support. A popular thin client approach is to use terminal servers, such as in the Citrix Winframe/Metaframe model. The overall WAN implications of the terminal server approach are discussed later in this chapter. Briefly, this approach uses a remote presentation technology in which the client processes are actually executed in a centrally located server, while keystrokes and mouse movements are transmitted across the network (LAN or WAN or dial-up) via some proprietary protocol (Citrix calls this ICA or Independent Computing Architecture®). Thus each client need not be equipped with application logic. It is a compelling way to reduce end user support complexity. It also has the potential to improve client/server application performance over a WAN by providing "LAN-like" performance. However, this approach does introduce additional performance issues, such as added bandwidth and sensitivity to latency, especially over packet networks like frame relay. We will discuss these issues in detail in Section 9.6. For an interesting dissertation on this issue, see Edlund [5].

The other approach to thin client is made possible by the proliferation and maturation of Web technology and Java. This approach combines the functionality of the application server found in 3-tier client/server architecture and the thin client model. The classic example of this architecture is Oracle NCA. The application software resides in middle-tier application servers that communicate on a LAN with the back-end database servers. End users with Java enabled web browsers download an applet from a central web server to obtain the user interface to the application.

Currently, SAP R3 is a widely deployed three-tier client/server application. There appear to be little or no WAN-related performance issues in SAP R3 implementations. PeopleSoft and Oracle implementations have been two-tier until recently, with somewhat mixed performance results over the WAN. Both of these vendors currently support three-tier architectures.

9.3 Client/Server Application WAN Traffic Characterization

Client/server application traffic is generated from end user transactions. A transaction can be defined as the process in which an end user submits a request and receives a screen update from the server. In the classical 3270 world, this transaction is simple; typically, it consists of two packets, one small packet sent to the host (say, 100 to 300 bytes), and one large packet received from the host (say, 700 to 1000 bytes). In the client/server world, the traffic characteristic of a transaction can be complex. Some transactions in a two-tier environment can consist of several hundred "ping-pong" packets between the client and server processes. Other transactions may be a one-way bulk data transfer from the server to the client. Transactions can be hybrid as well: a series of "ping-pong" packets, followed by one-way bulk data transfer, then followed by some more ping-pong packets, and so on.

Traffic in a three-tier implementation, as in SAP R3, is very much like 3270—small in, large out.

Thus some types of transactions (i.e., ping-pong) are latency sensitive, whereas others may be bandwidth sensitive, or a combination of both. The performance of latency-sensitive applications is relatively difficult to tune over the WAN because traditional tuning methods, such as bandwidth upgrades and prioritization, will not improve performance. In addition, in a geographically dispersed enterprise network, end users are likely to experience vastly dissimilar performance for the same application depending on where they are located.

9.3.1 Examples of Two-Tier Application Traffic Patterns

We provide some examples of transactions from some applications using a two-tier mode. These examples cover the following three cases: ping-pong, bulk data transfer, and hybrid.

9.3.1.1 Ping-Pong Transactions

Figure 9.3 is a sniffer protocol trace of an end user in Europe logging on to a sales-aid application residing in San Jose, California, over the global frame relay connection shown in Figure 9.4. The sniffer is attached to the same LAN segment as the server. The source and destination IP addresses have been deleted.

The client process has TCP port 1026, and the server has TCP port 2001. Notice the following:

- It takes 550+ transactions for the client to log on to the application.
- The round-trip latency over frame relay between Europe and San Jose is slightly greater than 300 msec (look at frames 14, 18, and 21).
- There is a one-way transfer from the server to the client mixed in with some ping-pong packets.

Clearly, the login time across a global connection for applications with this type of traffic pattern will be significant (in minutes rather than seconds).

Frame	Delta T	Destination Source		Summary
12	0.0405	TCP D=2001 S=1026	SYN	SEQ=68497 LEN=0 WIN=0
13	0.0041	TCP D=1026 S=2001	SYN	ACK=68498 SEQ=186752000 LEN=0 WIN=8192
14	0.3140	TCP D=2001 S=1026		ACK=186752001 WIN=2880
15	0.0552	TCP D=2001 S=1026		ACK=186752001 SEQ=68498 LEN=512 WIN=2880
16	0.0020	TCP D=2001 S=1026		ACK=186752001 SEQ=69010 LEN=32 WIN=2880
17	0.0278	TCP D=1026 S=2001		ACK=69042 SEQ=186752001 LEN=198 WIN=8192
18	0.3271	TCP D=2001 S=1026		ACK=186752199 WIN=2880
19	0.0142	TCP D=2001 S=1026		ACK=186752199 SEQ=69042 LEN=26 WIN=2880
20	0.0085	TCP D=1026 S=2001		ACK=69068 SEQ=186752199 LEN=17 WIN=8192
21	0.3171	TCP D=2001 S=1026		ACK=186752216 WIN=2880
22	0.0066	TCP D=2001 S=1026		ACK=186752216 SEQ=69068 LEN=24 WIN=2880
....				
220	0.0161	TCP D=1026 S=2001		ACK=74735 SEQ=186784047 LEN=512 WIN=8192
221	0.0323	TCP D=2001 S=1026		ACK=186782511 WIN=2880
222	0.0184	TCP D=1026 S=2001		ACK=74735 SEQ=186784559 LEN=512 WIN=8192
223	0.1143	TCP D=2001 S=1026		ACK=186783023 WIN=2880
224	0.0172	TCP D=1026 S=2001		ACK=74735 SEQ=186785071 LEN=512 WIN=8192
.....				
582	0.0085	TCP D=1026 S=2001		ACK=75511 SEQ=186873819 LEN=296 WIN=8192
583	0.2423	TCP D=2001 S=1026		ACK=186871771 WIN=2880
584	0.0488	TCP D=2001 S=1026		ACK=186872283 WIN=2880
585	0.0268	TCP D=2001 S=1026		ACK=186872795 WIN=2880
586	0.0287	TCP D=2001 S=1026		ACK=186873307 WIN=2880
587	0.0271	TCP D=2001 S=1026		ACK=186873819 WIN=2880
588	0.0171	TCP D=2001 S=1026		ACK=186874115 WIN=2880

Figure 9.3 Login trace of the sales-aid application.

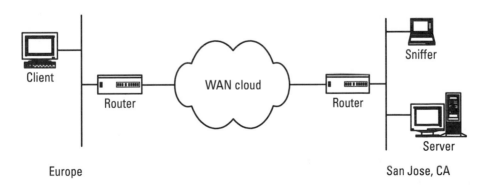

Figure 9.4 Reference connection for protocol trace of a sales-aid application.

In Figure 9.5, we present a trace of a sample application task that a user might invoke after logging into the application. Note the following characteristics of this application task:

- The transaction consists of 343 individual subtransactions.

- The application has a ping-pong characteristic. The sequence is a mixture of client-to-server packets, or server-to-client packets, and TCP acks, almost one at a time.

- The packet sizes (LEN) are fairly small—less than 300 bytes.

Frame	Delta T	Destination Source	Summary
1		TCP D=2001 S=1026	ACK=186875550 SEQ=76457 LEN=235 WIN=2880
2	0.0891	TCP D=1026 S=2001	ACK=76692 SEQ=186875550 LEN=479 WIN=8192
3	0.3758	TCP D=2001 S=1026	ACK=186876029 WIN=2880
4	0.0413	TCP D=2001 S=1026	ACK=186876029 SEQ=76692 LEN=240 WIN=2880
5	0.0406	TCP D=1026 S=2001	ACK=76932 SEQ=186876029 LEN=120 WIN=8192
6	0.3470	TCP D=2001 S=1026	ACK=186876149 WIN=2880
7	0.0362	TCP D=2001 S=1026	ACK=186876149 SEQ=76932 LEN=248 WIN=2880
8	0.0336	TCP D=1026 S=2001	ACK=77180 SEQ=186876149 LEN=168 WIN=8192
9	0.3241	TCP D=2001 S=1026	ACK=186876317 WIN=2880
10	0.0372	TCP D=2001 S=1026	ACK=186876317 SEQ=77180 LEN=248 WIN=2880
11	0.0749	TCP D=1026 S=2001	ACK=77428 SEQ=186876317 LEN=237 WIN=8192
12	0.3465	TCP D=2001 S=1026	ACK=186876554 WIN=2880
13	0.0383	TCP D=2001 S=1026	ACK=186876554 SEQ=77428 LEN=230 WIN=2880
14	0.0661	TCP D=1026 S=2001	ACK=77658 WIN=7962
15	0.4114	TCP D=1026 S=2001	ACK=77658 SEQ=186876554 LEN=512 WIN=8192
16	0.3087	TCP D=1026 S=2001	ACK=77658 SEQ=186877066 LEN=60 WIN=8192
17	0.0401	TCP D=2001 S=1026	ACK=186877066 WIN=2880
18	0.3066	TCP D=2001 S=1026	ACK=186877126 WIN=2880
19	0.0724	TCP D=2001 S=1026	ACK=186877126 SEQ=77658 LEN=512 WIN=2880
20	0.0035	TCP D=2001 S=1026	ACK=186877126 SEQ=78170 LEN=93 WIN=2880
.....			
339	0.0019	TCP D=1026 S=2001	ACK=91866 SEQ=186933092 LEN=29 WIN=8192
340	0.3605	TCP D=2001 S=1026	ACK=186932580 WIN=2880
341	0.0302	TCP D=2001 S=1026	ACK=186933092 WIN=2880
342	0.0018	TCP D=2001 S=1026	ACK=186933121 WIN=2880
343	0.0081	TCP D=2001 S=1026	ACK=186933121 SEQ=91866 LEN=20 WIN=2880

Figure 9.5 Transaction using multiple packet exchanges.

It is very clear that performance over the WAN for applications such as these that exchange numerous packets between the client and the server will be unsatisfactory.

9.3.1.2 Bulk Data Transfer Transactions

In Figure 9.6 we show an example of another client/server transaction that contrasts with the transaction shown earlier. In this example the client TCP port is

Frame	Rel T	Destination Source	Summary
1	0.00000	TCP D=1527 S=1040	ACK=657541464 SEQ=1971777 LEN=1357 WIN=7767
2	0.19754	TCP D=1040 S=1527	ACK=1973134 SEQ=657541464 LEN=45 WIN=32768
3	0.19941	TCP D=1527 S=1040	ACK=657541509 SEQ=1973134 LEN=27 WIN=7722
4	0.26702	TCP D=1040 S=1527	ACK=1973161 SEQ=657541509 LEN=261 WIN=32768
5	0.37145	TCP D=1527 S=1040	ACK=657541770 WIN=7461
6	0.79395	TCP D=1527 S=1040	ACK=657541770 SEQ=1973161 LEN=131 WIN=7461
7	1.00494	TCP D=1040 S=1527	ACK=1973292 WIN=32768
8	2.91647	TCP D=1040 S=1527	ACK=1973292 SEQ=657541770 LEN=1460 WIN=32768
9	2.99540	TCP D=1040 S=1527	ACK=1973292 SEQ=657543230 LEN=588 WIN=32768
10	2.99589	TCP D=1527 S=1040	ACK=657543818 WIN=8760
11	3.18485	TCP D=1040 S=1527	ACK=1973292 SEQ=657543818 LEN=1460 WIN=32768
12	3.37353	TCP D=1040 S=1527	ACK=1973292 SEQ=657545278 LEN=1460 WIN=32768
13	3.37428	TCP D=1527 S=1040	ACK=657546738 WIN=8760
14	3.56219	TCP D=1040 S=1527	ACK=1973292 SEQ=657546738 LEN=1460 WIN=32768
15	3.56304	TCP D=1527 S=1040	ACK=657548198 WIN=8760
16	3.77217	TCP D=1040 S=1527	ACK=1973292 SEQ=657548198 LEN=1460 WIN=32768
17	3.88290	TCP D=1527 S=1040	ACK=657549658 WIN=8760
18	3.96095	TCP D=1040 S=1527	ACK=1973292 SEQ=657549658 LEN=1460 WIN=32768
19	4.08356	TCP D=1527 S=1040	ACK=657551118 WIN=8760
20	4.14966	TCP D=1040 S=1527	ACK=1973292 SEQ=657551118 LEN=1460 WIN=32768
21	4.15046	TCP D=1527 S=1040	ACK=657552578 WIN=8760
. . . . . .			
250	27.34931	TCP D=1040 S=1527	ACK=1973292 SEQ=657723398 LEN=1460 WIN=32768
251	27.45981	TCP D=1527 S=1040	ACK=657724858 WIN=8760
252	27.55929	TCP D=1040 S=1527	ACK=1973292 SEQ=657724858 LEN=1460 WIN=32768
253	27.66045	TCP D=1527 S=1040	ACK=657726318 WIN=8760
254	27.74793	TCP D=1040 S=1527	ACK=1973292 SEQ=657726318 LEN=1460 WIN=32768
255	27.74879	TCP D=1527 S=1040	ACK=657727778 WIN=8760
256	27.79640	TCP D=1040 S=1527	ACK=1973292 SEQ=657727778 LEN=3 WIN=32768
257	27.96150	TCP D=1527 S=1040	ACK=657727781 WIN=8757

Figure 9.6 Transaction using bulk data transfers, where Rel T is the relative time.

1040 and the server TCP port is 1527. Note the following characteristics of this transaction:

- Only TCP acknowledgment packets are sent in the client-to-server direction; there are no data packets.

- The TCP window size that one needs to consider for throughput calculations is that advertised by the client, which is approximately 8 Kbytes.

- The maximum segment size option in TCP is used; the server is able to send TCP segment sizes larger than 512 bytes.

9.3.1.3 Hybrid Transactions

The trace shown in Figure 9.7 shows a transaction with a hybrid nature from an application different from the ones just described. Note the following characteristics:

- There are three distinct phases for this transaction: an initial exchange of data between client (TCP port 1213) and server (TCP port 1521), a transfer of data from server to client (approximately 25,000 bytes), and a final exchange between the client and server.

- The are some significant delays in the midst of the transaction (frames 85 and 100) totaling about 4.5 sec. This is commonly referred to as *client build delay* where the client processes the data received. One should collect several samples to determine the variability of the client build delay.

9.3.2 Example of a Three-Tier Transaction

Figure 9.8 shows an example of three SAP R3 transactions. Note the following important characteristics of these transactions:

- This is a single transaction of a few packet exchanges between the client and server. This makes the application more WAN friendly.

- The inquiry is small and the response is large.

- The delta time between transactions is relatively long. This is the user think time between transactions. Think times are important because they directly affect the amount of traffic each user contributes to the WAN.

Frame	Delta T	Destination Source	Summary
Ping-Pong			
1	0	D=1521 S=1213 SYN	SEQ=6901576 LEN=0 WIN=8192
2	0.081	D=1213 S=1521 SYN	ACK=6901577 SEQ=507022000 LEN=0 WIN=49152
3	0	D=1521 S=1213	ACK=507022001 WIN=8760
4	0.009	D=1521 S=1213	ACK=507022001 SEQ=6901577 LEN=50 WIN=8760
5	0.18	D=1213 S=1521	ACK=6901627 WIN=49152
6	0.001	D=1521 S=1213	ACK=507022001 SEQ=6901627 LEN=260 WIN=8760
7	0.12	D=1213 S=1521	ACK=6901887 SEQ=507022001 LEN=8 WIN=49152
8	0.001	D=1521 S=1213	ACK=507022009 SEQ=6901887 LEN=50 WIN=8752
9	0.28	D=1213 S=1521	ACK=6901937 WIN=49152
10	0.001	D=1521 S=1213	ACK=507022009 SEQ=6901937 LEN=260 WIN=8752
11	0.08	D=1213 S=1521	ACK=6902197 SEQ=507022009 LEN=24 WIN=49152
12	0.002	D=1521 S=1213	ACK=507022033 SEQ=6902197 LEN=167 WIN=8728
13	0.089	D=1213 S=1521	ACK=6902364 SEQ=507022033 LEN=127 WIN=49152
14	0.009	D=1521 S=1213	ACK=507022160 SEQ=6902364 LEN=35 WIN=8801
15	0.08	D=1213 S=1521	ACK=6902399 SEQ=507022160 LEN=33 WIN=49152
. . . .			
Bulk Data			
57	0.099	D=1213 S=1521	ACK=6905207 SEQ=507023728 LEN=1460 WIN=49152
58	0.002	D=1213 S=1521	ACK=6905207 SEQ=507025188 LEN=587 WIN=49152
59	0	D=1521 S=1213	ACK=507025775 WIN=8760
60	0.012	D=1213 S=1521	ACK=6905207 SEQ=507025775 LEN=1460 WIN=49152
61	0.008	D=1213 S=1521	ACK=6905207 SEQ=507027235 LEN=1460 WIN=49152
62	0	D=1521 S=1213	ACK=507028695 WIN=8760
63	0.008	D=1213 S=1521	ACK=6905207 SEQ=507028695 LEN=1460 WIN=49152
64	0.008	D=1213 S=1521	ACK=6905207 SEQ=507030155 LEN=1460 WIN=49152
65	0	D=1521 S=1213	ACK=507031615 WIN=8760
66	0.062	D=1213 S=1521	ACK=6905207 SEQ=507031615 LEN=1460 WIN=49152
67	0.01	D=1213 S=1521	ACK=6905207 SEQ=507033075 LEN=1460 WIN=49152
68	0	D=1521 S=1213	ACK=507034535 WIN=8760
. . . .			
81	0.007	D=1213 S=1521	ACK=6905207 SEQ=507046215 LEN=1460 WIN=49152
82	0.168	D=1521 S=1213	ACK=507047675 WIN=8760
83	0.08	D=1213 S=1521	ACK=6905207 SEQ=507047675 LEN=85 WIN=49152
84	0.14	D=1521 S=1213	ACK=507047760 WIN=8675
Ping-Pong			
85	0.417	D=1521 S=1213	ACK=507047760 SEQ=6905207 LEN=18 WIN=8675
86	0.079	D=1213 S=1521	ACK=6905225 SEQ=507047760 LEN=14 WIN=49152
87	0.002	D=1521 S=1213	ACK=507047774 SEQ=6905225 LEN=136 WIN=8661
. . . . .			
117	0.079	D=1213 S=1521	ACK=6906565 SEQ=507049045 LEN=11 WIN=49152
118	0.179	D=1521 S=1213	ACK=507049056 WIN=7379

Figure 9.7 A hybrid transaction.

Frame	Delta T	Bytes	Destination Source	Summary
1		244	TCP D=3200 S=1104	ACK=1998545017 SEQ=20365418 LEN=172
2	0.050	1096	TCP D=1104 S=3200	ACK=20365590 SEQ=1998545017 LEN=1024
3	0.122	72	TCP D=3200 S=1104	ACK=1998546041 WIN=8712
4	2.748	139	TCP D=3200 S=1104	ACK=1998546041 SEQ=20365590 LEN=67
5	0.019	113	TCP D=1104 S=3200	ACK=20365657 SEQ=1998546041 LEN=41
6	0.138	72	TCP D=3200 S=1104	ACK=1998546082 WIN=8671
7	0.566	1280	TCP D=1104 S=3200	ACK=20365657 SEQ=1998546082 LEN=1208
8	0.135	72	TCP D=3200 S=1104	ACK=1998547290 WIN=7463
9	1.860	133	TCP D=3200 S=1104	ACK=1998547290 SEQ=20365657 LEN=61
10	0.051	72	TCP D=1104 S=3200	ACK=20365718 WIN=65340
11	0.087	772	TCP D=1104 S=3200	ACK=20365718 SEQ=1998547290 LEN=700
12	0.106	72	TCP D=3200 S=1104	ACK=1998547990 WIN=8712

Server = TCP port number 3200, Client = TCP port number 1104.

Figure 9.8 SAP R3 transactions.

- TCP port number 3200 is a common SAP port. In fact, SAP uses the range of TCP ports from 3200 to 3399.

Note that three-tier applications can also exhibit bulk data characteristics, such as the case when remote printing occurs over a WAN.

9.4 Data Collection

Network managers need to plan for bandwidth in the network in anticipation of a full-scale deployment of client/server applications. These applications may be in a test mode in a LAN environment or perhaps deployed over the WAN at a few sites. It is important to be able to estimate bandwidth requirements without having to resort to detailed tests and tools.

STEP 1: Collect Transaction Data Using Protocol Traces

This is an important first step. Protocol traces are absolutely necessary for bandwidth estimation and response time calculations. Sniffers can be deployed at the server location or client location or both. In most situations, a single LAN/WAN sniffer at the server side is used. The advantage of using two sniffers at both ends of the connection is that they can be used to isolate client and server delays and also typical think times between transactions. A set of transactions from a single user, preferably separated by pings to differentiate individual transactions, should be traced.

STEP 2: Summary Traffic Characteristics

Based on the data collected in step 1, the transaction should be classified as ping-pong, bulk data, hybrid, or three-tier inquiry-response. Additional information needed includes the following:

- Number of packets exchanged per transaction in the inbound and outbound directions;

- Number of bytes sent in any one direction (actually, the maximum of the inbound and outbound direction), if the transaction is predominantly bulk data transfer;

- Rough estimate for the average packet size for both directions;

- Receiver TCP window size advertisements, especially for bulk data transfer and hybrid transactions; and

- Client build delays and/or server delays, as discussed in the previous section.

A word of caution: If the traces are not collected properly, one can easily mistake user think time for client build delays. One way to avoid this situation is to use pings to separate individual transactions.

STEP 3: Other Relevant Information

Additional information needed for bandwidth estimation includes the number of users per location using this specific application, and the peak hour transaction rate.

If this information is not available, then an educated guess will suffice. For example, business processes may suggest that not more than 50% of all users are likely to be using this application at any given time, and perhaps 10 transactions/per user/per hour. Alternatively, if think time information is available for individual transactions (as for SAP R3), then one can estimate the transaction rate as follows: If T_H is the think time (in seconds), then the number of transactions per hour per user cannot exceed $3600/T_H$. If N is the number of active users in the peak hour, then the transaction rate is approximately $N \times 3600/T_H$.

9.5   Bandwidth Estimation Guidelines

We will provide some guidelines for bandwidth estimation for the four types of transactions just discussed.

9.5.1   Applications With a Ping-Pong Traffic Characteristic

These applications, as mentioned before, are two tier and exhibit a traffic characteristic similar to Figure 9.5. For such applications, one can adopt the approach described here. Let

> N_T = number of packets per transaction in a given direction
>
> S = bytes per packet in the given direction (including WAN overhead)
>
> D = one-way network latency (essentially propagation delay)
>
> T_H = user think time for the transaction + client build delay + server delay (in seconds)
>
> N_U = number of active users

We need to estimate bandwidth for the application in the inbound and outbound directions, for which the values of N_T and S can be different. However, the variables D, T_H, and N_U remain the same. The estimated bandwidth will then be

$$\text{max(Inbound estimate, Outbound estimate)}$$

In any given direction, the bandwidth estimate is given by

$$B = 8 \times N_U \times N_T \times S / (N_T \times D + T_H) \text{ bps}$$

The rationale for this formula is as follows. The time between successive transactions generated by a single active user is $N_T \times D + T_H$, since $N_T \times D$ is the total network latency for the N_T packets (implicitly assumes a one-to-one ping-pong exchange), and T_H is the time between transactions. During this time, that is, $N_T \times D + T_H$, the user will have submitted, in the specified direction, $N_T \times S \times 8$ bits of data. The factor N_U is then used because it is the number of active users.

Example

As an illustration, assume that an application, with a server in the United States and clients in Europe, has the following characteristics:

N_T = 100 packets inbound, 120 packets outbound

S = 60 bytes/inbound packet, 200 bytes/outbound packet (+ 48 bytes WAN overhead)[1]

T_H = 1 min (think time)

D = 150 msec (one-way latency)

N_U = 25 active users at a branch office

Then the estimated inbound bandwidth is

$$[25 \times 100 \text{ packets} \times 108 \text{ bytes/packet} \times 8 \text{ bits/byte}] / (0.15 \times 100 + 60) = 28 \text{ Kbps (approximately)}$$

The estimated outbound bandwidth is

$$[25 \times 120 \times 248 \times 8] / (0.15 \times 120 + 60) = 76 \text{ Kbps (approximately)}$$

Hence 76 Kbps would be the peak bandwidth usage between Europe and the United States for this application. However, if one were to engineer physical and logical links in the network to, say, 70% utilization, then the bandwidth required should be 76 Kbps/0.7 = 108 Kbps, or 128 Kbps (to use the next highest multiple of 64 Kbps).

9.5.2 What Happens When Think Times Are Not Available?

The preceding formula relies on a good estimate for think time between successive transactions. For instance, if the think time is 2 min instead of 1 min, the bandwidth required would essentially be halved, to 64 Kbps.

If think times are not easily available, then an estimate of L, the number of transactions per user during a peak hour, is needed. Assuming a uniform rate during the hour, the number of transactions per user per second can be calculated as $L / 3600$. Call this L^*. Then the estimated bandwidth will be

---

1. 48 bytes = 20 (TCP) + 20 (IP) + 8 (Link layer, approximately).

$$(8 \text{ bits/byte}) \times (N_T \text{ packets/transaction}) \times (S \text{ bytes/transaction}) \times (N_U \text{ users})$$
$$\times L^* \text{ transactions per user per second}$$

9.5.3 Bandwidth Estimation for Bulk Data Transfers

Bandwidth estimation for bulk data transfers needs a somewhat different approach. The reason is that one needs additional information: the end user tolerance for the transaction time. Without this information, bandwidth may be underestimated (slow transaction time) or overestimated. In addition, adding bandwidth beyond an upper threshold will not improve the transaction time. This threshold depends on the window size and the latency in the network. If latency in the network is large, or window size advertisement is small, then the desired transfer time will not be achieved.

If no information is available regarding the end user tolerance time, one can assume a "human factors" delay number, such as 2 to 3 sec for a screen update.

We will first provide a general formula, and then illustrate using an example. Let

$F =$ total number of bytes transferred in any one direction

$S =$ segment size

$W =$ receiver advertised window size

$W^* =$ optimal window size for the connection

$T =$ preferred end user response time (in seconds)

$D =$ one-way propagation delay

$O_f =$ protocol overhead factor $= (S + \text{Overhead bytes}) / S$

Using this notation, the following steps can be used to estimate bandwidth:

$$B = [F \times O_f \times 8 \text{ bits/byte}] / T$$

Tune, if necessary, W to exceed W^*, where $W^* = (S + \text{Overhead bytes}) + 2 \times D \times B$.

Example

Assume that a client/server application sends image files from the server to the client. The client and server are separated by about 1000 miles. Images can be requested in blocks of 1 Mbyte. The protocol is TCP/IP with a client-advertised window size of 4096 bytes and segment size of 1024 bytes. End users would like the images to appear on the screen in 10 sec or less.

Note that the overhead factor O_f is 1.05 [(1024 + 48) / 1024], or 5%. Propagation delay is likely to be 10 to 20 msec (depending on the WAN type—frame relay or private line—and issues such as routing).

Using the approach given earlier, the initial bandwidth estimate is

$$B = 1,024,000 \text{ bytes} \times 8 \text{ bits/byte} \times 1.05 / 10 \text{ sec} = 860 \text{ Kbps}$$

For the sake of simplicity, let us assume that the decision is made to use a full T1 (1536 Kbps) for this application, given the large file size and tight constraints on transfer time. The next step is to ensure that the window size W advertised by the client is at least as large as the optimal window size, W^*.

The optimal window size is calculated as

$$W^* = 1024 + 48 + 2 \times 0.02 \text{ sec} \times 1,536,000 \text{ bps} / (8 \text{ bits/byte}) = 8750 \text{ bytes}$$

It is clear that the current window size of 4 Kbytes will be too small to "fill the pipe." Thus, in addition to recommending a T1 line for this application, one also has to ensure that the window size W is increased to at least 8 Kbytes.

9.5.4 Bandwidth Estimation for Hybrid Transactions

If the transaction exhibits a hybrid traffic pattern consisting of packet exchanges interspersed with one-way data transfers, then the only approach would be to use a combination of the two methods suggested earlier. The first step would be to isolate portions of the transaction that exchange packets between the client and the server, from those that involve a one-way data transfer.

Next, aggregate the ping-pong portions and apply the formula from Section 9.5.1. One can then aggregate the one-way data transfer portions to determine the overall amount of data exchanged in any one direction and proceed exactly as in Section 9.5.3.

Example

Let us analyze the transaction shown in Figure 9.7. The first step would be to combine the two ping-pong phases—frames 4 to 56, and frames 85 to 118. Notice the one-for-one exchange in these phases. There are approximately 43 transactions from client to the server and the same number in the reverse direction. Let us assume that the average packet size is 130 bytes in the client-to-server direction and about 70 bytes in the reverse direction.[2] Let us also

---

2. This is based on a quick analysis (using a spreadsheet) of the original data.

assume the following parameters: $N_U = 50$ active users, $N_T = 43$ packets, $T_H = 60$ sec (think time plus client and server delays), and $D = 0.04$ (40-msec one-way latency). Then the estimated bandwidth in the inbound direction for the ping-pong portion of the transaction is

$$B = 8 \times 50 \times 43 \times (130 + 48) \, / \, (43 \times 0.04 + 60) = 49 \text{ Kbps}$$

Conservatively engineered, this would translate to a bandwidth of 64 Kbps. We still have to account for the bulk data transfer portion of the transaction, which is approximately 25,000 bytes of actual data (not accounting for protocol overhead). Notice that the segment size is 1460 bytes. This implies that the protocol overhead is about 3%. Hence the actual amount of bulk data sent is likely to be $25,000 \times 1.03 = 25,750$ bytes. To transfer these data over a connection with 64-Kbps bandwidth, the estimated time is 25,750 bytes/(8000 bytes/sec) > 3 sec. Thus, 64 Kbps is likely to be insufficient bandwidth for this application with the given usage characteristics. A better choice would be 128 Kbps.

Readers should note the following:

- The preceding analysis demonstrates an approach to estimate bandwidth for hybrid applications. Real-life situations may require fine-tuning of the approach. For instance, some transactions may not be an exact ping-pong exchange. There may be a few instances where two or three packets flow in one direction, but there may not be a large enough numbers of packets to classify the transaction as bulk data. In this case, the previous formula will slightly overestimate the bandwidth required.

- The transaction studied here refers to "loading a form." If users typically load the form once in a business day, then it may not be appropriate to use this transaction to estimate bandwidth. It may be better to use transactions in which users input data using these forms.

- Note that if many users were to load reports simultaneously, then the response time will suffer because there are likely to be several bulk data transfers occurring at once.

9.5.5 Example of an SAP R3 Application

Look at the transaction shown in Figure 9.8. The following traffic assumptions can be made:

N_T = three inbound packets, two outbound packets (characteristic of a three-tier application)

S = 70 bytes/packet inbound, 672 bytes/packet outbound (including 48 bytes overhead)[3]

D = 30 msec one-way

N_U = 50 active users

T_H = 60 sec of think time

Clearly, the outbound bandwidth will be bigger than the inbound bandwidth. Hence, let us estimate bandwidth in the outbound direction:

$$50 \text{ users} \times 2 \text{ packets} \times 672 \text{ bytes} \times 8 \text{ bits/byte} / (2 \times 0.03 \text{ sec} + 60 \text{ sec}) = 9 \text{ Kbps, approximately}$$

Again, based on engineering the link at 70% utilization, the actual bandwidth estimate will be 9 Kbps/0.7 = 13 Kbps, approximately.

If a separate PVC for a frame relay connection is being considered, then a 16-Kbps CIR would be sufficient for this application.

Let us study the bandwidth formulas in a little more detail. The following observations can be made:

- Notice how the think time dominates the denominator, especially in the case of SAP R3. This is consistent with the observation that if a user has a long think time (say, a complex screen data input),[4] the number of transactions submitted by the user per unit time will be relatively small.

- For a two-tier application that exchanges several packets, the formula assumes a one-for-one exchange of packets between the client and server. Protocol traces may show instances where many packets are sent in one direction. As long as the number of such packets is small (say, two or three) *and* the number of such instances in a transaction is small (say, less than five), then the bandwidth formula can be used. Otherwise, the transaction needs to be classified as hybrid. The approach in Section 9.5.4 can be used.

---

3. In this example, the second transaction, which appears to be "worst case", is considered. In reality, one would have to scan individual transactions and build best case and worst case scenarios.

4. For instance, consider human resource applications where the user has to input name, address, social security number, emergency contact numbers, and so on.

- Just as think times and propagation delays inversely affect bandwidth requirements, so do client and server processing delays. These can be significant. For a two-tier application, the delay can be at the client—sometimes called "client build delay"—and at the server. The formula given ignores client and server delays. To that extent, the bandwidth requirements forecasted by the formula would be conservative.

9.5.6   An Approach to Computing Response Times

Computing application-level response times is usually a complex task, especially if it involves background load. The task is especially complex in the case of two-tier client/server applications. However, it is useful and relatively easy to calculate baseline end user response times under the assumption of unloaded network resources. These baseline response times are useful in troubleshooting WAN performance problems for client/server applications.

The same parameters used for estimating bandwidth requirements (see Section 9.5.3) can be used for computing response times.

Instead of providing general formulas, we illustrate the concept using the examples from the previous section. Consider the example in Section 9.5.1. Assume that the connection between the United States and Europe is frame relay with the line speeds given in Figure 9.9.

Under a no-load scenario, and assuming full bursting on the PVC, one can use the following approach to estimate response time:

$$\text{Inbound} = [8 \times 108 \text{ bytes} / 64 \text{ Kbps}] + 0.15 + [8 \times 108 \text{ bytes} / 256 \text{ Kbps}] = 0.166 \text{ sec}$$

$$\text{Outbound} = [8 \times 248 \text{ bytes} / 256 \text{ Kbps}] + 0.15 + [8 \times 248 \text{ bytes} / 64 \text{ Kbps}] = 0.188 \text{ sec}$$

$$\text{Approximate total response time} = 100 \times 0.166 \text{ sec} + 120 \times 0.188 \text{ sec} = 39 \text{ sec}$$

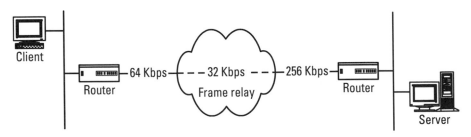

Figure 9.9 Reference connection for a client/server application.

The following issues are important and impact the applicability of this approach for some cases:

- We assumed a one-for-one exchange. However, note that there were 120 transactions outbound for every 100 transactions inbound. Hence, on an average, there are 20% more transactions outbound than inbound. The assumption of a one-for-one exchange overestimates the response time because if several packets are sent together in the outbound direction, then the total transfer time for these packets is affected only once by propagation delay. One way to address this issue is to increase the outbound packet size by 20%, while reducing the outbound packet count to 100.

- To calculate the transfer time in this new model, use the following approach outbound response per packet:

$$= [8 \times 248 / 256,000 + 0.150 + 8 \times 248 \times 1.2 / 64,000] = 0.195 \text{ sec}$$

Note that the first packet size is 248 bytes, not 248×1.2 bytes, because of the pipelining effect. Hence the new response time is

$$100 \times 0.166 \text{ sec} + 100 \times 0.195 \text{ sec} = 36 \text{ sec}$$

- The fine-tuning approach given above should be used cautiously. The basic assumption is that the transaction is ping-pong in nature, albeit a few packets being sent together in one direction, typically two or three packets. If a large number of packets are sent in one direction, then the transaction needs to be classified as hybrid.

- The effect of 25 active users and potential other background load has been ignored. As mentioned before, this is a complex exercise in queuing theory and beyond the scope of the book.

9.5.7 Response Times for Bulk Data Transfer Transactions

The approach in Section 9.5.3 for estimating bandwidth requirements can be used without any change for estimating response times.

Consider the example in that section: Clients request 1-Mbyte images over a frame relay connection. The one-way latency through the connection is 20 msec. Assume the frame relay connection has port speeds of 512 and 128 Kbps and a PVC at 64-Kbps CIR at the two locations, respectively. The

TCP segment size is 1024 bytes and the protocol overhead is 48 bytes, a protocol overhead of approximately 5%. The optimal window size that should be advertised by clients in New York is given by

$$W^* = \text{Round-trip delay for a packet} \times \text{Slowest link} = \{[8 \times 1072 \, / \, 512,000 + 0.02 + 8 \times 1072 \, / \, 128,000] + [8 \times 48 \, / \, 128,000 + 0.02 + 8 \times 48 \, / \, 512,000]\} \times 128,000 \text{ bps} = 2040 \text{ bytes}$$

Under this assumption, the response time will be

$$\text{File size} \, / \, \text{Throughput} = 1,024,000 \text{ bytes} \, / \, [(16,000 \text{ bytes/sec}) \, / \, 1.05] = 67 \text{ sec}$$

9.5.8 Response Times for Hybrid Transactions

As indicated in Section 9.5.4, the transaction needs to be broken down into its ping-pong and bulk data transfer components. Then the approach in Sections 9.5.1 and 9.5.3 can be applied. The overall response time is the sum of the response times for the two components.

9.6 The Thin Client Solution

The attraction of thin client technology is that it can dramatically reduce end user hardware, software, and ongoing support requirements. This is because applications are deployed, managed, and executed completely on the centrally or regionally deployed servers, and not on the client PCs. As mentioned in the introduction, there are two approaches to thin client technology in the context of client/server networking: a remote presentation approach as implemented in Citrix Winframe/Metaframe, and the Java-based network computing (NC) approach. We will discuss these approaches in some detail in this section. We will show that, although the thin client approach is compelling, it does introduce a new set of performance issues that needs to be considered.

9.6.1 The Remote Presentation Approach

In this approach, terminal servers are typically deployed on the same LAN as the database server (see Figure 9.10). The terminal server runs the client software and communicates with the database server on the local LAN. Communication between the remote client and the terminal server occurs in a manner similar

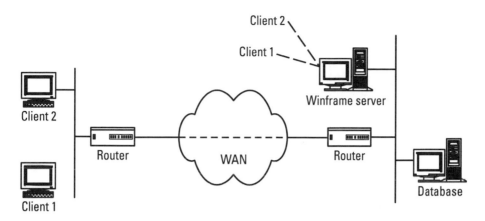

Figure 9.10 Remote presentation architecture for two-tier applications.

to X-Windows: Clients only send mouse movements and keystrokes across the WAN. Thus remote presentation promises to deliver "LAN-like" performance.

It is important to note that the communication between the end user and the terminal server approach is echoplex in nature in that keystrokes from the client are echoed back from the terminal server. Mouse movements are acknowledged by the server, but not echoed. Thus, the user experience is highly sensitive to round-trip delay. As round-trip delay increases, a larger percentage of users may be dissatisfied with performance due to slow terminal and mouse movement response.[5] Additionally, low delay variability is an important factor for end user perception of good performance. We will discuss these issues in more detail later in this section.

9.6.1.1 "LAN-Like" Performance: A Quantitative Perspective

Let us study this issue in a little more detail. For the sake of illustration, we will consider two applications. Application A is a two-tier client/server application with a ping-pong characteristic, and application B is a bulk data transfer from a server to the client. We will also consider two WAN scenarios: a 56-Kbps link and a T1 WAN link.[6]

---

5. Human factors issues are important here. Users accustomed to "local" performance may be dissatisfied with slow keyboard response and mouse movements. Irate users will then tend to retype, hit Enter keys several times, or move the mouse rapidly in frustration. This will only make the situation worse because more data will be sent into the network.

6. For the sake of simplicity, we will consider only private lines here. The conclusions are valid for frame relay as well.

We make the following assumptions:

- Application A: 50 packets outbound with 250 bytes average, 50 packets inbound with 100 bytes average, one-for-one exchange. Packet size includes protocol overhead.
- Application B: User views a 1-Mbyte document remotely.
- One-way propagation delay is 30 msec.
- WAN links: 56 Kbps, T1 (1536 Kbps).
- Unloaded WAN links.
- A screen update is 2000 bytes (actually, in many situations, screen updates may involve much smaller packets).
- TCP segment size is 512 bytes (about 10% overhead) and the window size is large enough to get the maximum throughput possible.

We will estimate the performance for these applications in native mode and using the terminal server approach.

Application A: Native Mode

Estimated screen response on a 56-Kbps unloaded WAN link:

$$[(8 \times 250 \text{ bytes} / 56,000 + 0.03) + (8 \times 100 \text{ bytes} / 56,000 + 0.03)] \times 50 = 5.5 \text{ sec}$$

Estimated screen response on a T1 link (192 bytes/msec) unloaded WAN link:

$$[(8 \times 250 \text{ bytes} / 1,536,000 + 0.03) + (8 \times 100 \text{ bytes} / 1,536,000 + 0.03)] \times 50 = 3.1 \text{ sec}$$

Terminal Server

Estimated screen response on a 56-Kbps unloaded WAN link:

$$8 \times 2000 \times 1.1 / 56,000 + 0.06 = 0.38 \text{ sec}^7$$

Estimated screen response on a T1 unloaded WAN link:

---

7. One needs to be a little more careful here, because the inquiry from the client will take some time to reach the server on a 56 Kbps link. However, this delay is likely to be small and it is very hard to make any assumptions about packet sizes in a terminal server environment.

$$8 \times 2000 \times 1.1 \ / \ 1,536,000 + 0.06 = 0.07 \text{ sec}$$

Thus it is clear why the terminal server approach will provide "LAN-like" performance over the WAN. The most significant component of the delay in the native form is propagation delay, which is totally eliminated with terminal servers.

Application B: Native Mode

Note that the entire file first needs to be downloaded before it is viewed. Hence, the estimated screen response on a 56-Kbps unloaded WAN link is

$$8 \times 1,024,000 \text{ bytes} \times 1.1(\text{Protocol overhead}) \ / \ (56,000 \text{ bps}) = 2 \text{ min } 41 \text{ sec}$$

Estimated screen response on a T1 link unloaded WAN link:

$$8 \times 1,024,000 \text{ bytes} \times 1.1(\text{Protocol overhead}) \ / \ 1,536,000 = 6 \text{ sec}$$

Terminal Server

Assuming a 2000-byte screen update, the numbers are the same as before, that is, 380 and 70 msec, respectively, for a 56-Kbps and a T1 link.

Thus, even in the case of bulk data transfers, a terminal server will provide better screen response times. Of course, as the user scrolls through the document, additional data will be sent from the server to the client. Nevertheless, some important performance issues over WAN exist that need to be discussed.

9.6.1.2 WAN Bandwidth Issues

It is not surprising that additional bandwidth is required with terminal servers because keystrokes and mouse movements are sent across the network on TCP/IP connection. However, it not straightforward to estimate the additional bandwidth requirements (compared to using applications in the native mode) because of two factors:

- "Administrative" work, such as number of colors used, screen resolution, and mouse movements, impacts bandwidth.

- The nature of the application itself impacts bandwidth. For instance, a graphics-intensive application is likely to be more bandwidth intensive compared to a word-processing application where the user merely types with very little scrolling. These two scenarios can be considered as extremes. Within this range, there are many situations that have to be considered.

As such, it is impossible to provide applicable guidelines for bandwidth.[8] However, it is useful to list some of the factors that influence bandwidth consumption.

Number of Colors Used
For some graphics-intensive applications, using a smaller number of colors (16 colors instead of 256 colors) can make a significant difference in the bandwidth.

Compression
It is hard to predict the exact compression ratios achievable, but compression of data between the client and the terminal server is definitely recommended.

Screen Resolution
An 800×600 screen resolution should provide reasonable resolution. Higher resolutions will impact bandwidth negatively, although its effect may be somewhat less when compared to using a larger number of colors and/or disabling compression.

Blinking Cursors
Blinking cursors will add bandwidth to the network. One study, conducted by the authors, estimates that approximately 800 bps of traffic per user is generated via a blinking cursor. Much of the bandwidth consumption is due to the protocol overhead that TCP/IP and the WAN imposes on small packets (e.g., 48 bytes of overhead for about 5 bytes of data for a blinking cursor).

While 800 bps per user may not seem significant, consider a 100-person remote office, 50 of whom leave for lunch during midday with their PCs on-line. This would mean 50×800 bps = 40 Kbps of precious WAN bandwidth!

Mouse Movements
Typically, mouse movement generates more traffic in the client-to-server direction than in the reverse direction, the latter traffic in this case being TCP level acknowledgments. This is in keeping with the observation that mouse movements are not echoed by the terminal server, but are acknowledged. With

---

8. It has been mentioned several times in journals and white papers that one should plan to add 20 Kbps per user while deploying new applications with Citrix Winframe/Metaframe access. From the authors' experience with laboratory testing and real-life experience, this appears to be an overestimate in many cases.

compression turned on, mouse movements are expected to contribute about 4 to 5 Kbps per user.

Keyboard Data Entry

Data entry from remote users also adds bandwidth to the network. At a minimum, each character results in 49 bytes of data sent to the server and back from the server over a WAN: 1 + 40 (TCP/IP) + 8 (WAN). On a LAN the overhead is likely to be more. If a user types at the rate of five characters per second (an experienced typist), the bandwidth added in both directions is

$$5 \text{ packets/sec} \times 49 \text{ bytes} \times 8 \text{ bits/byte} = 1960 \text{ bps per user!}$$

To refine this analysis, one needs to better understand the number of bytes transferred with each keystroke between the client and the server. It is possible that more than 1 byte is exchanged per packet.

Application Characteristics

Application characteristics contribute significantly to network bandwidth in a terminal server environment. While simple typing of characters, mouse movements, and so on, contribute perhaps 5 Kbps per user with compression, graphic-intensive applications can contribute more than 50 Kbps in the server-to-client direction per user, especially with 256 colors. Compression is absolutely essential in such cases, as well as a lower number of colors (16).

Studies conducted by the authors in using Winframe for two-tier versions of client/server applications from Oracle and PeopleSoft show that these applications will require 5 to 10 Kbps of bandwidth per user. However, we hasten to add that these are overall estimates and each situation should be analyzed separately.

9.6.1.3 Other WAN Performance Issues

We also need to discuss other performance-related issues such as the impact of network latency and background load on end user response times, and differences between private lines and frame relay in the context of terminal servers.

Impact of Network Latency

Network latency is primarily a function of propagation delay, backbone network loading (in the case of public data networks like frame relay), and packet processing delays. Network latency is a very important consideration because of the echoplex nature of the interaction between the remote users and the terminal server. Since each character is echoed from the server, the round-trip latency in the network needs to be consistently small. To see what this delay should be, consider an end user typing at a rate of three or four characters per second. Since the interval between keyboard entries is 250 to 350 msec (approximately), the round-trip network latency should also be within that range. Otherwise, characters will not appear on the screen as fast as they are entered.

Round-trip network latencies of 250 to 350 msec are achievable in many public data networks. For example, the round-trip delay on a domestic U.S. frame relay network is usually less than 100 msec in the worst case. However, for international connections, it is not uncommon for latencies to exceed 350 msec.

Note that delays[9] due to background load in the network have not been considered in the previous discussion. Thus one can restate the WAN delay requirement for keyboard entries as follows:

> Overall network delay, as measured by a small ping (say, 100 bytes) between a client and server, should be consistently below 350 msec for satisfactory echoplex response times.

An immediate corollary from this guideline is that the terminal server approach may not be suitable for some global connections, and definitely unsuitable for satellite connections.

Impact of Background Load

As discussed before, background load has the potential to increase delays in the network beyond a point at which keyboard entries and mouse movements will appear to be slow to the end user. Routers can potentially prioritize terminal server traffic using the specific TCP port number used between the terminal server and the client (ICA uses port number 1494). Note, however, that this prioritization only protects terminal server sessions against nonterminal server sessions. It will not address prioritization within applications using the terminal server, because all of these sessions may use a common TCP port number. For instance, a user accessing a graphics-intensive application could overwhelm other users accessing data entry applications.

Suitability of Frame Relay for the Terminal Server Solution

The following issues should be kept in mind when supporting applications using remote presentation over frame relay:

---

9. We differentiate between latency and delay in a network. Loosely, latency is the sum of propagation delay and switch processing delays in the backbone network. As such latency is independent of factors such as packet size, line speeds, line utilization, and router delays. Overall end-to-end delay is the sum of network latency and delays due the factors just mentioned.

- As a general rule, private line networks have lower latency and more consistent delay characteristics than frame relay, which is a packet network. Therefore, delay and loading in frame relay networks need to be monitored carefully. It is also advisable to overengineer frame relay ports and PVCs.

- Frame relay traffic discrimination methods should be used to ensure good performance for terminal server users. One method is router prioritization. As previously discussed in the book, router prioritization is not always effective in frame relay. A separate PVC for the terminal server traffic may be needed.

- Global frame relay connections may not be appropriate for terminal server traffic.

9.6.2 The Network Computing Approach

The network computing architecture combines the functionality of the application server found in three-tier client/server architecture and the thin client model. The classic example of this architecture is Oracle NCA. The application software resides in middle-tier application servers that communicate on a LAN with the back-end database servers. End users with Java enabled Web browsers download an applet from a central Web server to obtain the user interface to the application.

There are a number of WAN performance issues with the network computing approach.

- Java applets for the user interface can be fairly large. A few megabytes of data transfer is not unusual. This will add traffic to the WAN each time a new version of the application is released.

- In the initial stages of software deployment, incremental changes to the graphical user interface is most likely a rule than an exception. These incremental changes will add bandwidth. This issue can be overcome somewhat by maintaining a local forms server.

- Field validation may actually require the transmission of packets to the central server. This has implications regarding network bandwidth and latency sensitivity, somewhat similar to the echoplex delay phenomenon of remote presentation.

- HTML documents (forms, etc.) are downloaded using the HyperText Transfer Protocol (HTTP). Unlike Telnet and FTP, whose WAN performance characteristics are relatively straightforward, HTTP's data transfer characteristic is somewhat more complex, as discussed in Chapter 7. There are two versions of HTTP in use today—HTTP/1.0 and HTTP/1.1. HTTP/1.0 has some inefficiencies over the WAN. It relies on a separate TCP connection to download each image in the HTML document. The performance overhead of establishing and tearing down TCP sessions can be quite significant. In addition, a penalty is paid in the use of TCP slow start in terms of round trip delays. To address these issues, browsers usually open multiple (typically four) parallel TCP sessions with HTTP/1.0 servers to download images. While this addresses the latency issues, it does introduce bursty traffic. HTTP/1.1 supports a persistent TCP session, although some browsers appear to prefer multiple TCP sessions even while communicating with HTTP/1.1 servers. In addition, few web servers appear to support HTTP/1.1.

9.7 Summary

This chapter presents a detailed discussion of the performance issues in supporting client/server applications over a WAN. We discuss the differences between two-tier and three-tier applications. We demonstrate that two-tier applications are generally not "WAN-friendly," whereas three-tier applications tend to perform better over a WAN. Because planning for bandwidth upgrades is an important task prior to full deployment of client/server applications in an enterprise network, we provide guidelines for data collection and simple formulas for estimating bandwidth. Finally, we discuss the thin client approach to supporting client/server applications, and highlight the advantages and performance potential pitfalls in using this approach over wide-area networks.

References

Several articles in trade journals discuss the issue of client/server application performance. We mention five below. The authors are not aware of any books that specifically deal with this topic.

[1]   Comer, D. E., *Computer Networks and Internets*, 2nd ed., Englewood Cliffs, NJ: Prentice-Hall, 1999.

[2]   Jessup, T., "WAN Design with Client-Server in Mind," *Data Commun. Mag.*, *August 1996.*

[3] Bruno, L., "Tools that Troubleshoot Database Transactions," *Data Commun. Mag.*, August 1996.

[4] Robertson, B., "Those Application Networking Nightmares," *Network Computing*, August 1996.

[5] Edlund, A., "How Thin Clients Lead to Fat Networks," *Business Communications Review*, Vol. 28, Number 7, 1998.

10

WAN Design and Performance Considerations for SNA Networks

10.1 Introduction

It seems like a long time ago, but in the early to mid-1980s there were few WAN architectures to choose from other than IBM's SNA (Systems Network Architecture). Today, just over a decade later, the buzzwords in the networking industry are not SNA, but TCP/IP, Internet, intranets, frame relay, VPNs, and so on. From the early skepticism of "SNA and IP—Oil and Water" to "My users need web access to 3270 applications," the facelift from SNA networking to TCP/IP appears to be more or less complete. However, there is one important difference: mainframes and AS/400s are still the anchors of many large corporations and legacy SNA applications[1] have a significant presence in the financial, retail, insurance, and government markets. It is the access to SNA applications that has undergone a massive facelift, not the applications themselves. Thus the issue of supporting these applications in a TCP/IP environment is very important.

SNA applications are typically transaction oriented and therefore require high availability and fast and consistent response times. These requirements have not changed in the migration from traditional SNA networking to TCP/IP. It has, however, given rise to a different set of issues that are, in many ways, more challenging. Perhaps the biggest concern is that mixing SNA

---

1. Although SNA and legacy applications have different connotations, SNA being an access method to legacy applications, we will refer to mainframe and AS/400 applications as "SNA applications."

applications and multiprotocol LAN-oriented traffic will compromise performance. Although these concerns apply for other applications (e.g., SAP R3), there are some SNA-specific design and performance issues, such as the migration from multidrop SNA networks to router networks, the role of routers in connecting Front-end Processor (FEP)-to-FEP links, and the design of the data center, at which all the SNA traffic in a large network typically converge.

Our objective in this chapter is twofold. First, we want to review the ways in which traditional SNA networks have migrated to TCP/IP networks and, second, we want to discuss the issue of SNA performance in a multiprotocol network. As in the rest of the book, we will discuss these issues in the context of leased line and frame relay WANs. In Section 10.2, we briefly review the various transport methods for SNA applications—encapsulation, using IP gateways, TN3270, native frame relay support using RFC 1490, and Web access. In Section 10.3, we discuss the data center architecture from a WAN perspective, which is an important issue for large SNA networks. In Section 10.4, we discuss overall performance issues in supporting SNA applications in an IP network. Specific topics include the migration of traditional SNA multidrop networks to router networks, SNA over frame relay, and traffic discrimination.

The reader is referred to Guruge [1] for an excellent in-depth treatment of these and related topics.

10.2 SNA Transport Methods: A Review

Broadly speaking, there are five different ways to transport SNA applications over a WAN[2]:

- Encapsulate SNA in TCP/IP (data link switching);
- Use centralized SNA gateways (emulation);
- SNA directly over frame relay (RFC 1490);
- Translate SNA into TCP/IP (TN3270); and
- Web access to SNA applications.

10.2.1 TCP/IP Encapsulation: Data Link Switching

Data link switching (DLSw) is a standard method for transporting SNA and NetBIOS across IP internets using TCP/IP encapsulation (see Figure 10.1).

---

2. Strictly speaking, all of these options except frame relay can be used over a campus network as well.

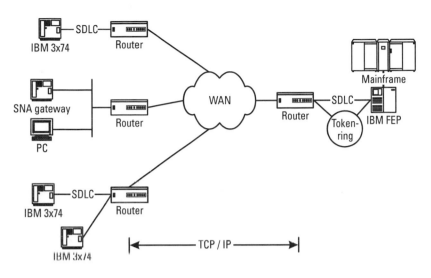

Figure 10.1 Data link switching.

Prior to DLSw several proprietary TCP/IP encapsulation schemes were followed by router vendors. The term *DLSw* was coined by IBM to refer to their implementation in the IBM 6611 router. The specifications were submitted to the IETF and DLSw became an informational RFC 1434. However, RFC 1434 did not address some key issues like interoperability between vendors, flow control, prioritization, and other aspects of SNA and NetBIOS support. Later versions of DLSw address these issues. The current version of the standard is RFC 2166.

DLSw terminates the SNA link layer (SDLC or LLC2) locally at the routers, encapsulates the SNA data in TCP/IP, and transports it over an IP backbone (leased lines, frame relay, ATM) to the destination router (its peer). The destination router "decapsulates" the TCP/IP headers and adds SNA link layer overhead (SDLC, LLC2) and sends the frame to the destination SNA devices.

DLSw describes a switch-to-switch protocol (SSP). Some of its main functions are:

- Interoperability between vendors through a function known as *capabilities exchange.*

- Establishment of an end-to-end connection between end devices, for example, between a 3174 controller and a 3745 FEP.

- Protocols to locate resources in the network, analogous to explorer frames in source route bridging. DLSw uses a protocol called "canureach/icanreach" to locate resources in the network.

- End-to-end sessions are broken up into three parts: two separate data link connections between the DLSw switch and the local SNA device, and a TCP session between the DLSw switches. Circuit IDs are used to identify an end-to-end session.

- In addition to providing local LLC2 acks, and preventing LLC2 administrative messages from going across the WAN, DLSw also terminates the routing information field (RIF) in the token ring MAC layer header. This addresses some hop count limitation with source route bridging.

- DLSw has a somewhat large overhead—16 bytes for DLSw and 40 bytes for TCP, plus link layer overhead (PPP, frame relay), which is usually 8 bytes. To minimize TCP overhead, DLSw allows multiple DLSw packets to be encapsulated in a single TCP segment.

DLSw has become popular in recent years and is supported by most major router vendors. The link layer termination and broadcast control features of DLSw make it an attractive alternative for SNA to IP migration, although, strictly speaking, SNA is not eliminated over the WAN. DLSw does not distinguish between leased lines, frame relay or ATM; it is IP traffic and therefore can use any transport medium.

10.2.2 Emulation Using SNA Gateways

SNA gateways have been in use for many years. These gateway servers are placed either at the branch offices or at the central site. Client PCs with 3270 emulation connect to these gateways using LAN protocols such as Novell IPX or TCP/IP. SNA gateways look exactly like controllers to the SNA host, and as such, DLSw is used when the gateways are deployed at the branch locations (see Figure 10.1). The SNA gateways can also be centralized, eliminating SNA from the WAN and limiting it to communications between the gateway server and the host/FEP [see Figure 10.2(a)].

There are many SNA gateway vendors—Microsoft, Novell, and others—and 3270 emulation vendors for client PCs—Attachmate, WRQ, and others.

The advantage of using SNA gateways with 3270 emulation for the client/PCs is that they provide a full range of 3270 functions, unlike TN3270, which we discuss later.

Some of SNA gateways can also be channel attached to the mainframe. The IBM 3172 and the Cisco Channel Interface Processor (CIP), are examples of SNA gateways that can be channel attached to the mainframe. See Figure 10.2(b).

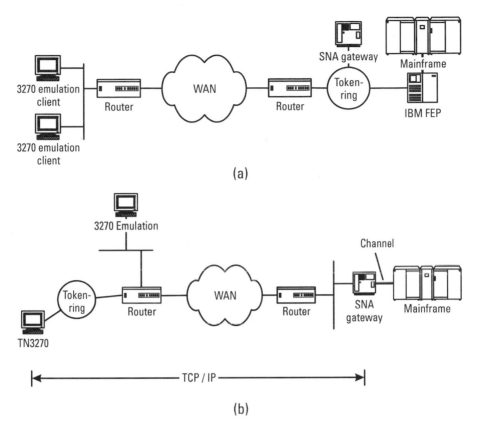

Figure 10.2 (a) Centralized SNA gateway with FEP and (b) centralized SNA gateway channel attached to the host.

10.2.3 Direct Encapsulation Over Frame Relay: RFC1490

RFC 1490 specifies a standard way to encapsulate multiple protocols over a frame relay PVC. The objective is to enable interoperability of different vendors' equipment over frame relay. RFC 1490 has taken on a life of its own in the context of SNA. Although RFC 1490 specifies a multiprotocol encapsulation standard, there is no mention of SNA, APPN, or NetBIOS/NetBEUI in the RFC. The use of RFC 1490 in the SNA context is specified in the Frame Relay Forum FRF.3 document, which was authored by IBM.

In the context of SNA, RFC 1490 specifies how SNA traffic is carried directly over frame relay without TCP/IP encapsulation. Without TCP to guarantee end-to-end delivery, one needs a reliable protocol to carry SNA. In RFC 1490 this is accomplished by LLC2 (logical link layer type 2), since frame relay is an unreliable link layer protocol.

To support frame relay within the context of an SNA network, IBM incrementally introduced frame relay features into its product lines. With NCP 6.1[3] or higher, two 3745 FEPs are able to communicate directly over a frame relay network. With NCP 7.1 or higher, one can attach a 3174, a FRAD, or a similar device with a frame relay interface directly into a 3745 FEP using RFC 1490 encapsulation. There are two ways in which this can be achieved: boundary network node (BNN) and boundary access node (BAN), the latter requiring NCP 7.3. In a nutshell, BNN describes connectivity between serially attached SNA devices (such as a 3174) and a FEP over frame relay [see Figure 10.3(a)]. BAN specifies this connectivity for LAN attached devices such as controllers behind a router/FRAD [see Figure 10.3(b)]. BAN can also be used when the FEP is token ring attached to a central site router. In a very real sense, RFC 1490 for SNA specifies a standard way to support source route bridging over frame relay.

10.2.4 SNA Translation: TN3270

TN3270 is an alternate way to transport SNA over TCP/IP. TN3270 is available on almost all client platforms supporting TCP/IP. It uses the standard telnet operations under TCP/IP to provide a basic set of functions for carrying 3270 traffic. These functions are a subset of the functions provided by a full-featured 3270 emulation device.

TN3270 operates on a client/server paradigm of computing, with the TN3270 server being resident in the host or on in a separate device having a gateway into the mainframe (see Figure 10.4). If the mainframe is used as the telnet server, then clearly the mainframe must be TCP/IP enabled. Thus the host becomes an TCP/IP host, with clients using TN3270 to access host applications.

TN3270 has had some limitations arising from the fact that it is not a full function 3270 protocol. For instance, host-based printing is not supported, nor are certain keys important in the 3270 world supported. These have been more or less addressed in TN3270E, which is still referred to at times as TN3270.

10.2.5 Web Access to Mainframe Applications

With the proliferation of Web browsers, it is not surprising that Web access to mainframe applications has become an important issue (see Figure 10.5). In a sense, this is similar to the TN3270 approach in that the goal is to use the "thin

---

3. NCP (Network Control Program) is the software that runs the FEP.

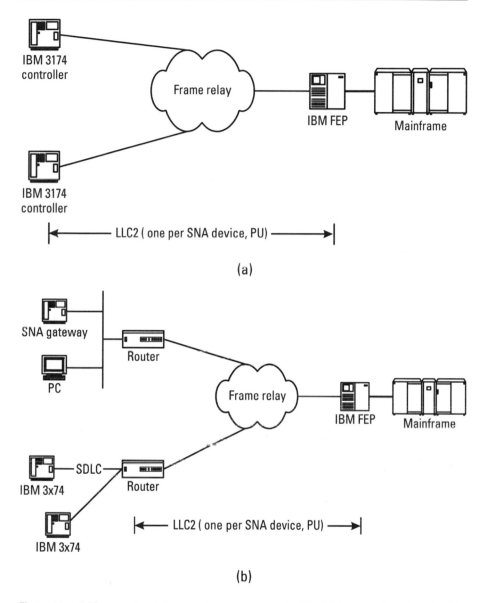

Figure 10.3 RFC 1490 direct frame relay connection to FEP (a) in BNN format and (b) in BAN format.

client" approach—TCP/IP and Web browsers have become mainstays in client PCs. There are two ways to enable end users with Web browsers to access 3270 applications: HTML conversion and by downloading 3270 Java applets for a "green screen." With HTML conversion, the 3270 screen coming from

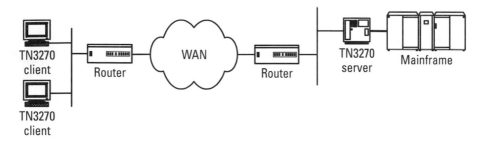

Figure 10.4 SNA Translation; TN3270 via an external TN3270 server.

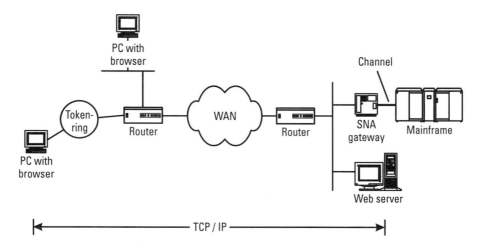

Figure 10.5 Web access to mainframe via an SNA gateway.

the mainframe is converted to standard HTML format. With Java, an end user wishing to access the mainframe first downloads a 3270 Java applet from a central Web server. From that point on, the governing protocol across the WAN is TN3270(E).

In summary, these are fundamental architectural decisions that must be made when migrating SNA networks to a multiprotocol environment. Very often, a single solution will not fit the bill. For instance, a network manager migrating from a multidrop SNA network to an IP-based WAN (via PCs at the remote locations and SNA gateways at the data center) may still need to support FEP-to-FEP links, either on SDLC links, directly via frame relay, LLC2 connections via source route bridging, or SDLC tunneling via routers.

The DLSw and centralized gateway options are popular choices. RFC 1490 does not appear to have the same popularity as DLSw. To a somewhat lesser extent, network managers have replaced central FEPs with channel attached gateways with or without TCP/IP on the mainframe. Web access technology is being used to a somewhat limited extent, and is still evolving.

10.3   Data Center Architecture Issues for Large SNA Networks

For large SNA networks (typically ranging from hundreds of locations to thousands of locations, all connected to a few central sites), the data center architecture is an important issue. We briefly discuss some of these issues next.

Discard or Retain the Data Center FEP?

Technology exists today that enables the data center FEP to be replaced by routers that are channel attached to the mainframe. It is not clear that channel attached routers are the right alternative for large SNA networks. Many small SNA networks (say, a 100 sites) have successfully replaced the FEP by channel attached routers.

How Many Parallel Routers Are Needed at the Central Site?

Routers have some hard limits in terms of PVCs per port and per box, the number of DLSw peers, the number of sessions, and so on. The limits depend on many factors, chief among them being port densities, router memory and utilization, broadcast traffic, overall user traffic, specific DLSw implementations, and so on. As an example, a 1000-node SNA network migrating to a router-based network with DLSw is likely to require 4 or 5 high-end routers to terminate DLSw peers. This assumes that a single router can support 200 to 250 DLSw peers. (Some router vendors claim to be able to support upwards of 300 peers.) If router redundancy or protection from single points of failure is required, then 8 to 10 routers may be required.

This does not include the consideration of the number of PVCs that can be terminated on a serial port on a router. Some large SNA networks are architected with DLSw peer routers positioned behind frame relay WAN routers, that is, the WAN routers do not terminate the TCP/IP sessions on which DLSw rides.

How Much Bandwidth Should Be Allocated at the Central Site?

This is also a somewhat complex issue. One needs to make some assumptions regarding volume of SNA and non-SNA traffic, concurrency (how many sessions from remote locations will be active simultaneously?), and CIR to port oversubscription. While SNA traffic volumes are easy to obtain, often non-SNA applications may not be fully deployed. Therefore, educated guesses or overengineering may be the only alternatives. It is imperative in such cases to actively monitor port/PVC utilization as the applications become widely deployed. Utilization must be maintained at a reasonable level, say, less than 70%. As for oversubscription in frame relay, there are no absolute rules, except that utilization levels, time-of-day, and time zone characteristics may allow a 2-to-1 oversubscription.

Redundancy Issues

What is the backup strategy for WAN links, frame relay switches, routers, and SNA gateways? How about ISDN dial backup? Are SONET rings at the central site locations viable?

The fundamental issue with redundancy and backup strategies is cost—the level of redundancy desired in the network versus the cost to provide it. Redundancy levels are strongly dependent on business drivers; a securities trading company is likely to need more levels of redundancy than, say, a manufacturing company. Most network managers are concerned about routers and WAN links into the central site supporting many remote locations (especially for frame relay), because these are single points of failure. One possible backup strategy for data center frame relay links and routers is to set up dual routers with diversely routed access links into the frame relay ports. One router can be designated as the "SNA router" (primary) and the other as the "non-SNA router" (secondary). The second router can back up the first[4] (see Figure 10.6). Frame relay networks can be configured so that PVCs to the secondary router carry actual traffic, as opposed to backup PVCs which are inactive until a failure occurs. Backup PVCs are more relevant in the context of disaster recovery, which is discussed next. Access at the remote locations can also be protected via dial backup.

Most large networks with multiple data centers/hub locations have self-healing SONET rings at the these locations.

Disaster Recovery and Data Center Backup Strategy

Many SNA networks have data center disaster recovery support from Comdisco, Sunguard, or IBM Sterling Forest. When disaster strikes—such as a fire in the building—the connections will need to "swing" from primary to the backup data center within a relatively short time. Frame relay is particularly attractive in this context because the connections are logical, and moving the PVCs from one location to another is easily accomplished (see Figure 10.7). It is also important to test this capability periodically.

---

4. The ability to accomplish this without manual intervention depends on the router.

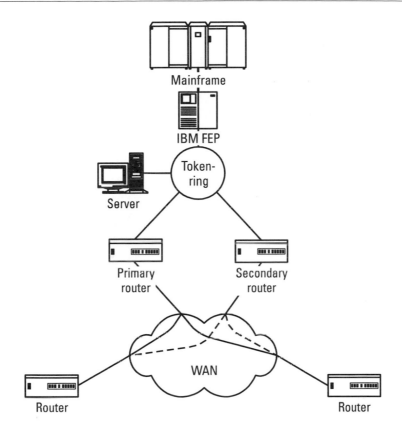

Figure 10.6 Data center redundancy.

When Does ATM at the Central Site Make Sense?

There are two reasons for considering ATM at the hub sites: (1) A collection of hub sites requires high-bandwidth connectivity among themselves, in addition to bandwidth required for connecting remote sites. (2) In a many-to-one network, aggregate bandwidth required at the central site may require several T1s. In this case, it may be beneficial to use DS-3 ATM for access consolidation [see Figures 10.8(a) and (b)].

If ATM is used to connect existing hub sites, it makes sense to use the same ATM ports to support remote frame relay sites as well [Figure 10.8(a)]. In the latter case [Figure 10.8(b)] where ATM is used purely for access consolidation, one needs to consider whether or not there is sufficient traffic to keep a 45-Mbps ATM facility sufficiently utilized. One must also balance the cost savings due to access consolidation with the cost of ATM interfaces on routers. Other factors are important such as the operations expense and complexity of

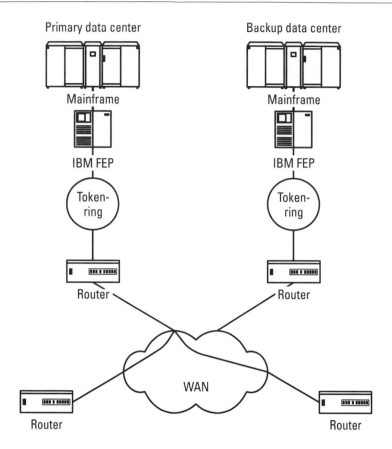

Figure 10.7 Disaster recovery.

supporting an additional technology. Hence, the cost/performance/functionality trade-offs must clearly favor an ATM solution at the data center.

10.4 Quality of Service Issues for SNA

Although the term "mission critical" is overused, SNA applications truly fall into that category. Many businesses depend on mainframe and AS/400-based transactions for some of their basic needs: reservations, order entry, and so on. Hence there is very little tolerance for network downtimes and poor response times.

Traditional SNA networks, whether they are multidrop, FEP-to-FEP or LAN-based, can be tuned to provide consistent response times for end users. When these networks migrate to router-based multiprotocol networks, SNA

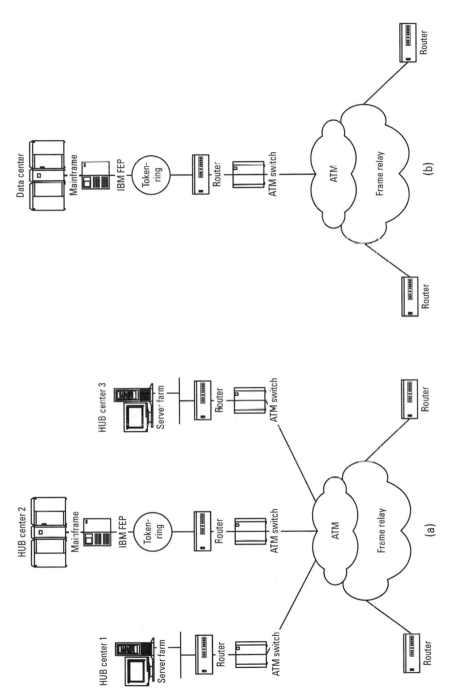

Figure 10.8 ATM at the data center (a) with frame relay at the remotes and (b) for access consolidation.

performance can be compromised. We discuss some of these issues below and provide some guidelines to help minimize the impact that a multiprotocol network may have on SNA applications.

10.4.1 Delay Trade-Offs in SNA Migration to IP

Consider a 56-Kbps multidrop SNA connection with five geographically separate 3174 controllers attached to a 3745 FEP[5] [see Figure 10.9(a)]. Consider now the connection in Figure 10.9(b) where DLSw is used over frame relay with 56-Kbps remote port speed and T1 port at the data center.

The following trade-offs affect SNA response times over the WAN. The factors that contribute to an increase in SNA response times include these:

- *Store-and-forward delays.* Since the router is a store-and-forward device, the insertion delay for an SDLC frame on the 56-Kbps link between the controller and the router in Figure 10.9(b) will be the same as that for the frame between the controller and the FEP in Figure 10.9(a). However, in the DLSw scenario, there is an additional 56-Kbps insertion delay. If the controller was token ring attached instead of SDLC attached, then the 56-Kbps insertion delay is eliminated. There is also an additional delay on the T1 port, but this is negligible.

- *Protocol overhead.* In traditional SNA networks, such as multidrop, the protocol overhead is relatively small. For example, a 100-byte SNA inquiry requires about 15 bytes of overhead.[6] When this SNA inquiry is transported using DLSw, the overhead is 73 bytes.[7] Thus the overhead is significantly higher.

Factors that contribute to a decrease in SNA response times include:

- *Reduced polling overhead.* Because the router polls the controller behind it, rather the remote FEP in the multidrop case, the polling delay will be significantly reduced. In addition, the router only polls one device

---

5. These arguments can be applied to multidrop AS/400 networks with 5 × 94 controllers.

6. 15 bytes = 3 bytes (Request/response header or RH) + 6 bytes (Format identifier type 2 or FID2) + 6 bytes (SDLC).

7. 3 bytes (RH) + 6 bytes (FID2) + 40 bytes (TCP/IP) + 16 bytes (DLSw) + 8 bytes (FR) = 173 bytes.

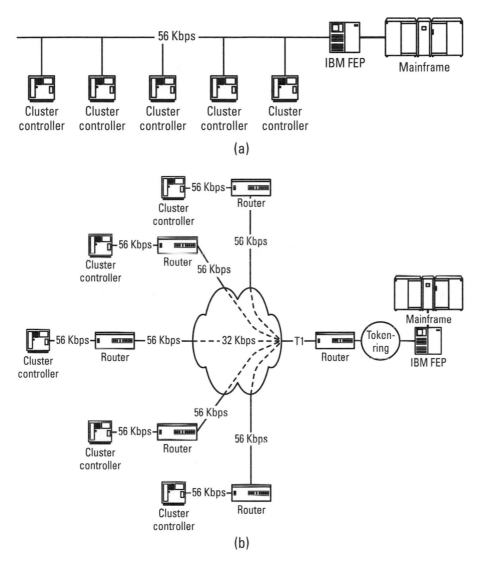

Figure 10.9 (a) A multidrop 56-Kbps SNA connection with five drops. (b) Migration of multi-drop to DLSw.

in Figure 10.9(b), as opposed to the FEP polling five controllers in Figure 10.9(a).

- *Reduced load.* Instead of five controllers contributing traffic to a 56-Kbps link in Figure 10.9(a), each controller has a dedicated 56-Kbps link.

In reality the nature of 3270 transaction traffic is that inquiries are small (say, less than 100 bytes) and responses from the host are large (say, 700 to 1000 bytes), and SDLC links are rarely utilized more than 40%. Hence, SNA response time usually improves when migrating from a multidrop with many controllers (say, more than 4). For a point-to-point connection between a single controller to the FEP, the migration to DLSw will always result in increased response times.

If an SNA gateway is used instead of the controller, then the same arguments hold if the SNA gateway is local to the remote locations. If the SNA gateway is centrally located, then the protocol overhead would be somewhat smaller because of the fact that DLSw is not needed.

10.4.2 FEP-to-FEP Issues

Many SNA networks use remote FEPs for concentration of low-speed multidrop links into one or more high-speed links (transmission groups, or TGs). With all end locations needing multiprotocol support, it is debatable whether these remote FEPs have a role to play in the new network paradigm. Indeed, many network managers have replaced these FEPs with routers and realized cost savings.

However, this replacement needs to be done with care. In some situations replacing FEPs is not an option. This is because there are many types of FEP-to-FEP links, used for different purposes. Some remote FEPs actually concentrate traffic from large regional offices to the data center. Others connect data centers on high-speed links. Still others connect different SNA networks via SNI (SNA network interconnection) links. SNI links are FEP-to-FEP links between disparate SNA networks.

In general, high-speed FEP-to-FEP links (say, 256 Kbps and higher) between data centers and major network hubs are better left alone—replacing FEPs by routers or private line by frame relay is not advisable! This is because FEP-to-FEP links support many features that are near and dear to SNA network managers: TGs, class of service priorities, and so on. In addition, one can utilize these links at very high levels and yet provide good response times for interactive traffic. A multiprotocol router/FRAD really adds very little value to a SNA traffic when used to tunnel FEP-to-FEP traffic. Indeed, class of service (COS) priorities and transmission group support will be lost, not to mention a definite degradation in response times—two store-and-forward delay boxes are introduced where there were none before.

For SNI links, replacing the FEP or in any way altering configurations on the FEP may be out of the question because these FEPs are typically owned by the clients of the customers (such as a credit reporting agency interfacing with banks).

10.4.3 Traffic Discrimination

Chapters 4 and 5 discussed the different options for traffic discrimination when supporting a mixture of time-sensitive and bulk data transfer applications on the same WAN connection. We briefly review these methods in the context of SNA applications.

Router Prioritization for SNA

Routers provide different ways to prioritize one class of traffic over others going into a WAN link. For example, TCP/IP can be prioritized over Novell IPX, or SNA can be prioritized over TCP/IP FTP traffic. These prioritization schemes have essentially two flavors: strict queuing and bandwidth allocation. With strict queuing, there are no bandwidth guarantees for the lower priority traffic. Bandwidth allocation schemes ensure that lower priority traffic gets a specified minimum share of the total bandwidth.

Router prioritization schemes are useful, and sometimes indispensable, tools to ensure that SNA will "stay ahead" of the non-SNA traffic. However, the need to prioritize traffic at routers decreases as the line speed increases. Typically, router priorities provide the maximum benefit at relatively lower link speeds (say, 56/64 to 128 Kbps).

In the context of frame relay, router prioritization of SNA is not always effective. This is because of the "congestion shift" phenomenon, where the speed mismatch between frame relay port speed and the CIR on a connection causes congestion to occur at the network switch buffers. This is discussed in detail in Chapter 5.

Giving SNA Its Own PVC

By assigning SNA its own logical path in a frame relay network, one can address the congestion shift issue. Whether the congestion occurs at the network ingress or egress, a separate PVC will enable preferential treatment for SNA in the network. Explicit prioritization of the SNA PVC may be needed at the network egress, if the port speed there is relatively slow speed (say, 128 Kbps or lower).

However, this approach cannot be recommended as a general enterprise-wide QoS solution for SNA applications (or, indeed, for any other class of applications). This is not a scalable solution. For large SNA networks with hundreds of remote locations connected to a few central sites, doubling the number of PVCs makes the design, engineering, and ongoing network management more complex. While it might resolve specific SNA performance issues, it does not address the QoS needs for other applications. It should be limited to small networks (say, less than 25 sites).

Router-Based Traffic Shaping for Frame Relay

The congestion shift is caused by the router being "unaware" of the PVC's CIR. It seems reasonable to build intelligence in the router about this speed mismatch. Traffic shaping does precisely this. Consider a router at an SNA data center with a T1 frame relay port speed serving a remote branch office with a 64-Kbps port speed and a 32-Kbps CIR. If the router can be configured so as to not exceed a transmission rate of 64 Kbps on that PVC, and buffer the excess traffic, the congestion shifts back to the router, making the router prioritization schemes more effective. In a very real sense, traffic shaping makes the frame relay connection look like a private line connection with a 64-Kbps line speed. The difference is that each individual frame is transmitted at T1 rates, but the overall transmission rate over a period of time cannot exceed 64 Kbps.

This is an interesting perspective for SNA networks, but one has to consider the scalability of this solution, especially as it relates to the burden traffic shaping places on the central site router. Conservative engineering rules would be required regarding the number of PVCs that can be traffic shaped simultaneously on a router. For this reason, it cannot be recommended as a general QoS solution for SNA networks over frame relay.

Bandwidth Management via External Devices

At the time of this writing, bandwidth management devices (such as Xedia, Packeteer, and others) appear to hold promise in being able to provide QoS for a broad set of applications in a network in a relatively more scalable manner than routers. They offload the prioritization and traffic shaping function from the routers. In the context of SNA applications, these devices can be used to allocate desired levels of bandwidth to SNA traffic, and minimum and maximum bandwidth to background traffic. They use queuing or TCP rate control to manage the bandwidth allocated to each traffic class. In an SNA network, they can be placed at the data center or hub sites, or even at the branch offices, as in the case of the devices that use TCP rate control.

Bandwidth management and traffic shaping are discussed in more detail in Chapter 5.

10.5 Summary

In this chapter, we discussed the various performance issues that arise in supporting SNA applications over a wide-area network. Traditionally, SNA has been used to access mainframe or AS/400 resident applications. While corporations around the world still depend on these applications for their businesses,

TCP/IP is increasingly the preferred transport method over WANs for access to mainframes and AS/400s. Some methods, such as DLSw, rely on encapsulation of SNA within TCP/IP. This does not eliminate SNA over the wide area—it merely hides SNA protocols. RFC 1490 is a similar transport method that retains SNA over the WAN but uses LLC2 encapsulation rather than TCP/IP. Centralized SNA gateways, TN3270, and Web access to mainframes are methods that use TCP/IP but also eliminate SNA from the WAN. We also discussed the challenges in designing SNA networks, especially as related to the data center design. Finally, we presented a qualitative overview of performance issues for SNA over WANs, including the trade-offs in migrating from multidrop networks to router networks and methods to guarantee consistent performance for SNA applications.

Reference

[1]   Guruge, A., *Reengineering IBM Networks,* New York: John Wiley & Sons, 1996.

Part III
Case Studies

11

Case Studies

11.1 Introduction

This chapter contains some case studies that illustrate how the principles and methods described in the rest of the book can be applied to addressing real-world networking issues. All case studies are related to actual situations. The authors were personally involved in every one of these case studies, most of which are performance related. Needless to say, not all wide-area networking problems lend themselves to the type of analysis described in the case studies in this chapter. Many problems have little to do with application or protocol characteristics and everything to do with LAN broadcast issues, improper router hardware and software configurations, routing instability, bugs in router code, faulty TCP/IP stacks, carrier problems (such as faulty repeaters, access line problems, improperly configured WAN switches, less than optimal routing within the "cloud"), and so on. The approach and methods illustrated here are most relevant in situations where performance issues persist in spite of the fact that underlying LAN/WAN components are healthy, that is, low to moderate utilization of WAN connections, reasonable WAN latency, no packet drops, and so on.

The case studies are arranged by protocol/application as follows:

- *TCP/IP case studies:* validating frame relay network latency and throughput via pings; FTP throughput; sizing bandwidth for an intranet application;

- *Client/server applications case studies:* troubleshooting WAN response time for a sales-aid application; troubleshooting WAN response time for a custom client/server application; troubleshooting WAN performance issues for an Oracle Financials application using Citrix Winframe;

- *Novell NetWare case studies:* how Novell SAPs can impact performance; impact of packet burst; does more bandwidth mean worse performance?;

- *SNA-related case studies:* frame relay network design for SNA migration; mixing SNA and TCP/IP on a frame relay network; and

- *Estimating WAN bandwidth impact of network management via SNMP.*

11.2 TCP/IP Case Studies

The case studies presented in this section deal primarily with TCP/IP-based application performance issues over WANs. Bandwidth sizing, response time, and throughput are common themes in these case studies. Recommendations for tuning TCP/IP parameters for optimal performance are also discussed. Although Section 11.3 also deals with TCP/IP-based applications, the emphasis there is more on the nature of client/server applications.

The reader is referred to Chapters 6 and 7 for detailed information on TCP/IP performance issues.

The first case study in this section is somewhat different; it illustrates how a simple ping calculation revealed serious routing problems in a carrier's frame relay network.

11.2.1 Validating Network Latency and Throughput

This case study illustrates how simple "back-of-the-envelope" calculations can uncover serious latency issues in the network. Faced with end user complaints about poor response times, the first call that a network manager makes is usually to the carrier's NOC (network operations center) to troubleshoot the circuit(s) in question. These simple calculations give the network manager a tool to verify claims made by the carriers about circuit delays in their networks.

11.2.1.1 Background and Context

A network health check was conducted for a large multinational corporation based in Pittsburgh, Pennsylvania. The network is a mixture of private lines and frame relay with Cisco routers. Several aspects of the WAN, such as peak and average resource utilization (links, ports/PVCs, router CPU, memory, buffers, and so on) using Concord Network and Router Health® tools, router configurations, bandwidth impact of new applications, latency and throughput characteristics, were studied.

A few key frame relay locations were isolated for the latency and throughput study. The approach, as discussed in Chapter 6, was to use small and large router-to-router pings across the WAN, and to compare the observed ping delays against calculated ping delays.

One connection chosen for study was between Pittsburgh and Sidney, Ohio, which are a few hundred miles apart (see Figure 11.1). Concord reports showed that the traffic between these locations is fairly light, and that the routers are healthy as well.

11.2.1.2 Issue

The observed delay for a 100-byte ping between the two WAN routers was observed to be 76 msec. For a large ping (18,024 bytes[1]) between the two routers, the observed delay was 728 msec. The issue is whether or not these delays reflect any unusual latency and throughput problems in the connection.

11.2.1.3 Analysis

Let us first compare the observed ping delay (76 msec) for a 100-byte ping with the expected delay for that connection. Please refer to Chapter 6 for details on calculation of ping delays.

Assume that Pittsburgh, Pennsylvania, and Sidney, Ohio, are separated by about 300 miles (airline). If we further assume a two-hop connection (three WAN switches in the cloud), and that switch latency is about 2 to 3 msec (usually is) we can estimate the expected ping delay for a 100-byte packet as follows:

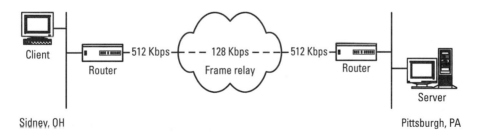

Figure 11.1 A test connection.

---

1. This is the largest ping size configurable on Cisco routers.

Frame size = 100 + 20 + 8 + 8 = 136 bytes

Network latency = Propagation delay + switch latency = 2×300 miles $\times$ 0.01 msec/mile + 2×3 switches $\times$ 3 msec = 24 msec

Expected ping delay = $2 \times 136 \times 8 \, / \, 512{,}000 + 0.024$ sec = 28 msec (approximately)

Observed ping delay = 76 msec!

Why the difference? Before we answer the question, let us calculate the expected ping delay for the large ping packet.

The default serial line MTU on a router is usually 1500 bytes. When an 18,024-byte ping is sent to the far-end router, the near-end router will fragment the ping into 1500-byte IP datagrams, with the first packet containing the ICMP header. Thus the ping will be fragmented as follows:

$$18{,}024 = (1500 + 20 + 8 + 8) + (1500 + 20 + 8) + \ldots + (1500 + 20 + 8) + (24 + 20 + 8) = 18{,}396 \text{ bytes}$$

with 13 datagrams in all. Hence the ping delay, using the network latency calculated above, is given by

$$(1536 \times 8 \, / \, 512{,}000 + 0.012 + 18{,}396 \times 8 \, / \, 512{,}000) \times 2 = 0.647 \text{ sec}$$
(best case, that is full bursting)

$$(18{,}396 \times 8 \, / \, 128{,}000 + 0.012 + 52 \times 8 \, / \, 512{,}000) \times 2 = 2325 \text{ msec}$$
(worst case, that is no bursting at all)

Compare this with the observed ping delay for the large ping of 728 msec!

Hence the large ping delays match much more closely than the small ping delays. Indeed, if we were to use the observed ping delay of 76 msec in the calculation for the large ping delay (to reflect the actual latency in the network), then the expected ping delay should be increased by 52 msec (76 − 24). Hence the revised expected large ping delay will be 647 + 52 = 699 msec, which is much closer to observed ping delay (728 msec).

The immediate conclusion, therefore, is that the PVC is able to burst to full port speed, while the latency in the network is too high. The reason that high latency does not affect the large ping delay is because of the bulk data transfer nature of the large ping (which is precisely the reason for choosing the largest ping allowable on the router).

In trying to analyze the reason for the long latency, the carrier's frame relay network operations center claimed that the network latency is 23 msec (compare with the 28 msec that we calculated with just basic information about the connection!). Further investigation revealed that the carrier measures network latency between the frame relay switches at the ends of the PVC connection.

Given that the connection is lightly loaded, the only explanation for the long latency in the network is excessive backhaul.

Frame relay carriers advertise frame relay points of presence, but may not have physical frame relay access switches at all points of presence. Thus the carriers sometimes need to backhaul circuits to the nearest frame relay switch. Backhaul can also occur when the nearest switch is congested.

As it turned out, the frame relay carrier performed a rather long backhaul of access circuits at both Pittsburgh and Sidney. Pittsburgh was back-hauled approximately 600 miles and Sidney was back-hauled approximately 1400 miles. Hence the total unaccounted miles in the preceding calculation is 2000 miles. Thus the unaccounted delay is 2 × 2000 × 10 msec/1000 miles = 40 msec, a significant amount!

Note that the unaccounted delay resolves the discrepancy in the latency almost perfectly: 28 msec + 40 msec = 78 msec (versus 76 msec observed)!

The complex issue is that a 76-msec round-trip delay between two frame relay points of presence is within what the carrier would call "delay objectives." However, "objective" delays may not be the same as "reasonable" delays as this example shows.

11.2.2 TCP Bulk Data Transfer

A manufacturing company based in Dallas, Texas, has a requirement to periodically send a 640-Mbyte file to clients in the United Kingdom. The file needs to reach the clients within a 3-hour window. A frame relay connection exists between Dallas and the United Kingdom. The network manager wants to know how much bandwidth over the WAN should be allocated for this application.

There are several ways to answer this question. Perhaps the easiest way (and perhaps the most common approach) is

Bandwidth required = File size / Required transfer time = (640 × 1,024,000 bytes × 8 bits/byte) / (3 × 3600 sec) = 485 Kbps

This is a perfectly reasonable approach. However, there are three issues with the formula:

- It ignores protocol overhead.

- The TCP segment size for this application should be larger than the default 512 bytes in order to maximize throughput. In particular, this means that the MSS option should be invoked when the TCP session is set up. This issue is also related to protocol overhead.

- Perhaps most important of all, the formula assumes that window size is large enough so that the bottleneck is the WAN connection and not the window size. One needs to estimate the optimal window size and ensure that the window size advertised by the client is at least as big as the optimal window size.

Let us study these issues a little more carefully. The protocol overhead is a function of the segment size. The recommended TCP segment size for this transaction is 1460 bytes.[2] The protocol overhead is therefore (1460 + 48) / 460 = 1.03 or 3%. The file size needs to be multiplied by this factor in the preceding formula.

The more critical factor is the window size advertised by the client in the United Kingdom. If the window size is less than optimal for the connection, then the throughput (transfer time) will actually be lower (longer). The optimal window size can be estimated as follows. Assuming that a 512-Kbps WAN connection (CIR for frame relay or fractional T1 connection for private line) is used, the optimal window size is given

$$W^* = \text{Bandwidth} \times \text{Round-trip delay}$$

A simple estimate for the round-trip delay in this case is the latency between Dallas and the United Kingdom. Assuming a frame relay connection between the two locations, the latency is likely is to be about 300-msec round-trip.[3] Hence the estimated optimal window size is

---

2. Assume an Ethernet attached client and server.

3. For frame relay networks, the latency estimate using the 10 msec/1000 mile rule will be optimistic. For instance, the airline distance between Dallas, Texas, and the United Kingdom would be about 6000 miles. Hence the round-trip latency will be 120 msec. This will definitely be an underestimate because it ignores the fact that potentially three different frame relay networks, strung together seamlessly via NNIs would be needed for this connection: a domestic U.S. network, an international backbone network, and a local network in the host country. See discussion on international frame relay circuits in Chapter 5.

$$W^* = 64,000 \text{ bytes/sec} \times 0.3 \text{ sec} = 19 \text{ Kbytes}$$

If one considers the fact that most TCP implementations (including Windows NT) use default window sizes of 4 or 8 Kbytes, it is immediately clear that the throughput will be smaller.

Hence the final recommendation to the network manager ought to be

$$\text{Bandwidth required} = 485 \text{ Kbps} \times 1.03 = 499 \text{ Kbps or } 512 \text{ Kbps}$$

$$\text{TCP Maximum segment size} = 1460 \text{ bytes}$$

and the window size advertised by client should be greater than 19,200 bytes.

Other "what if" questions can be easily answered using these estimates. For instance, what if the window of transfer time is relaxed to 5 h instead of 3 h? How much less bandwidth is needed?

Note the linear relationships in the formula. If the transfer time requirement is 5 h, then the bandwidth required would be 60% less (512 Kbps × 3 / 5), or 307 Kbps, which, when rounded to the next highest multiple of 64 Kbps, is 320 Kbps. Correspondingly, the window requirement can be lowered to 11,520 bytes, or about 12 Kbytes.

11.2.3 Sizing Bandwidth for a TCP/IP Application

A financial services company would like to provide a web-based intranet service for their users. The web servers are centrally located with a few remote locations. Their current network is frame relay with predominantly 56-Kbps remote ports and 32-Kbps CIR. The network is fairly lightly loaded. The network manager is looking for some guidance on the number of users that can be supported on the existing network for this application.

Detailed usage statistics on the application is available primarily because of business requirements and educated guesses:

- Average bytes from client to host: 490 bytes; host to client: 98,000 bytes;
- Peak hour number of simultaneous users = 1/3 of total user population;
- Peak users send requests every 2 min;
- Among nonpeak users, 50% access the application four times a day, 25% two times a day, and 25% once a day.

11.2.3.1 Methodology

We need to estimate the number N_U of end users that can be supported at a remote location. The number of active users at the peak hour is 0.33 N_U. Assuming all requests occur in a peak hour, the total number of requests during the peak hour is

$$0.33 \times N_U \times 30 + 0.67 \times N_U \times [0.5 \times 4 + 0.25 \times 2 + 0.25 \times 1] = 12.65 \times N_U$$

(Note that 1 request every 2 min implies 30 requests in an hour for peak users.)

The bandwidth issue is clearly more important in the server-to-client direction than in the reverse direction. Therefore we will only size the application in the server-to-client direction. We also need to make some critical assumptions regarding the nature of the 98,000-byte data transfer. This data can be sent as a single bulk data transfer or could be sent in multiple transactions (as in a client/server application). Clearly, a sniffer trace is needed to characterize the transaction. In the absence of a sniffer trace, and given that a "quick-and-dirty" method is needed, let us assume that the entire response from the server occurs in a single bulk.[4]

Assume that the TCP segment size is 1460 bytes. The 98,000-byte bulk data transfer will consist of about 67 segments of size 1506 bytes. Hence the outbound data in a peak hour is

$$12.65 \times N_U \times 67 \times 1506 \text{ bytes} \times 8 \text{ bits/byte} = N_U \times 12{,}211{,}282 \text{ bits} = N_U \times$$
$$2.8 \text{ Kbps}$$

How many users can be supported on a 56-Kbps connection? If engineering rules call for loading the WAN link at no more than 70% utilization, then the number of users that can be supported is

$$N_U \times 2.8 \text{ Kbps} / 56 \text{ Kbps} = 0.7, \text{ or } N_U = 14 \text{ users}$$

11.2.3.2 Further Analysis

Very often, usage patterns are not available in as much detail as provided in the previous case study. The only available information is likely to be the number

---

4. Another important difference is whether or not the end-to-end protocol is HTTP, as is likely the case for Web-based applications. Although HTTP uses TCP for transport, there may be multiple parallel TCP sessions involved in the data transfer. As mentioned, only a protocol trace will enable one to perform a detailed analysis.

of users at a remote location, the frequency with which users request information from the server, and the expected response time. This information, together with some assumptions, can be used to estimate bandwidth requirements and response times. This approach is discussed next.

Let us assume that there are N_U users at a remote location, and that a typical user will access the information from the server once every 10 min during a peak hour. The expected response time is 5 sec. Again, assume that the data transfer occurs in a single chunk of 98,000 bytes. These assumptions can be verified by a sniffer trace.

An initial estimate for bandwidth can be obtained from the size of the data transfer and the expected response time. Assuming that the client window size is large enough, the approximate bandwidth required is

$$\text{Data transfer size / Response time} = [98,000 \text{ bytes} \times 8 \times 1.1(\text{overhead}) / (5 \text{ sec})] = 173 \text{ Kbps (approximately)}$$

Rounding off to the nearest multiple of 64 Kbps, the estimated bandwidth will be 192 Kbps.

With multiple users accessing the server, the 5-sec response time cannot be guaranteed. For instance, if two users were simultaneously downloading the 98,000-byte chunk of data, then each user will get approximately half the available throughput, or experience twice the response time, that is, 10 sec. But what are the chances of two or more users simultaneously accessing the system? If this is sufficiently small, say, less than 5%, then it may be of less concern. Given the number of users N_U, and the frequency with which they access the server, we can estimate the probability of multiple users.

From a single-user point of view, for this application, there are ON and OFF periods. The ON period is when the user is being sent the 98,000-byte data. The sum of the ON and OFF periods is 10 min. The fraction of the ON period is equal to

$$[98,000 \text{ bytes} \times 1.1/24,000] / 600 \text{ sec} = 4.5 / 600 = 0.0075$$

To estimate the probability of simultaneous users, one can use the following approach: Let Z be the random variable denoting the number of simultaneous users. The maximum value of Z is N_U and the minimum is 0. Z has a binomial distribution with parameters N_U and $p = 0.0075$.[5] It is known that Z

---

5. Think of Z as the number of heads in a coin tossing experiment where the coin has 0.007 as the probability of turning up a head, and there are N tosses.

can be approximated, for large values of N_U, by a normal distribution[6] (call it variable Z^*) with mean $N_U \times p$ and variance $N_U \times p \times (1 - p)$. Specifically the variable $(Z -$ mean)/standard deviation has a standard normal distribution (mean 0, variance 1).

The objective of the exercise is to choose that value of N_U so that $P[Z > 1]$ = 0.05. If the total number of users is less than N_U, then the chance of more than one simultaneous user (and thereby causing unacceptable response times) will be less than 5%. Now

$$P[Z > 1] = P[Z^* > (1 - N_U 0.0075) / \text{SQRT}(N_U \times 0.0075 \times 0.9925] = 0.05$$

From standard normal distribution tables, it is seen that

$$(1 - N_U 0.007) / \text{SQRT}(N_U \times 0.0075 \times 0.9925) = 1.645$$

Solving this quadratic equation, it is seen that N_U is approximately 30 users. With this many users, the chance of more than two simultaneous users is

$$P[Z > 2] = P[Z^* > \{(2 - 30 \times 0.0075) / \text{SQRT}(30 \times 0.0075 \times 0.9925)\}] =$$
$$P[Z > 3.756] < 0.001$$

Hence the conclusion is that, for this application, up to 30 clients can be supported for this application on a 192-Kbps connection, and that 95% of the time the response time will be within 5 sec, and between 5 and 10 sec about 5%. The chance of the response time exceeding 10 sec is less than one-tenth of 1%.[7]

11.3   Client/Server Application Case Studies

In this section we present three case studies involving two-tier fat client applications. The first two applications involve troubleshooting response time problems over a frame relay network. The third case study discusses how bandwidth

---

6. The approximation is valid for large values of N and small values of p.
7. Note that this analysis relates only to this application. Additional bandwidth considerations should be taken into account for other traffic. The value of this analysis lies in describing how the methodology can be applied to make quantitative judgments.

can be estimated for an application that contains dissimilar transactions and usage patterns.

The reader is referred to Chapter 9 for details on client/server performance issues. A case study of sorts for estimating bandwidth and response time for an SAP R3 application was presented in that chapter.

11.3.1 Troubleshooting Response Times for a Sales-Aid Application Over a Global Frame Relay Network

Case studies usually have "happy endings," especially in the context of troubleshooting, where a fundamental issue is uncovered. The case study presented here did not have a "happy ending." It involved troubleshooting response times for a sales-aid application over a global frame relay network. The network managers involved were adamant that the problem was caused by frame relay, and that private line was really the answer to resolving WAN response time issues, all evidence to the contrary notwithstanding.

11.3.1.1 Context

The company in question is based in the Silicon Valley and manufactures products for communication networks. Their frame relay network spans the globe with headquarters in San Jose, California, and locations in Europe, Asia, and Australia. The company had deployed the sales-aid application at all of their major locations around the globe. This application was crucial to the company's business. End users in the Europe, Asia/Pacific, and Australian locations were complaining about poor response times—screen update times of up to 2 min. End users use the application to load customer cases, check customer order status, and so on. The frame relay network was recently installed and it was felt that this application was unsuitable for frame relay, and that it would perform much better on leased lines.

11.3.1.2 Analysis

The first step was to isolate a few locations and collect sniffer traces of the application, isolating a single end user and the server. Given the logistics, it was decided that a sniffer could not be placed at the remote locations, and that a single sniffer at the San Jose location would be used. Figure 11.2 shows a reference connection between San Jose and The Netherlands.

The first step was to capture the login transaction shown in Figure 11.3. From the sniffer point of view, "Frame" refers to the frame numbers seen on the LAN, and "Delta" is the time differential between successive instances of the frames. The IP addresses are suppressed and replaced by "Server" and "Client."

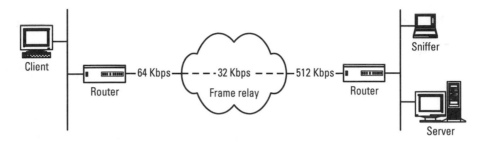

Figure 11.2 A reference connection.

Frame	Delta	Dest	Source		Summary
12	0.0405	Server	Client	TCP	D=2001 S=1026 SYN SEQ=68497 LEN=0 WIN=0
13	0.0041	Client	Server	TCP	D=1026 S=2001 SYN ACK=68498 SEQ=186752000 LEN=0 WIN=8192
14	0.3140	Server	Client	TCP	D=2001 S=1026 ACK=186752001 WIN=2880
15	0.0552	Server	Client	TCP	D=2001 S=1026 ACK=186752001 SEQ=68498 LEN=512 WIN=2880
16	0.0020	Server	Client	TCP	D=2001 S=1026 ACK=186752001 SEQ=69010 LEN=32 WIN=2880
17	0.0278	Client	Server	TCP	D=1026 S=2001 ACK=69042 SEQ=186752001 LEN=198 WIN=8192
18	0.3271	Server	Client	TCP	D=2001 S=1026 ACK=186752199 WIN=2880
....					
588	0.0171	Server	Client	TCP	D=2001 S=1026 ACK=186874115 WIN=2880

Figure 11.3 Sniffer trace of a login transaction.

The following conclusions can be immediately drawn:

- There are approximately 600 packet exchanges in a more or less ping-pong fashion to log on to the application.

- A rough estimate of latency in the network is 300 msec (look at delta times for frames 14 and 18).

- To estimate round-trip latency from frame 14, look at frames 13 and 14. Frame 13 sends a SYN packet and frame 14 is the acknowledgment. This is like a ping of 48 bytes from the server to the client. To calculate network latency, we need to subtract the insertion delay, which is equal to $2 \times (48 \times 8 / 64{,}000 + 48 \times 8 / 512{,}000) = 13$ msec. Hence the network latency estimate is $314 - 13$ msec, or about 300 msec.

- Of course, this includes the client processing delay (not the server delay—because the sniffer is next to the server), but we can assume that this is minimal. The 300-msec latency for a United States to

Europe frame relay connection is not unusual. In a similar way, we can calculate the round-trip latency from frame 18. Frame 18 is an acknowledgment from the client for a 198-byte (data) packet from the server. We need to add 48 bytes overhead. As in the previous calculation, we need to subtract the insertion delay, which is $(246 \times 8 / 512,000 + 246 \times 8 / 64,000 + 48 \times 8 / 64,000 + 48 \times 8 / 512,000)$ sec = 41 msec.

- Hence, the estimated latency is 327 − 41 msec = 286 msec.
- The time to log on to the application will be at least 300 exchanges × 300-msec round-trip delay = 90 sec or 1.5 min.

Once the user has logged on to the application, he/she is ready to "load a small customer case." The protocol trace captured for this application task is shown in Figure 11.4.

Note the following immediate aspects of this application task:

- Total number of packet exchanges for the task is about 350.
- Total transaction time is about 76 sec (look at frame 349).

Frame	DeltaT	Rel T	Dest	Source		Summary
1	0.0000		Server	Client	TCP	D=2001 S=1026 ACK=186875550 SEQ=76457 LEN=235 WIN=2880
2	0.0891	0.0891	Client	Server	TCP	D=1026 S=2001 ACK=76692 SEQ=186875550 LEN=479 WIN=8192
3	0.3758	0.4649	Server	Client	TCP	D=2001 S=1026 ACK=186876029 WIN=2880
4	0.0413	0.5062	Server	Client	TCP	D=2001 S=1026 ACK=186876029 SEQ=76692 LEN=240 WIN=2880
5	0.0406	0.5468	Client	Server	TCP	D=1026 S=2001 ACK=76932 SEQ=186876029 LEN=120 WIN=8192
....						
122	0.3793	10.1881	Server	Client	TCP	D=2001 S=1026 ACK=186896007 WIN=2880
123	**27.85**	**38.03**	Server	Client	TCP	D=2001 S=1026 ACK=186896007 SEQ=83867 LEN=125 WIN=2880
124	0.2886	38.3266	Client	Server	TCP	D=1026 S=2001 ACK=83992 WIN=8192
. . .						
208	0.0275	45.4920	Server	Client	TCP	D=2001 S=1026 ACK=186911917 WIN=2880
209	**16.03**	**61.52**	Server	Client	TCP	D=2001 S=1026 ACK=186911917 SEQ=87293 LEN=143 WIN=2880
. . .						
349	0.3198	**75.86**	Client	Server	TCP	D=2001 S=1026 ACK=186933203 WIN=2880

Figure 11.4 Sniffer trace of user application.

- Although the entire trace is not shown, the transaction consists, for the most part, of ping-pong exchanges.

- The network latency can be calculated from frames 2 and 3; insertion delay is $(527 \times 8 / 512{,}000 + 527 \times 8 / 64{,}000 + 48 \times 8 / 64{,}000 + 48 \times 8 / 512{,}000) = 81$ msec, and $376 - 81 = 295$ msec.

- Note the spikes in the delta T numbers in frames 123 and 209. This holds the answer to the response time problem. The spike in delays shown in these frames appears to be "client build delays"—the client acknowledges the previous segments from the server (frames 122 and 208), then waits for 28 and 16 sec, respectively, before submitting the next request. One could construe these spikes as user think times. However, there is no end user interaction with the application. (If this was indeed the case, then there is an error in the characterization of the application.)

- The two spikes (44 sec) contribute more than half of the application response times.

- There is no evidence of frame drops in the network (which would be indicated by TCP retransmissions).

- The spikes may not be application related, but caused by the particular TCP/IP stack used by the clients.

Further investigation showed that these delay spikes occurred for other application tasks and at every other location in Europe and Asia/Pacific. In all cases, the delay spikes were at the client side; the client would send back a TCP acknowledge for the previous segment from the server, and essentially go to "sleep" for several seconds.

While the analysis is not conclusive, it clearly shows that there is no evidence of any sort that frame relay is unsuitable for this application, and that private lines would yield significantly better performance. The network manager must first understand the reason for the delay spikes. As mentioned earlier, there may be several reasons for the delay spikes: a local LAN issue, TCP/IP stack problems, client build delays, and so on.

However, it is abundantly clear that this application is not "WAN-friendly," especially over global connections. For the application task named "loading small customer case," the response time can be as large as 175×0.3 sec $= 52$ sec (assuming a complete one-for-one packet exchange with network latency of 300 msec). In reality, the packet exchanges are likely to be less than completely one for one. However, packet size, line speed, and background load will affect the response times.

11.3.2 Troubleshooting Response Times for Custom Client/Server Application Over a Frame Relay Network

This case study is in the context of a company in the entertainment industry. A custom application was developed to provide some essential functions such as contractual issues, artist profiles, and production schedules. Remote access to these functions was a prime consideration when the application was being developed. Unfortunately, many months of effort were put in to the development of the application with absolutely no consideration for how the application would perform over the WAN, until a few months before the application was to be deployed in production mode.

The company had a corporate frame relay network over which end users in New York City (NYC) accessed the servers in Los Angeles (LA). End users were complaining about long response times making the application almost unusable; the acceptable response time is about 2 to 3 sec versus average response times of 20 to 30 sec experienced by users. The application had many subtasks, of which a few are mentioned here. A sniffer trace shown in Figure 11.5 (with a WAN sniffer at the New York City location) revealed the following characteristics of the subtasks.

Frame	Delta	Dest	Source		Summary
1	0	Server	Client	TCP	D=1521 S=1213 SYN SEQ=6901576 LEN=0 WIN=8192
2	0.001	Client	Server	TCP	D=1213 S=1521 SYN ACK=6901577 SEQ=507022000 LEN=0 WIN=49152
3	0	Server	Client	TCP	D=1521 S=1213 ACK=507022001 WIN=8760
4	0.009	Server	Client	TCP	D=1521 S=1213 ACK=507022001 SEQ=6901577 LEN=50 WIN=8760
5	0.18	Client	Server	TCP	D=1213 S=1521 ACK=6901627 WIN=49152
6	0.001	Server	Client	TCP	D=1521 S=1213 ACK=507022001 SEQ=6901627 LEN=260 WIN=8760
7	0.12	Client	Server	TCP	D=1213 S=1521 ACK=6901887 SEQ=507022001 LEN=8 WIN=49152
8	0.001	Server	Client	TCP	D=1521 S=1213 ACK=507022009 SEQ=6901887 LEN=50 WIN=8752
9	0.28	Client	Server	TCP	D=1213 S=1521 ACK=6901937 WIN=49152
10	0.001	Server	Client	TCP	D=1521 S=1213 ACK=507022009 SEQ=6901937 LEN=260 WIN=8752
...					
83	0.08	Client	Server	TCP	D=1213 S=1521 ACK=6905207 SEQ=507047675 LEN=85 WIN=49152
84	0.14	Server	Client	TCP	D=1521 S=1213 ACK=507047760 WIN=8675
85	0.417	Server	Client	TCP	D=1521 S=1213 ACK=507047760 SEQ=6905207 LEN=18 WIN=8675
86	0.079	Client	Server	TCP	D=1213 S=1521 ACK=6905225 SEQ=507047760 LEN=14 WIN=49152
....					
115	0.085	Client	Server	TCP	D=1213 S=1521 ACK=6906550 SEQ=507048982 LEN=63 WIN=49152
116	0.001	Server	Client	TCP	D=1521 S=1213 ACK=507049045 SEQ=6906550 LEN=15 WIN=7390
117	0.079	Client	Server	TCP	D=1213 S=1521 ACK=6906565 SEQ=507049045 LEN=11 WIN=49152
118	0.179	Server	Client	TCP	D=1521 S=1213 ACK=507049056 WIN=7379

Figure 11.5 Sniffer trace of the Load Application.

11.3.2.1 Load Application

Note the perfectly ping-pong nature of the application task. There appears to be spikes of about 420 msec and 180 msec at frames 85 and 118. These can be attributed to client delay (recall that the sniffer is physically colocated with the client). Note also that the round-trip latency in the WAN is about 80 msec (consistent with a frame relay connection between NYC and LA). Thus overall performance of this application can be estimated roughly as

$$60 \text{ round-trips} \times 80 \text{ msec} + 420 \text{ msec} + 180 \text{ msec} = 5.4 \text{ sec}$$

In reality, the response time will be more because of packet insertion delays and background load. In addition, one needs to use more samples, spread through the business day, to correctly estimate the client delay. However, this is a good start. Even if client delays are eliminated, the time to load the application remotely will be at least 4.8 sec.

All other application tasks show the same behavior. For instance, look at the trace of another function shown in Figure 11.6. Notice the perfect ping-pong characteristic of the application. If we assume 400 packets make up the transaction, then the minimum response time for the application task will be $200 \times 0.08 = 16$ sec, again not including background load and packet insertion delays. (The spike of 23 sec in the first frame, 1084, is really the time that this user took between the application tasks.) Thus a response time of a half-minute or more for this important task, during periods of peak load, is to be expected.

Frame	Delta	Dest	Source		Summary
1084	23.161	Client	Server	TCP	D=1521 S=1213 ACK=507118428 SEQ=6998993 LEN=15 WIN=7748
1085	0.081	Server	Client	TCP	D=1213 S=1521 ACK=6999008 SEQ=507118428 LEN=11 WIN=49152
1086	0.006	Client	Server	TCP	D=1521 S=1213 ACK=507118439 SEQ=6999008 LEN=18 WIN=7737
1087	0.08	Server	Client	TCP	D=1213 S=1521 ACK=6999026 SEQ=507118439 LEN=14 WIN=49152
1088	0.001	Client	Server	TCP	D=1521 S=1213 ACK=507118453 SEQ=6999026 LEN=132 WIN=7723
1089	0.079	Server	Client	TCP	D=1213 S=1521 ACK=6999158 SEQ=507118453 LEN=35 WIN=49152
. . .					
1458	0.002	Client	Server	TCP	D=1521 S=1213 ACK=507155995 SEQ=7051194 LEN=151 WIN=8489
1459	0.083	Server	Client	TCP	D=1213 S=1521 ACK=7051345 SEQ=507155995 LEN=42 WIN=49152
1460	0.001	Client	Server	TCP	D=1521 S=1213 ACK=507156037 SEQ=7051345 LEN=17 WIN=8447
1461	0.084	Server	Client	TCP	D=1213 S=1521 ACK=7051362 SEQ=507156037 LEN=63 WIN=49152
1462	0.137	Client	Server	TCP	D=1521 S=1213 ACK=507156100 WIN=8384

Figure 11.6 Sniffer trace of another client/server application.

11.3.2.2 Resolution

What is the resolution to the problem of poor response times? Clearly, it is fruitless to change any network parameters—increase bandwidth, prioritize the application, and so on. It may be possible to make some changes to the application itself, but time and resource constraints may be significant enough to make this option not viable. The best option, it appears, is to support this application via Citrix Winframe/Metaframe or a similar thin client solution. Citrix Winframe/Metaframe is typically used to minimize end user support requirements (as discussed in Chapter 9), but can also be used solely for the purpose of improving WAN performance of client/server applications.

It is clear that if the performance of the application over the WAN was a priority item during application development, many of the subsequent problems just prior to full deployment could have been overcome. Perhaps an application rewrite would have been feasible, or even the possibility of using a different application platform.

11.3.3 Troubleshooting Performance Problems for an Oracle Financials Application Using Citrix Winframe Over a Global Frame Relay Network

In Chapter 9 we discussed the performance advantages and disadvantages in using the remote presentation thin client approach to client/server applications. One of the important issues in this regard is the latency factor and how it affects the screen response time of end users. Since remote presentation uses echoplex (that is, keystrokes are echoed from the central server), the end user's perception of application performance is closely tied to network latency and to background traffic load. Therefore, networks with inherently long latencies, such as global frame relay networks, pose some serious issues for data-entry applications using the remote presentation approach. The case study discussed next illustrates this point.

The context of the case study is a global frame relay network with several locations in Europe and other countries and a central hub site in the United States (see Figure 11.7). As seen in the figure, there are three hub locations in Europe. End users in Scandinavian countries access the server farm in the United States through multiple hops through frame relay and leased lines. In particular, a section of the user population use an Oracle Financials application (two-tier) using a Citrix Winframe thin client approach. The Citrix Winframe approach was adopted, among other things, to make transaction response time for native-mode Oracle Financials application over the global connection perform better.

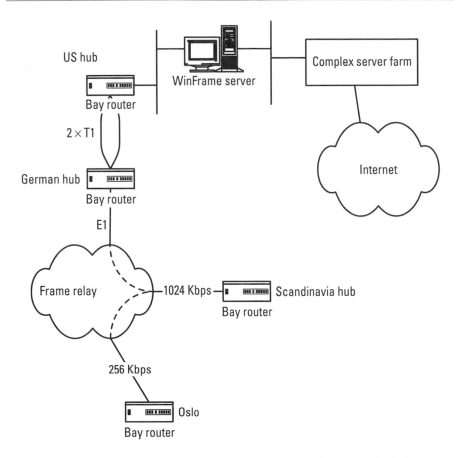

Figure 11.7 Reference connection connectivity between Scandinavia and the U.S. hub.

As the company quickly discovered, the Citrix Winframe approach solved some issues with respect to poor response times for Oracle applications, but raised a few others as well. End users in Scandinavia, using the Oracle application for data-entry purposes (think of time and expense entry), were quite unsatisfied with the response times and there was a significant impact on productivity. Sometimes the performance became so bad as to require session reinitiation.

The challenge was to uncover the source of the problem and recommend ways to resolve the performance issues for the Oracle Financials application.

11.3.3.1 Analysis

An in-depth analysis of the performance problem revealed the following:

- The performance problem persisted in spite of routers at the United States and European hub prioritizing Winframe traffic (via the appropriate TCP port number).

- Round-trip delays from the U.S. hub to Scandinavia were consistently higher than 300 msec. Recall that echoplex applications require response times lower than 300 msec for satisfactory end user perception of performance.

- Sniffer protocol traces of packets between the European hub and Scandinavian hub demonstrated that the congestion was due to Internet and intranet web traffic (emanating from the United States), Lotus Notes, and Novell SAP broadcasts. These traces also revealed that the back-end Oracle database server, Winframe server, and the remote clients were not adding significant delay.

- As further confirmation of the preceding point, it was noticed that a significant number of frames on the PVC between the European hub and the Scandinavian hub were marked with the FECN/BECN bit.

- In particular, the European hub router received frames with BECN and the Scandinavian hub router received frames with FECN. The frame relay switch terminating the connection from the European hub router showed a significant amount of data in its PVC buffer. This clearly shows that there is excessive traffic coming from the United States destined toward the Scandinavian hub.

Based on this analysis, one can immediately come to the conclusion that the performance problem is caused by two factors: (1) excessive bursty traffic in the direction from the United States to Scandinavia and (2) Winframe's echoplex nature and the need to provide a consistently low round-trip delay.

11.3.3.2 Recommendations

There are several approaches to addressing this performance issue, from a "band-aid" solution to a longer term strategy that addresses quality of service for applications in general. Some of these recommendations are as follows:

Provide Local Internet Access

Because Internet traffic appears to be contributing to the performance issue, it makes sense to provide Internet access at the European and Scandinavian hub sites instead of back-hauling all the European users to the firewall in the United States. This will relieve the congestion caused by Internet traffic on the PVC between the European hub and the Scandinavian hub, and thereby improve Winframe response times.

Separate PVC for Winframe Traffic

Another option, which can be used in conjunction with the recommendation above, is to separate the Winframe traffic between the two European hubs on a separate PVC. Since the congestion is in the direction of the European hub to the Scandinavian hub, occurring specifically on the PVC buffers at the ingress frame relay switch, a separate PVC for Winframe will insure that Winframe traffic does not get "stuck behind" Lotus Notes, Internet, and Novell SAP broadcasts in the same buffer (a separate PVC buffer would be required). This PVC should also be prioritized in the network to address potential congestion issues at the frame relay port on the egress switch.

Rearchitect the Network for Winframe Traffic

The current network architecture, consisting of hybrid private lines and frame relay with multiple hubbing locations, may be cost effective, but the performance penalty is clear. For instance, look at a packet traveling from Oslo to the U.S. hub. On a round-trip, this packet will go through 10 insertions on serial lines, four network latencies over the frame relay network in Europe, and potentially four delays through two PVCs. This is certainly not optimal for latency-sensitive traffic, especially in the context of a global connection. While Internet and Lotus Notes may tolerate this latency, Winframe cannot.

If Winframe is the only latency-sensitive traffic between Scandinavia and the U.S. hub, it makes sense to provide separate PVCs between individual locations in Scandinavia directly to the U.S. hub on a separate frame relay port. This will eliminate the latency due to multiple hops and the impact of background traffic. Of course, this comes at an extra cost. As for sizing these PVCs, one can use an estimate of 7 to 10 Kbps per user for Winframe traffic (this is usually the case for applications like Oracle Financials; graphics-intensive applications need significantly more bandwidth with Winframe).

Traffic Shaping

A longer term strategy, which will address the Winframe performance issues as well lay the groundwork for an overall QoS framework in the network, is to deploy bandwidth management/traffic shaping devices at selected locations. These devices are typically placed behind WAN routers and manage the user-defined allocation of bandwidth to individual traffic streams using a combination of techniques such as class-based queuing, weighted fair queuing, TCP rate control, and rate throttling. Bandwidth can be allocated at a very granular level, such as traffic to and from an individual URL, individual users, specific servers, and so on.

In its simplest form, the router at the European hub can be configured to shape traffic on the PVC to the Scandinavian hub to a specific rate, say, the CIR. This will address the Winframe performance issue because all the queuing due to background traffic will take place at the router buffers making the router prioritization schemes more effective. Of course, this will also mean that the PVCs cannot burst.

A more complex but effective implementation would be to place bandwidth management devices at strategic locations in the network (essentially locations that are congestion points) and allocate bandwidth granularly to specific traffic classes. At the time of this writing, products from vendors such as Xedia, CheckPoint, and Packeteer appear to be capable of performing this task.

11.4 Novell Networking Case Studies

The following case studies deal with two main issues regarding Novell NetWare performance over WANs: SAP broadcasts and packet burst. The authors have not investigated the performance issues with respect to NetWare 5 where TCP/IP is the dominant protocol. However, preliminary indications are that these issues will not arise.

The reader is referred to Chapter 8 for details on Novell performance issues.

11.4.1 How Novell SAPs Can Impact Performance

This case study deals with how SAP broadcasts and improper router configurations can lead to disastrous results. More importantly, it demonstrates how numerical methods discussed in this book can be applied effectively to troubleshoot and explain WAN problems.

The company in question had an SNA multidrop network and a Novell/IP network in parallel on leased lines. The Novell/IP network was based on Bay routers and limited to a handful of locations in the network. The rest of the locations (about 50 in all) only had SNA presence. With the coming of public frame relay services and router support for SNA using DLSw, the company decided to consolidate its SNA and LAN networks into a frame relay network. Clearly, the SNA applications were the most important, with a 2- to 3-sec end user requirement being an absolute requirement.

When the network was migrated to frame relay, SNA response times were unacceptably large—spikes of 10- to 15-sec delays for SNA transactions were not uncommon. There was FECN and BECN indications in the network, as seen by the routers.

The company's worst fears about SNA over frame relay were realized.

11.4.1.1 Resolution

The frame relay network is as shown in Figure 11.8. As mentioned before, most of the locations are SNA only, with a few controllers attached to the router via SDLC; DLSw was used to support SNA over the WAN. Clearly, SNA response times will increase in the migration from multidrop to frame relay, due to increased overhead, store-and-forward delays, and more latency through frame relay, in spite of local polling. However, these factors cannot explain the fact that the response times have increased more than fivefold, from 2 to 3 sec to 15 sec or more. The initial conclusion was that there must be something unusual happening in the network—due to the service provider's network (dropped packets), router issues, or other traffic in the network. The FECN and BECN indication was further proof that the network was unstable. In particular, the router at the head end was receiving BECN frames and the router at the far end was receiving FECNs. This was a clear indication that there was excessive traffic in the direction from the data center to the remote locations. This could not be due to SNA alone, since SNA 3270 traffic is usually low volume. However, since the PVC carried only SNA traffic—the SNA-only locations were not even configured for LANs—it could not be occurring due to other protocol traffic. Hence, it would appear that either the frame relay network was severely congested and dropping packets, forcing frequent retransmissions, or the routers were causing the problem. Or so it seemed.

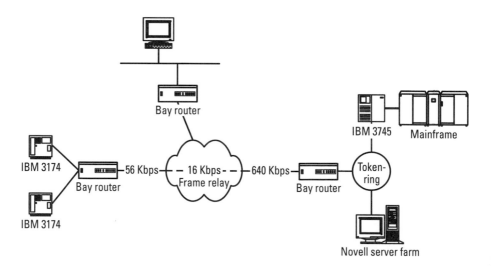

Figure 11.8 A frame relay network showing SNA-only and LAN attached locations.

The first step in troubleshooting is to capture a protocol trace of the packets traversing the specific SNA-only PVC in question. This is shown in Figure 11.9. This is a trace of the packets traversing on the PVC between an SNA-only location and the data center. The sniffer trace is an output from the packet capture facility of the Bay router at the data center.

The DLSw packets are readily seen—TCP port numbers 2067 and 2065. But why send IPX RIP packets and IPX SAP packets to a location that does not even have a LAN?

It turned out that the routers were configured in a mode that essentially assumed that the frame relay network was a LAN, using group-mode PVCs. When the router was reconfigured to direct-mode PVCs, that is, recognizing that each individual connection constitutes a network in its own right, the SNA response time problem disappeared!

11.4.1.2 Rationale

Why did the change from group-mode to direct-mode PVCs resolve the problem so dramatically? The Bay router at the head end essentially assumed that all routers in the group needed to hear SAP and RIP broadcasts every 60 sec, including the SNA-only routers.

We can do more—we can explain exactly why SNA response times were large and why FECNs and BECNs were being sent by the network.

Recall that Novell RIP and SAP broadcasts are sent out every 60 sec by default. The trace revealed 87 SAP broadcasts and 7 RIP broadcasts. To calculate the amount of data sent by the head end router every 60 sec, recall from Chapter 8 that a full SAP packet is approximately 500 bytes (7 services of 64 bytes each plus overhead), and that a RIP broadcast is approximately 450 bytes (50 networks, 8 bytes per network plus overhead). Hence every 60 sec the head-end router will send $87 \times 500 + 7 \times 450 = 46,650$ bytes. Now this amount of data will take $46,650 \times 8 / 640,000 = 0.58$ sec to be sent over the 640-Kbps frame relay port. But, since the maximum throughput of the connection is 56 Kbps, only 7000 bytes/sec $\times$ 0.58 sec = 4060 bytes can be cleared from network buffers. Hence the remaining amount, $46,650 - 4060 = 42,590$ bytes (approximately) will be in the network buffers, a definite reason for FECN and BECN indication. Furthermore, a SNA response frame sitting behind this amount of RIP/SAP broadcast packets will have to wait at least $42,590 \times 8 / 56,000 = 6$ additional seconds. In addition, there may be delays on the router serial port, which has to empty out large amounts of data (RIP/SAP broadcasts to all 50 locations).

Thus it is clear why SNA response times are large and also spiked—corresponding to the 60-sec broadcast pattern.

Flags	#	Delta T	Bytes	CumByt	Destination	Source	Summary
M	1		490	490	9C414E02.FFFF.	9C414E02.Wllf.	SAP R AS0037558US0AIC1, AS0119877AND001F, AS0243781HNAFCDE7, ...
	2	0.0612	490	980	9C414E02.FFFF.	9C414E02.Wllf.	SAP R ASHP02082MTRFCDE1, ASHP02100FTWFCDE2, ASHP03488GR..
	3	0.0678	490	1470	9C414E02.FFFF.	9C414E02.Wllf.	SAP R ASHP05995HNAFCDE3, ASHP06515US05FS3, ASO449649UG04IS, ...
	4	0.0556	490	1960	9C414E02.FFFF.	9C414E02.Wllf.	SAP R DSAWATCH_2FA79C22000000000184EA_GREISM01, KIT1, KIT1, ...
	5	0.1342	490	2450	9C414E02.FFFF.	9C414E02.Wllf.	SAP R SMM1600DD0122848B, SMM1600DD012284A1, SMM1600DD01228...
	6	0.0748	490	2940	9C414E02.FFFF.	9C414E02.Wllf.	SAP R SMM1600DD012286FB, UB00DD01229EFC, AND001F, ...
	7	0.0208	56	2996	[156.65.79.1]	[156.65.78.13]	TCP D=2067 S=2065 ACK=1002478020 WIN=16000
	8	0.0460	362	3358	9C414E02.FFFF.	9C414E02.Wllf.	SAP R AND001F, 9C418014_ EDAGATE,...
	9	0.0516	442	3800	9C414E02.FFFF.	9C414E02.Wllf.	IPX RIP response: 26 networks, 00000002 at 3 hops, 0000000C at 4 hops, ...
	11	0.0770	442	4684	9C414E02.FFFF.	9C414E02.Wllf.	IPX RIP response: 26 networks, 8DC21200 at 4 hops, 8DC2120F at 5 hops, ...
	12	0.0744	442	5126	9C414E02.FFFF.	9C414E02.Wllf.	IPX RIP response: 26 networks, 9C413000 at 3 hops, 9C413001 at 3 hops, ...
	13	0.0456	442	5568	9C414E02.FFFF.	9C414E02.Wllf.	IPX RIP response: 26 networks, 9C41E029 at 3 hops, 9C41E21F at 3 hops, ...
	14	0.0505	56	5624	[156.65.78.13]	[156.65.79.1]	TCP D=2067 S=2065 ACK=176804132 WIN=16000
	15	0.0018	114	5738	9C414E02.FFFF.	9C414E02.Wllf.	IPX RIP response: 9 networks, CA157B00 at 7 hops, CA157B01 at 7 hops, ...
	16	0.0213	514	6252	[156.65.78.13]	[156.65.79.1]	TCP D=2065 S=2067 ACK=111030989 SEQ=1002482374 LEN=458 WIN=128
	17	0.0061	558	6810	[156.65.78.13]	[156.65.79.1]	TCP D=2065 S=2067 ACK=111030989 SEQ=1002482832 LEN=502 WIN=128
	18	0.0021	174	6984	[156.65.78.13]	[156.65.79.1]	TCP D=2065 S=2067 ACK=111030989 SEQ=1002483334 LEN=118 WIN=128

Figure 11.9   Protocol trace of SNA-only PVC traffic.

11.4.2 Comparing Leased Line and Frame Relay Performance for Novell File Transfers

While trying to establish frame relay performance benchmarks for their applications, one company encountered the following problem: Novell file transfers over frame relay were showing consistently worse performance compared to an "equivalent" private line. This case study explains why such a performance difference is to be expected and, in doing so, illustrates the power of "back-of-the-envelope" calculations in trying to establish performance benchmarks where several assumptions have to be made.

11.4.2.1 Description of the Issue

The frame relay connection in question has a 256-Kbps frame relay port on both ends with ACC routers and a zero CIR in between (see Figure 11.10). Novell NCOPY (server-to-server) file transfers were used to measure performance. The performance benchmarks compared this frame relay connection with a 56-Kbps private line. The NCOPY file size is 7,026,040 bytes. Payload compression on the routers on private lines could not be turned off. Frame relay round-trip latency was measured at about 80 msec.

Here are the performance benchmarks: (1) frame relay (without router payload compression): 5 min 40 sec; and (2) private line (56 Kbps) (with router payload compression): 4 min 45 sec.

Prior experience with router payload compression over private line has led the company to expect a 4:1 compression ratio.

The question is why private line performance with 56 Kbps is better than frame relay performance, although frame relay bandwidth (256 Kbps) is more than four times that of private line. Since the company wanted to replace their private line infrastructure with frame relay, the related question was whether or not tuning parameters are available to make frame relay performance equal to that of private line.

Other relevant information includes the fact that the frame relay network is carrying very little background traffic and is healthy.

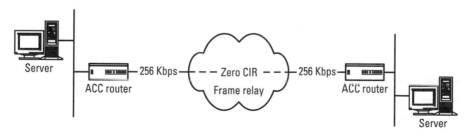

Figure 11.10 Reference connection for Novell file transfer issues.

11.4.2.2 Resolution

A protocol trace of the Novell file transfer revealed that packet burst was being used and that 10 frames were being sent in a single burst. See Chapter 8 for details on Novell packet burst. The frame size was 1024 bytes.

First let us try to validate the frame relay performance number. Because packet burst uses bursts of 10 frames with a payload of 1024 bytes, it is clear that approximately 686 bursts would be required to complete the file transfer of 7,026,040 bytes (file size/burst size). The frame size is 1024 + 36 (Pburst) + 30 (IPX) + 8 (FR) = 1026 + 72 = 1098 bytes.

Packet burst delays the transmission of successive frames within a burst by the IPG. Let us assume that the IPG is zero for this exercise (it is usually a few milliseconds). The following calculation shows how the file transfer can be calculated:

$$\text{Transfer time for a single burst} = (10 \times 1098 \times 8 / 256,000 + 0.04 + 1098 \times 8 / 256,000) + (0.04 + \text{small ack delay}) = 0.457 \text{ sec}$$

$$\text{Total transfer time} = 686 \times 0.457 = 5 \text{ min and } 13 \text{ sec.}$$

Compare this with actual observed value of 5 min and 40 sec!

Projected performance over a 256-Kbps private line (assume latency is equal to 50-msec round-trip) can be calculated as follows:

$$\text{Transfer time per burst} = 10 \times 1098 \times 8 / 256,000 + 0.05 = 0.393 \text{ sec}$$

$$\text{Transfer time for the file} = 0.393 \times 686 = 4 \text{ min } 30 \text{ sec}$$

Hence frame relay performance is about 1 min slower than private line. Why? This is because of the way packet burst sends multiple data packets in a burst and waits for an acknowledgment, and the fact that frame relay has a second insertion delay and a somewhat higher latency is larger than private line. The extra insertion delay over frame relay "costs" about $686 \times 1098 \times 8/256,000$ sec = 24 sec. The additional latency over frame relay "costs" an additional 686×0.03 sec = 21 sec—a total of about 45 sec. If this were a TCP/IP bulk data transfer with sufficiently large windows, one should expect the same file transfer times between the frame relay connection above and a 256-Kbps lease line.

How can one optimize frame relay performance for this Novell application? There is not much one can do; perhaps using a larger frame size and/or increased burst size would minimize the difference somewhat, but the difference between private line and frame relay will still remain.

One can even validate the leased line performance numbers with payload compression, using some reasonable assumptions. File transfer time over a 56-Kbps leased line can be calculated as follows:

$$\text{Transfer time for a burst} = 1098 \times 8 \times 10 \; / \; 56{,}000 + 0.05 \text{ msec} = 1.62 \text{ sec}$$

$$\text{Total transfer time} = 686 \times 1.62 = 1111 \text{ sec} = 18 \text{ min } 31 \text{ sec}$$

One can compare this to the observed leased line numbers with compression (assuming a 4-to-1 compression ratio):

$$\text{Observed leased line 56-Kbps file transfer time with compression} = 4 \text{ min}$$
$$45 \text{ sec} = 285 \text{ sec}$$

$$\text{Observed (expected) file transfer time over a 56-Kbps leased line without}$$
$$\text{compression} = 285 \text{ sec} \times 4 = 1140 \text{ sec!}$$

Compare this with the calculated value of 1111 sec!

11.4.3 A Paradox: Increasing Bandwidth Results in Worse Performance

Another company testing frame relay performance characteristics found that when they bought additional bandwidth from the carrier, the performance of their Novell applications actually degraded! The frame relay connection in question has a 56-Kbps port in Denver and a 384-Kbps port in Detroit with a 16-Kbps CIR. Again, NCOPY was used with a file size of 295,000 bytes. The frame relay file transfer time was 105 sec. In an effort to improve this performance, the company ordered an increase of the CIR to 56 Kbps. The file transfer time actually *increased* to 136 sec! Why?

It was not clear if packet burst was being used. However, the file transfer times hold a clue. If packet burst is assumed to be not present, then the file transfer time can be calculated as follows (assume a 30-msec one-way frame relay latency between Denver and Detroit).

$$\text{Number of packet transfers required} = 295{,}000 \text{ bytes} \; / \; (512 \text{ bytes/packet}) =$$
$$577 \text{ (approximately)}$$

Time for a packet transfer + ack time = $(578 \times 8 / 384{,}000 + 0.03 + 578 \times 8 / 56{,}000) + (68 \times 8 / 56{,}000 + 0.03 + 68 \times 8 / 384{,}000) = 166$ msec

where a 66-byte overhead and 2-byte acknowledgment are assumed. Hence file transfer time is $0.166 \times 577 = 96$ sec.[8] (Note that other delays, such as client, server, and router, have not been accounted for.)

Comparing the file transfer time to the actual observed value of 105 sec, it is clear that packet burst was not being used. If it were, then the file transfer time would have been much shorter.

What should happen to file transfer times with no packet burst when bandwidth is increased? Recall that without packet burst, Novell file transfers will use a stop-and-wait protocol, sending one packet of a 512-byte payload at a time. Such a file transfer method is bandwidth insensitive, and hence the file transfer time should be unaffected by additional bandwidth. The reason why the file transfer time increased has to do with how frame relay carriers deal with requests for increased bandwidth for PVCs. Because PVCs are logical entities, it is likely that the frame relay carrier chose a different path to be able to accommodate the request for additional bandwidth. If this path happens to be longer than the original path, the network latency for that connection will increase and this will have a pronounced effect on the file transfer performance.

The resolution of the issue is to use packet burst. With this method, additional bandwidth will result in better performance.

11.5 SNA-Related Case Studies

The following case studies relate to concerns about supporting SNA over frame relay in a multiprotocol environment. The first case study shows how one can design a frame relay network to replace a multidrop SNA network. Although multidrop SNA networks are becoming less prevalent with the advent of multiprotocol networking and frame relay, the case study can be used as an example to demonstrate how bandwidth for a frame relay network can be estimated. The second case study deals with a company's concerns about the adverse effects of supporting mission-critical SNA and native TCP/IP traffic (e.g., FTP) on the same connection over frame relay, and how these concerns can be addressed.

---

8. The approach to calculate file transfer times is slightly different here. Instead of calculating round-trip delays, throughput, and then the file transfers, we have opted to calculate the round-trip delay to send a single packet and receive an ack, and then multiply by the number of packets in the file. This approach is valid here because of the stop-and-go nature of the file transfer.

11.5.1 Migration From SNA MultiDrop to DLSw

The multidrop network consisted of 12 major sites and other smaller sites. The company (a financial services organization) wished to migrate the major sites to frame relay, the primary motivation being cost reduction and the introduction of new TCP/IP applications. The 12 major sites had several colocated controllers (3x74) and smaller sites multidropped through these major sites, as shown in Figure 11.11.

The multiplexers concentrate two or three low-speed lines (9.6 or 19.2 Kbps) into a higher speed (56 Kbps) to the host (muxes at the host are not shown). Each multidrop connection could carry several 3174 controllers. The company wished to replace the multiplexers with Bay routers and upgrade the 3174 controllers with a token ring attachment. The multidrop network would

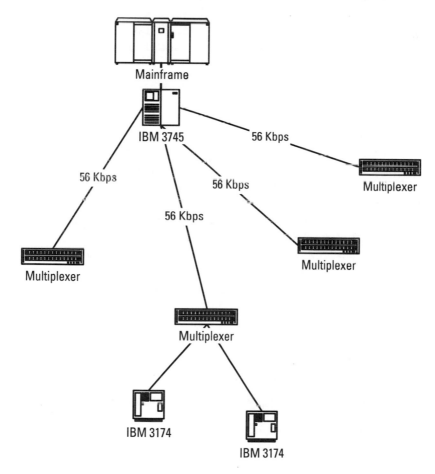

Figure 11.11 SNA multidrop.

then be replaced by a public frame relay service. The main issues are to size the frame relay ports and PVCs adequately and provide appropriate guidelines to ensure that SNA response times will not suffer in the migration.

It might seem that a simple replacement of the time-division multiplexer in the multidrop network with a statistical multiplexer is enough to guarantee better response times for SNA. After all, isn't it true that in the multidrop network, the 9.6-Kbps link cannot use the 56-Kbps bandwidth even if the latter is available, whereas with statistical multiplexing, the entire 56 Kbps is available to the SNA traffic? Actually, this is not true. It ignores the fact that the statistical multiplexer, such as a router, is a store-and-forward device—it stores and forwards SDLC frames, with some encapsulation. This store-and-forward nature introduces additional delays so that, in fact, the overall delays may actually increase. (It really depends on related aspects such as local polling, encapsulation overhead, and so on; see Chapter 10.)

However, since the decision was made to token-ring attach all the controllers, the preceding discussion is irrelevant. The proposed frame relay network would be as shown in Figure 11.12.

11.5.1.1 The Design Process

The company obtained transaction sizes and traffic volumes for their SNA 3270 applications. The transaction consisted of a 20-byte inquiry, followed by a 700-byte response. The traffic volumes, in overall bytes per working day, are broken down by location and in the inbound and outbound direction (to and from the mainframe, respectively). Several assumptions have to be made to make use of the information provided. First, we need peak hour traffic information. Second, we need to know the degree of simultaneity of the traffic: How many users in how many locations use the mainframe applications simultaneously? In reality, however, this type of information is very hard to obtain unless a detailed traffic study is undertaken. Most companies have neither the time nor the resources to expend on such studies. More importantly, exhaustive traffic studies are often not needed to accomplish the goals at hand, that is, design a robust frame relay network that will deliver good SNA response times. Transaction sizes and overall traffic volumes, together with a reasonable set of assumptions, usually suffice.

To obtain peak hour traffic volumes, one could make the 80–20 assumption, that is, that 20% of all the day's activities occur during a peak hour. From this, the overall bandwidth requirements for each location in the inbound and outbound direction can be obtained. We can also make the worst case assumption that all locations are simultaneously active at their peak loads.

The next step is to account for additional protocol overhead due to DLSw encapsulation.

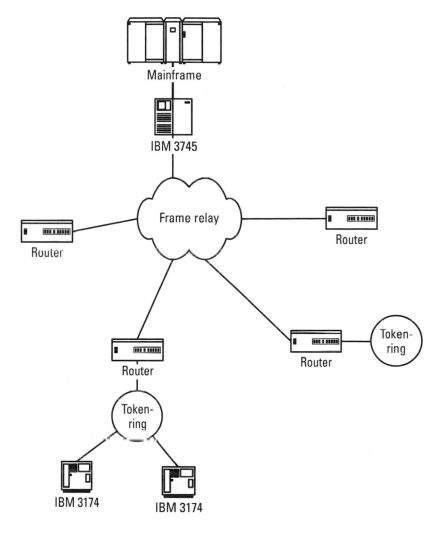

Figure 11.12 Frame relay replacement for the SNA multidrop with routers.

Inbound message size = 20 + 16 (DLSw) + 40 (TCP/IP) + 8 (FR) = 84 bytes

Outbound message size = 700 + 16 (DLSw) + 40 (TCP/IP) + 8 (FR) = 764 bytes

(An implicit assumption of no chaining or segmentation is made here.) Thus the peak hour inbound and outbound load will increase accordingly in the new network. This is shown in Table 11.1.

Table 11.1
Location-Specific Traffic and Design Information

Location	Load In/Out (Kbps)	Port Speed/CIR	Average Expected Response Time (sec)
Headquarters	10/352	768-Kbps port	—
Location 1	1.3/43.8	256/128	0.32
Location 2	16.6/58.1	256/128	0.34
Location 3	1.4/48.3	256/128	0.33
Location 4	1/35.4	128/64	0.53
Location 5	1.2/40.7	256/128	0.32
Location 6	0.94/32.8	128/64	0.5
Location 7	0.73/25.5	56/48	0.75
Location 8	0.5/16.9	56/32	0.85
Location 9	0.6/20.5	56/48	0.64
Location 10	0.4/12.8	56/32	0.72
Location 11	0.3/10.6	56/32	0.67
Location 12	0.2/6.8	56/16	1.1

The next step is size the PVC speeds for the remote locations. This will give us an indication of the port speeds at these remote locations. Once the remote port speeds and CIRs to the central site are specified, it is then easy to size the port speed at the data center.

Look at Table 11.1. To size a CIR for a location, we merely look at the outbound direction because the traffic in this direction is dominating. For instance, for location 5, the outbound traffic volume is 40.7 Kbps. What should be the CIR allocated for this PVC? There are several approaches. A risk-averse approach would be to use a rule that PVCs should not be loaded at higher than 60% utilization. A more reasonable approach would be to assume some amount of bursting in the network and size the CIR at, say, 100%. The former approach was used here, given the company's concerns about SNA. Hence the CIR allocated to this PVC is 128 Kbps (assuming that the carrier allocated CIRs in 64-Kbps increments). Similarly other CIRs are calculated. Port speeds are chosen to be the next higher DS-0 (64-Kbps) multiple value. To calculate the port speed at the central site, add all of the CIRs and roughly estimate the port speed. In this case, the sum of the CIRs is 848 Kbps. It seems appropriate to allocate 768 Kbps, or half of a T1 connection, to the central site. Table 11.1 shows these values in the first three columns.

Now we have a frame relay design that incorporates core characteristics of the company's SNA traffic. We should next estimate end user response times in the new network.

11.5.1.2 Response Time Calculation

We will illustrate the methodology for location 1. Assume that the network latency is 60 msec one way. Notice that the inbound utilization on the links (ports and PVCs) are almost zero. Hence inbound response time is

$$84 \times 8 / 128,000 + 0.06 + 84 \times 8 / 768,000 = 0.066 \text{ sec}$$

Outbound response time for location 1 can be calculated as follows.[9] Notice that the central port (768 Kbps) will be approximately 50% utilized (add all the outbound peak traffic volumes divided by 768 Kbps), and that the PVC to location 1 is about 35% utilized (44 Kbps/128 Kbps). Assume the worst case scenario that the PVC is not bursting above its CIR. Notice that, due to the underlying assumption of a closed-loop system, there can be no queuing occurring at the remote port. Hence, using an M/M/1 formula (see Chapter 3, Section 3.2), the outbound response time is

$$[764 \times 8 / 768,000] / (1 - 0.5) + [(764 \times 8 / 128,000) / (1 - 0.35)] + 0.06 + 764 \times 8 / 256,000 = 0.173 \text{ sec}$$

Hence total response time (network portion) is $67 + 173 = 240$ msec. Note that this ignores router delays and FEP delays. Thus, it is likely that end user response times are in the 300-msec range. When the network was actually turned up and tested, SNA response times were actually averaging 320 msec!

11.5.2 Insuring SNA Performance in the Presence of TCP/IP Traffic

The following case study illustrates the issues pertaining to mixing mission-critical SNA applications and "not-so-critical" TCP/IP applications on the same WAN connection.

The network in question supports the data communication needs for a large brokerage firm in New York City. There are 300 branch offices in the United States connected to the headquarters office in New York via frame relay. Large branches have a 256-Kbps/128-Kbps port speed and CIR

---

9. These calculations are approximate. It also assumes a closed-loop or leaky bucket implementation at the entrance to the frame relay cloud, essentially modeling the PVC as a serial line.

combination. Smaller branches have a 128-Kbps/64-Kbps port speed and CIR combination. There are multiple T1s at the headquarters location. Cisco routers are used in the network, which carries SNA (DLSw) traffic as well as some native TCP/IP applications, such as IP-based market data feeds (approximately 12 Kbps to each branch) and large FTPs from the mainframe to the branches, mostly during off hours.

Business requirements dictate that SNA traffic get consistent response times in the 1- to 2-sec range. Off-hour TCP file transfers from the mainframes (approximately 10 Mbytes to each branch) have a 3- to 5-h window of completion (to all locations). Although TCP file transfers occur mostly during offhours, it cannot be guaranteed that end users will not access files from the mainframe during the day. Hence the concern about mixing SNA and TCP/IP. The questions to be answered are:

- Will router priorities be sufficient to ensure good response times for SNA?
- Is a second PVC for SNA an absolute requirement?
- What about traffic shaping options at the headquarters location?

11.5.2.1 Resolution

The first step is to investigate whether or not the SNA transactions can be supported when TCP traffic is not present. The next step is to ensure that the TCP requirements (3- to 5-h window of completion for transfers) can be satisfied. The final step is to investigate what is likely to happen to SNA response times in the presence of TCP file transfers, and to recommend methods to address any issues.

It is clear that the SNA response times will be less than 1 sec. This does not require detailed analysis. Note that the SNA 3270 traffic is "small in, large out"; say, a few hundred bytes in and a thousand or so bytes back from the host. Given the relatively large ports and PVCs in the network to most locations, it is not unreasonable to expect that, in the absence of TCP traffic, SNA transactions will have acceptable response times.[10]

---

10. A more exact analysis will require information on bytes in and bytes out per transaction and number of users at a location, and user think times (or transaction rate). It is probably not worth the effort. For instance, with a 10-sec think time (a very busy user) between submitting successive inquiries, and a 1000-byte response time from the host, a single user will account for 100 bytes/sec of bandwidth. On a connection with 64-Kbps CIR, one can support more than 50 simultaneous users and still keep the link from being more than 70% utilized. Simple analyses of this nature can often be used to determine if there are likely to be issues with meeting given response time objectives.

Let's look at the TCP requirements now. Assume that a 10-Mbyte file needs to be sent from the host to the remote branch. How long will it take and what is the optimal window size? On a 128-Kbps connection (large branch), the transfer time is likely to be at most

$$1,024,000 \times 8 \times 1.05 \;/\; 128,000 = 11 \text{ min (5\% protocol overhead assuming}$$
$$\text{a 1024-byte segment size)}$$

For a 64-Kbps connection, the transfer time will likely be double that for a 128 Kbps, or 22 min.

The optimal window size for the transfers to the large branch can be calculated thus:

$$\text{Maximum bandwidth} = 256 \text{ Kbps (or 32 bytes/msec)}$$

$$\text{Expected round-trip latency } R_d \text{ (U.S. connections)} = 100 \text{ msec}$$

Hence optimal window size $W^* = 3200$ bytes (or approximately three 1024-byte segments).

Optimal window size for the smaller branches (128 Kbps) equals approximately 1500 bytes. Hence a window size of 1500 bytes is likely to provide a throughput of approximately 128 Kbps.

Because a T1 can hold about twelve 128-Kbps streams, one can achieve about 12 parallel TCP file transfers on a single T1. Thus, in the extreme case of all file transfers occurring on a single T1 link, the window of completion for 300 branches can be estimated as

$$(300/12) \times 11 \text{ min} = 300 \text{ min or 5 h}$$

Because there are multiple T1s at the data center, it is clear that the TCP requirements pose no problems from the WAN point of view. Indeed, it is an open question whether or not the mainframe will be able to "pump" as much TCP traffic as to be able to keep the WAN links busy.

What is the impact of TCP FTPs on SNA response times? Let us look at a small branch location with a 64-Kbps CIR. We assume the worst case scenario where the best throughput on the connection is 64 Kbps. How many parallel TCP sessions can be supported on this PVC before SNA response times become adversely affected? To answer this question, assume that the remote servers advertise a 1500-byte window. Because of the speed mismatch between a T1 frame relay port and a 64-Kbps PVC, buffering will occur at the network buffers (usually at the ingress or egress). If TCP FTP and SNA are the only users of this PVC, then the maximum data buffered will be $N \times 1500$ bytes (N is the number of parallel file transfers). These buffers can be emptied at a minimum rate of 64 Kbps (CIR). Hence the time to empty the buffer is $N \times 1500/8000 = N \times 0.187$ sec. If this has to be less than 1 sec, then the number of parallel transfers should be limited to five (1/0.187 is approximately 5).

Thus, here are our overall conclusions:

- SNA applications by themselves will perform adequately over this network.

- TCP FTP requirements of being able to transfer 10-Mbyte files to 300 locations within a 3- to 5-h window can be satisfied with one or more T1 links at the data center.

- Optimal window size (receiver advertised) should be set to 1500 bytes.

- SNA response times, when mixed with TCP FTPs, will certainly degrade. However, the response times may be acceptable provided precautions are taken to not have more than five parallel file transfers on a connection to a remote branch office. We will not address the question of whether or not this can actually be accomplished in mainframes (or other platforms). The statement is a design goal.

- Setting the window size to the optimal value is absolutely critical, because larger window sizes will not improve throughput, but will degrade SNA response times.

- Priority queuing mechanisms at the router will have little or no effect on SNA response times. Most of the queuing will occur in the network, not at the routers.

- A separate PVC for SNA may address some of the congestion problems in the network, but it would be too unwieldy to manage 600 PVCs in the network.

- One approach would be to use traffic shaping at the central site, either invoking traffic shaping on the routers themselves or using an external traffic shaping device. In its simplest form, the traffic shaping mechanism can limit the traffic rate on a PVC to not exceed the CIR rate. This will prevent congestion in the network buffers. More complex traffic shaping schemes, such as allocating specific bandwidth to SNA and FTPs, can also be invoked.

- The company should investigate externally deployed bandwidth management devices like Xedia and Packeteer.

11.6 Quantifying the WAN Bandwidth Impact of SNMP Polling

This case study deals with studying the bandwidth impact of multiple SNMP stations managing a large corporate data network. The analysis is fairly straightforward. However, data collection regarding the number of bytes exchanged during SNMP queries is an important component of the analysis. Several assumptions also need to be made.

The corporate data network has four HP OpenView® network management stations that monitor all aspects of the Cisco router network, which is partly frame relay and partly leased line. These management stations are centrally located at a data center, which is served by multiple T1 facilities. The question is this: What is the average load generated by the four network management stations?

We first list some assumptions made in the analysis and an assessment of how critical these assumptions are. Next, we list the input parameters such as polling frequencies and objects polled. We then discuss the bandwidth impact analysis. The last section deals with observations, conclusions, and recommendations.

Key Assumptions

Most of the assumptions are related to the polling data (SNMP Get requests and responses) generated by management stations. Specific assumptions are:

- SNMP V2 is not used.
  - SNMP V2 has many features that do not pertain to WAN bandwidth, but one feature in particular is useful—*bulk requests*—where a single SNMP Get request can collect a number of variables.
- Four network management stations are employed. They are all polling the *same objects* at the same frequencies.
  - This was the way it was set up in this particular customer network.
- The corporate WAN consists of 152 routers and 500 interfaces, including serial lines, subinterfaces, and Ethernet interfaces.
- There are also 19 hub routers in the network with 192 interfaces.
- No data were available on the traffic generated due to polling Cisco MIBs for IP, IPX, AppleTalk, and DECnet statistics. It is a safe

assumption that the data generated are less than that generated by polling full MIB2 interface statistics.

Objects Polled and Frequencies

The following objects are polled at the stated frequencies:

Output utilization	All router ports every 5 min
Full MIB 2 interface data	Only hub router ports every 15 min
Full MIB2 interface data	All router ports every 1 h
Protocol data using Cisco MIBs	All routers every 1 h
System data	All routers every 90 sec

Traffic Generated by SNMP Polling

The following information was gathered from a sniffer analysis of traffic generated by Node Manager release 5.0.1 running on HP UX® Version 10.2, polling a Cisco 7513 router with 24 interfaces.

The SNMP request is transmitted from the HP OpenView Network Node Manager 5.01 management station and receives a SNMP reply from the Cisco 7513 router device with 24 interfaces (mgmt.mib-2.system.sysUpTime: 48-byte SNMP get and approximately 53-byte reply).

CPU utilization as formulated by CiscoWorks for "CPU Load" using SNMP Get requests/replies as follows:

SNMP get of … cisco.2.1.57.0 (local.lsystem.avgBusy1)	51 bytes
SNMP reply (normally in 52–55 byte range)	52 bytes
SNMP get of … cisco.2.1.58.0 (local.lsystem.avgBusy5)	51 bytes
SNMP reply (normally in 52–55 byte range	52 bytes
SNMP get of … cisco.2.1.56.0 (local.lsystem.BusyPer)	51 bytes
SNMP reply (normally in 52–55 byte range)	52 bytes

Output utilization is measured using Node Manager's installed SNMP Data Collection routine, "if%util," which computes percent of available bandwidth utilization on an interface. The variables requested to obtain the preceding values were:

46 items requested (SNMP get) from sysUpTime ... ifSpeed = 775 bytes

46 item response (SNMP reply) from sysUpTime ... ifSpeed = 878 bytes

28 items requested (SNMP get) from sysUpTime ... ifSpeed = 487 bytes

28 item response (SNMP reply) from sysUpTime ... ifSpeed = 545 bytes

The 46 and 28 enumerated item totals were taken from the packet capture with no additional description identifying the particular variables polled except for the following: "sysUpTime ... ifSpeed."

The MIB-2 interface tree (mgmt.mib-2.interfaces.*) SNMP data request consists of a single value for "ifNumber" and 22 "ifTable.ifEntry" variables for each interface on the router.

SNMP get of "ifNumber"	50 bytes
SNMP reply (normally in 52–55 byte range)	49 bytes

All of the following "ifTable.ifEntry" SNMP Get-Nexts and replies should be multiplied by the number of interfaces on the router (for example, if there are 24 interfaces then you should multiply each of the SNMP MIB-object packet Gets and replies by that number, 24):

SNMP getnext of "ifIndex"	51 bytes
SNMP reply	52 bytes
SNMP getnext of "ifDescrip"	51 bytes
SNMP reply (normally in 62–64 byte range)	64 bytes
SNMP getnext of "ifType"	51 bytes
SNMP reply	52 bytes
SNMP getnext of "ifMTU"	51 bytes
SNMP reply	53 bytes
SNMP getnext of "ifSpeed"	51 bytes
SNMP reply (normally in 53–55 byte range)	54 bytes

SNMP getnext of "ifPhysAddress"	51 bytes
SNMP reply	57 bytes
SNMP getnext of "ifAdminStatus"	51 bytes
SNMP reply	52 bytes
SNMP getnext of "ifOperStatus"	51 bytes
SNMP reply	52 bytes
SNMP getnext of "ifLastChange"	51 bytes
SNMP reply (normally in 52–55 byte range)	55 bytes
SNMP getnext of "ifInOctets"	51 bytes
SNMP reply (normally in 52–55 byte range)	54 bytes
SNMP getnext of "ifInUcastPkts"	51 bytes
SNMP reply (normally in 52–55 byte range)	54 bytes
SNMP getnext of "ifInDiscards"	51 bytes
SNMP reply (normally in 52–55 byte range)	52 bytes
SNMP getnext of "ifInErrors"	51 bytes
SNMP reply (normally in 52–55 byte range)	52 bytes
SNMP getnext of "ifInUnknownProtos"	51 bytes
SNMP reply (normally in 52–55 byte range)	52 bytes
SNMP getnext of "ifOutOctets"	51 bytes
SNMP reply (normally in 52–55 byte range)	54 bytes
SNMP getnext of "ifOutNUcastPkts"	51 bytes
SNMP reply (normally in 52–55 byte range)	52 bytes
SNMP getnext of "ifOutDiscards"	51 bytes
SNMP reply (normally in 52–55 byte range)	52 bytes
SNMP getnext of "ifOutErrors"	51 bytes
SNMP reply (normally in 52–55 byte range)	52 bytes
SNMP getnext of "ifOutQLen"	51 bytes
SNMP reply (normally in 52–55 byte range)	52 bytes
SNMP getnext of "ifSpecific"	51 bytes
SNMP reply (normally in 52–55 byte range)	52 bytes

Bandwidth Impact Analysis

In general, for MIB-2 interface variables, it is reasonable to assume that approximately 52 bytes are generated in a SNMP request (either Get or Get-Next) and an SNMP response per variable. Protocol overhead characters for SNMP requests and responses are approximately 40 bytes, including UDP/IP overhead, SNMP overhead, and link overhead.

Based on the assumptions listed and the traffic profile shown above, the bandwidth impact can be assessed as follows:

- Output utilization for all router ports every 5 min:

46 items requested	775 bytes
46 items responded	878 bytes
28 items requested	487 bytes
28 items responded	545 bytes

$$\text{Bandwidth utilization} = 500 \text{ interfaces } (775 + 878 + 487 + 545 + 160) / 5 \times$$
$$60 = 37.9 \text{ Kbps}$$

- Full MIB-2 interface polling on hub routers every 15 min. The first poll for every hub router retrieves the number of interfaces, and then every interface is polled. Every interface contains 21 variables. There are 192 interfaces.

$$\text{Bandwidth utilization} = \{19 \times (100 + 40) + 192 \text{ interfaces} \times$$
$$[21 \times (52 + 40) \times 2]\} / 15 \times 60 = 6.6 \text{ Kbps}$$

- Full MIB-2 interface polling on all routers every 1 h. The first poll for every router retrieves the number of interfaces, and then every interface is polled. Every interface contains 21 variables. There are 500 interfaces.

$$\text{Bandwidth utilization} = \{152 \times (100 + 40) + 500 \text{ interfaces} \times$$
$$[21 \times (52 + 40) \times 2]\} / 60 \ 60 = 4.3 \text{ Kbps}$$

- Protocol statistics every 1 h on all routers. Because only one variable per protocol is polled, it is safe to assume that the total amount of traffic generated is significantly less than polling MIB interface data on very router every hour (that is, bandwidth utilization is less than 4.3 Kbps!).

- System data for all routers on a 90-sec interval. Polling system data for sysUpTime generates 50 bytes of request and response on an average.

Bandwidth Utilization = 152 routers × (50 + 40) × 2 = 2.4 Kbps

Hence the overall bandwidth utilization due to a single network management station is

37.9 + 6.6 + 4.3 + 4.3 + 2.4 = 56 Kbps (approximately)

Therefore the overall bandwidth utilization due to four parallel network management stations is

4 × 56 Kbps = 224 Kbps!

The overall observation is that the bandwidth impact of SNMP polling is minimal relative to the total bandwidth available at the data center.

Appendix A
Queuing: A Mathematical Digression

In this brief digression, we go into a little more of a mathematical discussion of queuing than is generally required. We begin by separately discussing the nature of the input and the service processes and then their combination to determine the queuing delay estimation process.

Consider the times of arrivals (or arrival epochs) of the cars to the tollbooth (refer to the discussion in Section 3.2.5 of the main text) taken as randomly distributed points on a line (see Figure A.1).

We denote the arrival points as T_0, T_1, ..., T_n and the interarrival times, that is, the time intervals between the arrival points, as a_0, a_1, ..., a_{n-1}. Each car brings with it an amount of work, that is, the amount of time the toll taker requires to service the car, which is denoted as s_0, s_1, ..., s_{n-1}. The total work in the system at the time of an arrival is the waiting time for that arrival.

From Figure A.1, it is evident that the waiting time is both a function of the arrival process and the service time process. A method to characterize the arrival process is to define the distribution for the interarrival times. A method to characterize the service process is to define the distribution of the service times for the individual arrivals.

The various types of queuing systems can be categorized in terms of (1) the input process, (2) the service process, and (3) the number of servers. Queuing theory then strives to determine the properties of systems combining specified input processes with specified service time distributions. Queuing models are categorized by specifying these processes in a shorthand notion. The shorthand notation is summarized as

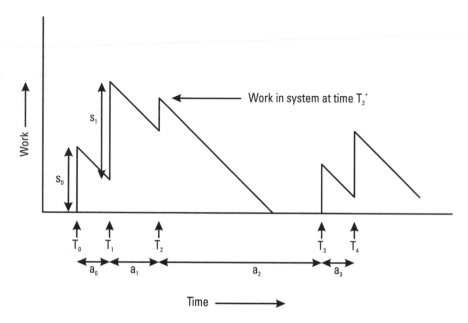

Figure A.1 Time line showing the arrival epochs and the interarrival times for an arrival process.

$$(\text{Input process}) \,/\, (\text{Service distribution}) \,/\, (\text{Number of servers})$$

Prominent queuing models are:

- *M/M/1:* Poisson input, exponential service time distribution, single server;
- *M/G/1:* Poisson input, general service time distribution, single server;
- *M/D/1:* Poisson input, deterministic service time distribution, single server; and
- *M/M/s:* Poisson input, exponential service time distribution, *s* servers.

A Poisson input process is one in which the interarrival times between incoming packets are distributed according to an exponential distribution [1]. In the preceding list, we just identified three types of service time distributions: exponential, general, and deterministic. Exponential service times are distributed according to an exponential distribution. General services times are unspecified. That is, the results for a system labeled as "general service time distribution" will apply to all systems independent of the service time statistics (assuming, however, that the specific service times of successive packets are not correlated in any way). A deterministic service time implies that the service time of every packet is identical.

Let us define some quantities characterizing queuing systems, then we will summarize some interesting results for the queuing models identified above. We define:

- τ = the average service time for a packet;
- σ^2 = the variance of the service time for a packet;
- λ = the average arrival rate of packets;
- $\rho = \lambda\tau$ = the average load on the system;
- $E(Q)$ = expected number of packets in queue; and
- $E(W)$ = expected queuing time for packets in queue.

Queuing theory has derived the following results for the M/G/1 queuing system (Remember that this refers to a system where the packets arrive according to a particular random arrival process referred to as a *Poisson process;* the service time of a given packet is randomly distributed according to a general distribution and there is a single server serving the queue.):

$$E(Q) = \frac{\rho^2}{2(1-\rho)}\left(1+\frac{\sigma^2}{\tau^2}\right) \tag{A.1}$$

$$E(W) = \frac{\rho\tau}{2(1-\rho)}\left(1+\frac{\sigma^2}{\tau^2}\right) \tag{A.2}$$

With this relatively simple formula, we can demonstrate the behavior discussed for our tollbooth example. First, the expected waiting times and number of customers in queue approach infinity as $p \to 1$. This simply states the fact that if the traffic becomes too large, then the lines at the tollbooth grow too large. (Just try commuting into Manhattan through the Lincoln Tunnel during morning rush hour.) We say that queuing systems are stable, that is, queue lengths and waiting times are finite, for loads in the range from 0 to 1 and are unstable, that is, waiting times and queue lengths grow infinitely large, for $p \geq 1$. Second, we see that the queuing delays are directly proportional to the service times. This is a simple reflection of the fact that the queuing delays on slower communications links will be longer than on faster communications links for a fixed load. Third, we see from the above formulas that

$$E(Q) = \lambda E(W) \qquad (A.3)$$

This simple expression, relating the mean values of queue length, waiting time, and arrival rate is known as Little's theorem and is valid for all systems (having finite means). A more general representation of Little's law is

$$N = \lambda \times D \qquad (A.4)$$

where N is the number in the system, λ is the rate through the system, and D is the delay in the system. This is a very general formula. We have already seen it applied to queues.

Another application of Little's law is the derivation of the relationship between throughput, window size, and round-trip delay in windowing systems. If we equate the window size W to N, the throughput X to λ, and the round-trip delay R_d to D in (A.4), we get

$$X = W / R_d \qquad (A.5)$$

This formula is derived in Section 3.5 of the main text.

Now consider the relatively simple exponential service time distribution, that is, $\tau^2 = \sigma^2$. This simplifies (A.1) and (A.2) to:

$$E(Q) = \frac{\rho^2}{(1-\rho)} \qquad (A.6)$$

$$E(W) = \frac{\rho}{1-\rho}\tau \qquad (A.7)$$

which are the queuing formulas for the M/M/1 queuing mode. Notice that these are extremely simple formulas and therefore are extremely useful aids to estimating ballpark queuing delays in network. One easy point to remember for simple estimations is that the queuing time for a system at 50% load is simply equal to the service time, that is, $E(W|\rho = 0.5) = \tau$. For example, if a communications facility is running at 50% utilization and the average transmission delay is 8 msec (e.g., the transmission delay for a 1500-byte packet on a

1.5-Mbps facility), then an estimate for the queuing delay is 8 msec. If you remember nothing else from this section on queuing delays, remember this.

What happens if the packet sizes are uniform? For deterministic service time distributions, $\sigma^2 = 0$. Therefore, from the previous M/G/1 results, we have the following results for the M/D/1 queuing model:

$$E(Q) = \frac{\rho^2}{2(1-\rho)} \tag{A.8}$$

$$E(W) = \frac{\rho\tau}{2(1-\rho)} \tag{A.9}$$

These also are extremely simple expressions. Note, however, that the waiting times and queue lengths for the M/D/1 are a factor of 2 less than the corresponding times and queue lengths for the M/M/1 systems for the same system loads. This is a dramatic demonstration of the effect that the service time distribution has on the system waiting times, which we discussed earlier.

Let's revisit the toll booth example in Section 3.2.5. If every car carried tokens (and did not require change), then the service times for each car is rather deterministic, and, from personal experience, the lines at the tollbooths are relatively short. However, if every car carries only cash and some require a fair amount of change (a lot of people carrying $20 bills for a 35-cent toll), then the lines become relatively long at the tollbooths for the same average traffic load. This is easy to see by comparing the M/D/1 formulas with the previous formulas for M/M/1 systems.

For exponential service times, Poisson arrival process, and s servers, the expected waiting time takes a more complex form, but can still be represented analytically [1]. Figure A.2 shows the effect of adding multiple servers on the queuing delay. This figure demonstrates the advantage of adding multiple tollbooths to our commuter model. The more servers the better, from the customer perspective (however, there is an expense associated with adding more servers). Notice the dramatic decrease in delay as the system goes from a single server to two servers and then to four servers. The decrease in queuing delays by doubling the number of servers clearly cuts the delays by significantly more than half (while maintaining the average load per server the same).

Finally, we wish to briefly discuss a formula related to the multiple server situation. This is the *erlang loss model*. This model is useful in estimating the blocking probability in systems where there are multiple servers, s, but no buffering. That is, if all s servers are busy, the new arrival is discarded, or blocked. If a server is available, then the arrival is assigned to a server. The erlang loss

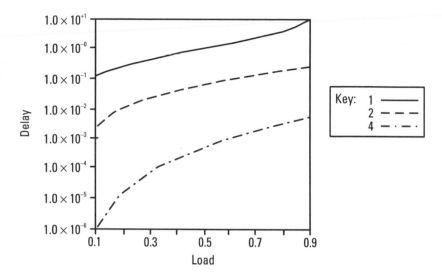

Figure A.2 The effect of adding servers to the expected waiting times in a queue.

model states that the probability of a dial user finding all s modems busy is given by:

$$S(s,a) = \frac{a^s / s!}{\sum_{k=0}^{s} a^k / k!}$$

(A.10)

This formula is extremely useful in determining the number of dial-in ports on a modem pool (as just mentioned), estimating whether the trunk bandwidth is sufficient to minimize the probability of being in an overload situation, or determining whether a sufficient number of client tokens is available for a given software package resident on a network server.

Reference

[1]   Cooper, R. B., *Introduction to Queuing Theory*, New York: Elsevier, 1981.

Appendix B
Throughput in Lossy Environments

B.1 Introduction

In this appendix, we consider the effects of transmission errors and packet losses on end-to-end throughput modeling. Transmission systems and data networks are not perfect systems. As a result not all data packets transmitted are received by the corresponding protocol layer in the receiver. Our discussion of throughput modeling in the main text of the book ignores the effects of packet losses on performance. In most enterprise networks, this is a reasonable assumption. For this reason, we chose to address the topic of packet loss and its impact on performance in this appendix.

Several factors contribute to packet losses:

- *Transmission errors.* When a transmitter sends a bit of information down a transmission path, it is not always interpreted correctly by the receiving entity on the transmission path. For this reason, protocol suites include error detection or error correction capabilities. Predominantly, error detection methods are employed in most modern WAN networking protocols, due to the relatively low bit error probabilities on transmission systems. These protocols, when detecting an error in a packet, discard the entire data packet. This appears to the higher layer protocols as a packet loss.

- *Buffer overflows.* Data communications equipment that employ statistical packet multiplexing maintain finite data buffers. Due to the statistical nature of the packet loads, there are occasions when the amount of data to be buffered exceeds the size of the data buffers and packet losses occur. To detect packet losses, protocol suites employ packet sequencing by numbering the packets in the order in which they are sent. A lost packet will result in the receiver seeing a packet number or several numbers skipped over.

- *Transmitter time-outs.* When a transmitting host sends a packet, it sets a timer. When the transmitter receives the acknowledgment for the packet, it terminates the timer. If the value of the timer exceeds a threshold, that is, the *transmitter time-out value,* the transmitter assumes that the packet was lost in transmission, and the transmitter enters into an error recovery state. Time-outs may occur due to packet losses or due to excessively long queuing delays in transit.

- *Out-of-sequence receptions.* Some network technologies ensure that the order of the packets sent to the network is the same as the order of the packets delivered by the network. Some networks do not maintain packet sequencing, most notably IP router networks. Some transport protocol implementations may discard out-of-sequence packets instead of storing them and reordering the incoming packet stream. In this case, out-of-sequence packets will affect the performance of throughput systems in a fashion similar to other packet loss mechanisms.

We now discuss each of these mechanisms in order in more detail.

B.2 Transmission Errors

The simplest model of transmission errors is one that assumes random, uncorrelated bit errors. This *random bit error* model simply assumes that for each bit of information transmitted, the probability of that bit being received in error is b_e, and that this is independent of the reception quality of any of the previous bits transmitted. The utility of this model is the simplicity of expressions attainable for quantities of interest. The quantity of most interest to us is the probability of a packet, M bits in length, being received in error; that is, at least one of the M bits is in error. Assuming independent bit errors, then the probability of a packet of length M bits having no bits received in error, $P\{\text{good}|L_p = M\}$, is

$$P\{\text{good}|L_p = M\} = (1 - b_e)^M \qquad \text{(B.1)}$$

Therefore the probability of a packet of length M bits having at least one bit received in error, $P\{bad|L_p = M\}$, is

$$P\{bad|L_p = M\} = 1 - (1 - b_e)^M \qquad (B.2)$$

which is simply $1 - P\{good|L_p = M\}$. This represents an extremely simple relationship between the bit error rate, the packet length, and the probability of a packet in error. However, the simplicity of this relationship is also its downfall.

In reality, bit errors are highly correlated. That is, if a bit is in error, then the probability that the following bit is also in error is much higher than if the preceding bit was not in error. This positive correlation between successive bit errors causes the preceding expression for the probability of a packet error to overestimate this quantity for a fixed bit error rate. However, there is value in engineering in having expressions that are conservative in nature.

E. N. Gilbert [1] proposed a model to describe and parameterize the error statistics measured on transmission systems. This model captures the fact that successive bit errors are highly correlated. This model, known as the Gilbert model for transmission errors, is depicted in Figure B.1. Models of this type are also referred to as *burst error models,* and they more accurately estimate the behavior of transmission systems. This model makes the assumption that the transmission system exists in two states: a good state where the probability of bit errors is zero and a bad state where the probability of bit errors is relatively large. The model further allows for transitions to occur between the two states.

The transmission system errors, which are characterized by the Gilbert model, are as follows. The transmission system consists of a discrete stream of bit transmissions. For a given bit transmission, the probability of that bit being received in error is determined by the state the transmission system is in and the corresponding probability of bit errors in that state.

Between each bit transmission, a state transition may occur as determined by the various state transition probabilities, shown as P, $1 - P$, Q, and $1 - Q$ in

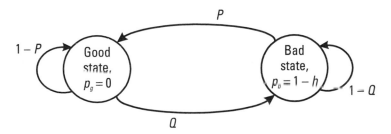

Figure B.1 Gilbert Model for transmission error statistics.

the figure. For example, if the system is in the good state, then the probability of a bit error is 0, the probability that the system will remain in the good state for the next bit is $1 - P$, and the probability that it changes to the bad state for the next bit is P.

The utility of this model is that one can derive expressions for the probability of a packet in error as a function of the bit error rate on the transmission system. The form of this expression is not as simple as that for the random bit error model. We illustrate the difference between the predictions of the independent bit error rate model and the Gilbert model in a plot. Figure B.2 shows the predictions of the random bit error model and the Gilbert model for the packet error probability as a function of the bit error rate on the transmission system.

An important point to notice in the plot is the fact that, for a given bit error rate, the burst models predict a much smaller packet error probability than the random bit error model. Also notice that as the bit error rate on the transmission facility increases, we see a rapid collapse of the ability of the facility to successfully transmit M bit packets in the neighborhood of bit error rates of one in 10^{-4}. For this reason, transmission systems aim to achieve error performance several orders of magnitude better than this.

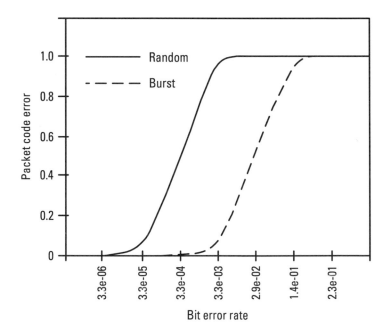

Figure B.2 A plot showing the effects of random bit errors versus burst errors on the probability of packet loss.

Typically, transmission systems' error statistics are characterized by the bit error rate (as already discussed) and the severely errored seconds (SES). Typical fiber transmission systems show a bit error rate of roughly 3×10^{-7}, which is much below the bit error rates at which the transmission systems show the performance collapse seen in Figure B.2. The SES is essentially the number of transitions to the bad error state per day. These states occur a few times a day, and last roughly 4 to 5 sec each. We mention this because it is common for long-haul transmission carriers to quote the SES value of their transmission facilities, in addition to the bit error statistics. The SES value gives an indication of the burstiness of the errors. For a given bit error rate, the fewer the SES, the greater the errors occur in bunches, and the better the throughput performance is of the transmission system.

To summarize the discussion of transmission errors, engineers have proposed several models to characterize the nature of these errors. The simplest is the random bit error model. It assumes that the bit errors are totally random. The value of this model is that it yields simple expressions for the quantities of interest, primarily the packet error rate. This model also tends to be very conservative. An improvement is found in burst error models, primarily the Gilbert model. These models are more accurate in that they assume that the bit errors occur in bunches. This is typically the case in today's long-haul transmission facilities. However, these models yield rather complex expressions for directly measurable quantities, such as the packet error rates. For most real-life modeling, the bit error rate model is probably sufficient to get an idea of whether the error rates on a given transmission system are the source of performance troubles.

B.3 Buffer Overflows

In networks that perform statistical multiplexing, for example, frame relay, ATM, and IP, packets are queued in data buffers until the resource in contention is available. This resource may be a transmission line, a switch backplane, internal transmission bus, or a CPU performing packet forwarding. In any case, a data buffer is used to hold the packet until the resource is available, and this data buffer is finite in size. Occasionally, the contention for this resource is too high, and the data buffer overflows, causing packets to be discarded.

In our discussion on queuing systems in the main text, we assumed that the data buffers were infinite in size, and therefore, the only interesting question was, what is the packet delay in the infinite queue? Given a finite buffer, queuing analysis also addresses the question of estimating the probability of packet loss. However, unlike asking the question of the mean queuing delay of a system, determining the probability of packet loss is equivalent to determining a high percentile of a complicated distribution. Also, the predictions on packet loss from queuing analysis are extremely sensitive to the detailed nature of the packet arrival process. For these reasons, we do not suggest a reliance on these queuing models for the purposes of engineering systems. However, we do wish to discuss some of the principles with a discussion of an example system. So let us digress for a minute into a bit of mathematical analysis.

Consider the M/M/1 queuing system with a finite buffer of m packets. Then, the probability of packet loss is akin to determining the probability of m packets in queue, given a Poisson arrival process and an exponential service distribution.[1] Defining p_m as the probability of m packets in queue, we have for the probability of packet loss in a finite queued M/M/1 system [2] the following expression:

$$p_m = \rho^m (1 - \rho) / (1 - \rho^{m+1})$$

(B.3)

where $\rho = \lambda / \mu$ is the system load. Figure B.3 shows a plot of the loss probability as a function of the buffer size derived from this equation. Evident from the plot is the fact that relatively small increases in the buffer size can have a dramatic effect on the loss probability at typical engineered loads, for example, $\rho = 0.7$. For example, the loss probability at this engineered load is roughly 10^{-1} for a buffer size of 5 packets. However, by doubling the buffer size to 10 packets, the probability of packet loss drops by more than an order of magnitude to less than 10^{-2}. Tripling the buffer size improves the loss performance by more than several orders of magnitude. Also, notice the strong dependence on the traffic load. In data networks, it is impossible to accurately estimate the load on given components in the network. For these reasons, it is impractical to utilize expressions like (B.3) for engineering purposes.

This estimate of packet loss is based on an M/M/1 model, which assumes that the packet arrivals are uncorrelated. In most instances, however, the packet arrivals to a buffer in a data network are highly correlated. For example, file transfers between end systems generate highly correlated packet arrivals. Therefore, this estimate is probably not accurate in most network engineering applications, even if you had accurate estimates of the traffic loads.

The probability of buffer overflows in data networks is one of the hardest measurables to predict a priori. The buffer overflow probability is extremely

---

1. The real question to be asked is what is the probability that a random packet arriving at the system will find the buffer full, that is, m packets in queue? However, for Poisson arrival processes, this is equivalent to the fraction of time the buffer is full, that is, p_m.

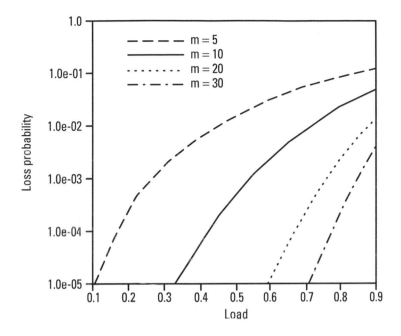

Figure B.3 The loss probability as a function of load for the M/M/1 model with finite buffer (here the buffer size, that is, *m*, is measured in terms of the number of data packets it will hold).

sensitive to the nature and load of the arrival processes, the size of the buffer in questions, and the details of the protocols and their interactions.

To demonstrate other complexities, a model that does not take into account feedback mechanisms between the network buffers and the end systems will fail in its prediction of today's packet networks' loss performance. For example, TCP relies on packet loss to dynamically adjust its running window. For these reasons, we spend very little time in this book quantifying this measure. In practice, the network engineer should rely heavily on actual packet loss measurements when engineering for buffer sizes and trunk speeds.

B.4 Transmitter Time-Outs

When a transmitter sends a data packet, it sets a timer if reliable communications are required. When the acknowledgment for the particular packet is received, the transmitter clears the timer value. In the event the timer exceeds a threshold, known as the *transmit time-out*, the transmitter assumes the packet was lost. It then enters an error recovery state, and retransmits the data packet.

However, packet losses are not the only reason for a time-out. Instead, the packet could have experienced extraordinary delays due to high buffer occupancies, for example, network congestion. Default transport level time-out thresholds are typically 1 to 3 sec. Therefore, if buffer delays reach these levels, timers will expire sending more traffic into the buffers due to retransmissions. This only exacerbates the buffer congestion problem. Some protocols, for example, TCP, implement congestion avoidance algorithms when time-outs occur to mitigate these concerns, as well as to dynamically adjust the transmit time-out value [3].

B.5 Out-of-Sequence Receptions

Some network protocols, for example, IP networks, do not guarantee packet sequencing. Others do, for example, the virtual circuit-based technologies like frame relay, X.25, and ATM. It is possible that a user of an IP network will receive packets out of sequence due to transitory routing patterns in the network. Under stable routing periods, the network generally develops single routing paths between any pair of end systems (assuming no load balancing is implemented[2]).

Single path routing will not result in out-of-sequence receptions. However, certain events in the network will trigger changes in the network routing patterns. These events may be caused by an equipment failure, transmission system failure, or planned network topology changes. A network relaxation time occurs following these events, during which the new routing patterns have not yet stabilized. During the relaxation period, routing loops or other routing patterns may develop that result in out-of-sequence packet deliveries. Given that these events are infrequent and that the network relaxation times are short, the probability of receiving packets out-of-sequence is small.

B.6 Impact of Packet Losses on Throughputs

We have discussed several causes of lost packets in data networks. We now wish to discuss the impact these packet losses have on the throughput of windowing systems. Windowing systems must have mechanisms built into them in order to detect packet losses. For the most part, two mechanisms are employed:

---

2. Some routing protocols, for example, Cisco System's EIGRP, are capable of supporting load sharing. However, they typically implement methods that prevent out-of-sequence receptions.

- *Packet sequence numbers.* Windowing systems include a sequence number on each transmitted packet. The receiver tracks the incoming sequence numbers, and if one or several numbers are skipped, the receiver assumes that packets were lost. The receiver then requests that the sender retransmit the packets associated with the missing sequence numbers.

- *Transmit timers.* The transmitter sets a timer on transmission of each packet. If the transmitter does not receive a packet acknowledgment before the timer expires, it will assume that the packet is lost and will retransmit the packet. This is known as the transmit time-out (as discussed earlier).

Independent of the mechanisms employed, packet losses will have a negative impact on the realized throughputs. The detailed relationship between packet losses and windowing throughput does depend on the nature of the windowing system and the recovery strategies it employs; for example, go-back-n versus selective repeats. The relationship between packet losses and system throughput also depends on the model for packet losses (similar to the problem of relating bit errors to packet losses). Due to the complexity of these issues and their relationships, we discuss only the simplest of cases. The qualitative impact on more complicated windowing systems is similar in nature.

One of the simplest systems to analyze is a simplex windowing system that implements a transmit time-out mechanism for packet loss detection in the presence of random, independent packet losses. Figure B.4 shows the throughput behavior of this system for certain specific packet loss sequences. In this figure, the first packet transmitted is not lost, and the transmitter receives an acknowledgment prior to the transmit time-out. For this, the round-trip delay is computed based on the delay components as discussed earlier. In this appendix, we indicate this as $R_d(0)$, where the 0 indicates no packet losses in the process of delivering the packet across the network. In the next packet transmission, the packet is lost, the transmit time-out expires, and the transmitter retransmits the packet. The retransmitted packet is then successfully received. Here, the round-trip delay, which we denote as $R_d(1)$ (where the 1 indicates a single packet loss), is $R_d(1) = R_d(0) + t_{out}$, where t_{out} is the value of the transmit time-out. In general, the round-trip delay, given k successive retransmissions, is $R_d(k) = R_d(0) + kt_{out}$. Assuming independent packet losses, the probability of k successive retransmissions is $p_{loss}^{k} \times (1 - p_{loss})$, where p_{loss} is the probability of a single packet loss. We can now write the average delay to successfully transmit a given packet as

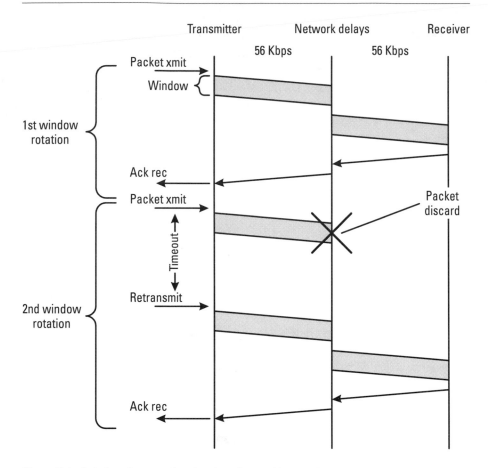

Figure B.4 A timing diagram showing the effects of lost packets on windowing behavior.

$$R_d = \sum_{k=0}^{\infty} \left(1 - p_{\text{loss}}\right) p_{\text{loss}}^k \times \left(R_d(k) = R_d(0) + kt_{\text{out}}\right) \tag{B.4}$$

or, performing indicated the summation,

$$R_d = R_d(0) + p_{\text{loss}} \times t_{\text{out}} / \left(1 - p_{\text{loss}}\right) \tag{B.5}$$

Therefore, the expected throughput is

$$X/X(0) = 1 / \left[1 + p_{\text{loss}} \times t_{\text{out}} / R_d(0) \times \left(1 - p_{\text{loss}}\right)\right] \tag{B.6}$$

where $X(0) = 8 \times N_w \times W / R_d(0)$ is the windowing throughput in a lossless environment, and the ratio of $X/X(0)$ is the expected drop in throughput due to packet losses.

In Figure B.5 we plot the expression given in (B.6) for throughput as a function of packet loss.

It is clear from this model that the system performance degrades rapidly for $p_{loss} > 0.01$. Therefore, if network measurements indicate packet losses near this threshold, the cause of the high losses should be investigated and resolved. Also, we have shown the effect for several different values of t_{out}. The time-out threshold must be set such that $t_{out} > R_d(0)$ to prevent timing out prematurely. But, t_{out} should not be too large to prevent wasting time in the event of an actual packet loss. Therefore, t_{out} is usually chosen to be roughly $3 \times R_d(0)$. There is no strict rationale for the factor of 3; it is simply larger than a single round-trip delay, but not a lot larger. Often, the estimate of the appropriate time-out period is based on delays typically found within a strict LAN environment. In LAN environments, typical round-trip delays are measured in tens or hundreds of milliseconds. If this is the case and you try to bridge multiple

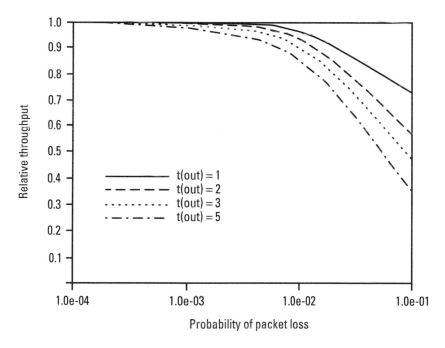

Figure B.5. Throughput degradation in the presence of packet losses for a simplex windowing model.

LAN segments over a WAN segment where round-trip delays are measured in seconds, the consequences can be disastrous. In these instances, the time-outs would need to be set to a much larger value than the default setting.

The expressions for other windowing systems are more complicated, but the behavior is similar to the preceding results.

References

[1]   Gilbert, E. N., "Capacity of a Burst Noise Channel," *Bell Syst. Tech. J.*, Vol .39, 1990, pp. 1253–1265.

[2]   Cooper, R. B., *Introduction to Queuing Theory*, New York: Elsevier, 1981.

[3]   Stevens, W. R., *TCP/IP Illustrated, Volume 1: The Protocols*, Reading, MA: Addison-Wesley, 1994.

Appendix C
Definitions

Several chapters in the latter part of the book develop and discuss simple formulas for estimating application performance. For the convenience of the readers, we list here the various terms and definitions used throughout those chapters.

Parameter	Definition
A	Size of the acknowledgment packet (bytes)
B_a	Available bandwidth of a link (bps)
C	Value of the CIR for the VC (bps)
C_v	Coefficient of packet variation
D_p	One-way latency for private line reference connection, for example, 0.030 sec (seconds)
D_f	One-way latency for frame relay reference connection, for example, 0.040 sec (seconds)
F	File size of data to be transferred (bytes)
K	Effective size of the token buffer in the leaky bucket ACF
L	Private line link speed (bps) (Note: Also used in Chapter 4 to represent the packet arrival rate.)
N_L	Number of private line links in the path
N_W	Number of packets in the transport window

Parameter	Definition
O	Protocol overhead (bits)
O_f	Protocol overhead factor $= 1 + O/S$
P_a	Port speed for the frame relay network on the client side of the connection (bps)
P_b	Port speed for the frame relay network on the server side of the connection (bps)
Q	Queuing delay (seconds)
R_d	Round-trip delay for a packet transmission and the return of an ack (seconds)
S	Segment (packet) size on the link (bytes)
T	Time to transfer the entire file (seconds)
T_A	Time to transmit ack on slowest link in path (seconds)
T_S	Time to transmit packet (segment) on slowest link in path (seconds)
T_S(average)	Average time to transmit a packet (seconds)
T_W	Time to transmit entire window on slowest link in path (seconds)
U	Utilization on the slowest link in the path
U_a	Utilization on the frame relay port on the client side of the network
U_b	Utilization on the frame relay port on the server side of the network
U_{CIR}	Utilization on the frame relay VC in the network
V	Variability in the packet transmit times
W	Window size in terms of the number of bytes
W^*	Optimal window size for the path (bytes)
X	Throughput for the reference connection (bps)
X_{max}	Maximum throughput for the reference connection (bps)

List of Acronyms

AAL ATM adaptation layer

ACF access control filter

API application programming interface

ARP address resolution protocol

ARPA Advanced Research Project Agency

AS autonomous system

ATM achronous transfer mode

BAN boundary access node

BECN backward explicit congestion notification

B-ISDN Broadband Integrated Digital Services Network

BNN boundary network node

CBR constant bit rate

CLLM consolidated link layer management

CIR   committed information rate

CPE   customer-premises equipment

CRC   cyclic redundancy check

DDCMP   digital data communications message protocol

DDS   Digital Data Service

DE   discard eligible

DLCI   data link connection identifier

DLSw   data link switching

DSP   Data Stream Protocol

DTE   data terminal equipment

EOB   end of burst

ERP   enterprise resource planning

FECN   forward explicit congestion notification

FEP   front-end processor

FIO   file input/output

FR   frame relay

FRAD   frame relay access device

FTP   file transfer protocol

GUI   graphical user interface

HTM   hybrid transfer mode

HTML   HyperText Markup Language

HTTP   HyperText Transfer Protocol

ICA   independent computing architecture

ICMP   Internet control message protocol

IDP   internetwork datagram protocol

IETF   Internet Engineering Task Force

ILMI   interim link management interface

IP   Internet Protocol

IPG   interpacket gap

IPX   internetwork packet exchange; Inter-Packet eXchange

ISDN   Integrated Digital Services Network

ITU   International Telecommunications Union

ITU-TS   International Telecommunications Union–Telecommunications Sector

IWF   interworking function

Kbps   kilobits per second

LAN   local-area network

LANE   LAN emulation

LIS   logical IP subnet

LLC   logical link control

LLC/SNAP   logical link control/subnetwork access point

LMI   link management interface

Mbps   megabits per second

MIB　management information base

MIR　minimum information rate

MTU　maximum transfer unit

NBMA　nonbroadcast multiple access

NCA　Network Computing Architecture

NCP　Network Control Program; network core protocol

NHRP　next hop resolution protocol

NIC　network interface card

NLM　network loadable module

NLPID　network layer protocol ID

NLSP　NetWare link services protocol

NNI　network-to-network interface

NOC　network operations center

NSAP　network services access point

OAM　operations, administration, and maintenance

OAM&P　operations, administration, maintenance, and provisioning

PAD　packet assembler/disassembler

PCM　pulse code modulation

PDU　protocol data unit

PLCP　physical layer convergence protocol

P-NNI　private network-to-network interface

PPP   point-to-point protocol

PTT   post, telegraph, and telephone authorities

PVC   permanent virtual circuit

QoS   quality of service

RDBMS   relational database management system

RED   random early detection

RIF   routing information field

RIP   routing information protocol

RMON   remote network monitoring

RSVP   resource reservation protocol

RTP   real-time transport protocol

SAP   service advertisement protocol

SAR   segementation and reassembly

SCR   sustainable cell rate

SDLC   synchronous data link control

SES   severely errored seconds

SIT   SAP information tables

SMTP   Simple Mail Transfer Protocol

SNA   System Network Architecture

SNAP   subnetwork access point

SPXP   sequenced packet exchange protocol

SQL   Structure Query Language

SSP   switch-to-switch protocol

STM   synchronous transfer mode

SVC   switched virtual circuit

TA   terminal adapter

TCP   Transmission Control Protocol

TDM   time-division multiplexing

TFTP   trivial file transfer protocol

TOS   type of service

UNI   user-to-network interface

VBR   variable bit rate

VC   virtual circuit

VCI   virtual circuit identifier

VLM   virtual loadable modules

VPN   virtual private networks

WAN   wide-area network

XNS   Xerox Networking System

About the Authors

Robert G. Cole holds a Ph.D. in Theoretical Chemistry from Iowa State University, and has held several post-doctoral positions at Yale, Boston, and Brown Universities prior to joining AT&T Bell Laboratories in 1985. He is currently a Technical Consultant in the Performance Analysis department, AT&T Laboratories. He was previously a Technical Manager of the IP Consulting Group within AT&T Solutions. His area of expertise is in performance and protocol analysis in router, frame relay, and ATM networks. He was an integral part of the AT&T development teams responsible for the rollout of the InterSpan frame relay service, the InterSpan ATM service and the WorldNet managed Internet service. He provided performance analysis and development support for the Lucent DataKit and BNS-1000 fast-packet switches, the BNS-2000 SMDS switch and the GlobeView 2000 ATM switch. He has published numerous articles in professional journals and holds several patents in the area of data communications and high-speed networking.

Ravi Ramaswamy holds a Ph.D. in Applied Probability from the University of Rochester. He is a Principal Consultant with AT&T Solutions. He joined AT&T Bell Laboratories in 1990, and prior to that he worked at BGS Systems Inc. His principal areas of expertise are wide-area network architecture, design, and performance engineering. He was responsible for supporting AT&T Frame Relay Service customers. Through the experience in supporting AT&T Frame Relay customers, he developed and delivered a 3-day course entitled "Multi-Protocol WAN Design and Performance Analysis" designed for AT&T's technical marketing organization. This course is now offered on a fee-basis for the industry at large. He delivered this course many times in the U.S., U.K., and Asia Pacific. A performance modeling tool, PerfTool, was developed based on the course material. Indeed, this book has its origins in this course and in Bob Cole's work on performance analysis. He has written many white papers on the topics discussed in this book. He has worked with many clients in designing and optimizing the performance of their networks.

Index

Access control filter (ACF), 23
    delaying, 25, 26
    input/output process from, 25
    output characteristics, 24
    schematic, 24
    strategy implementation, 23–24
    tag-and-send, 25, 26
Adaptive controls, 124–25
    explicit notification, 124
    mechanisms, 124
    See also Traffic discrimination methods
Address resolution protocol (ARP)
    defined, 42
    requests, 43
AppleTalk, 11, 81
Application analysis tools, 131
Application deployment, 132–35
    capacity management, 134–35
    characterization and, 132–33
    network redesign, 133–34
    traffic matrix, 133
Application programming interface
        (API), 118
Applications
    bandwidth sensitive, 101, 106–8
    client/server, 279–308
    ERP, 279
    latency sensitive, 101, 106–8
    sales-aid, 285, 343–46

time sensitivity, 104–5
    two-tier, 282, 284–89
Application server, 282
Asynchronous transfer mode (ATM), 2, 3
    adaptation layers (AALs), 34
    advantages/disadvantages, 37
    CBR support, 33
    cell stream, 34
    at hub sites, 321–22
    impetus, 12
    initial deployment of, 35
    internet example, 36
    IP on, 43–47
    layers, 33–34
    network deployment, 35
    SVC capabilities, 35
    switches, 32
    technology, 32
    trunking, 32
    VBR support, 33
    virtual connections, 35
ATM Forum, 36, 44

Background load, 306
"Back-of-the-envelope" calculations, 280
Backward explicit congestion notifications
        (BECNs), 21, 354, 355
    communication, 28
    DE bit marking support and, 170–71

Bandwidth
    allocation, 151
    allocation (large SNA networks), 319–20
    delay product, 212
    impact of SNMP polling, 369–74
    increasing results in worse
        performance, 359–60
    NetWare and, 256–60
    RIP, requirement, 258
    SAP, requirement, 260
    sizing, 150–64
    sizing case study, 339–42
Bandwidth estimation, 292–300
    bulk data transfers, 294–95
    hybrid transactions, 295–96
    ping-pong transactions, 292–93
Bandwidth management, 171–74
    external device techniques, 172–73, 328
    flexible approach, 173
    general techniques, 250
    SNA, 328
    traffic shaping vs., 173
Bandwidth sensitive
        applications, 101, 106–8
    defined, 108
    example trace, 109
    round-trip delay and, 108
    traffic volume, 109
Baud rate, 62–63
    bit rate relationship, 63
    defined, 62
Blinking cursors, 304
BNETX shell, 262, 264
Boundary access node (BAN), 316, 317
Boundary network node (BNN), 316, 317
Broadband Integrated Services Digital
        Network (B-ISDN) reference
        model, 33
Broadcast networks, 39, 42–43
Buffer overflows, 385–87
    defined, 381–82
    M/M/1 model example, 386, 387
    probability, 386–87
    See also Packet losses
Bulk data transfer transactions, 287–88
    bandwidth estimation, 294–95
    example, 287

response times for, 299–300
Burst error models, 383

Capacity management, 134–35
    strategy, 134
    tools, 131
Case studies, 331–74
    client/server
        application, 333–34, 342–53
    increasing bandwidth results in worse
        performance, 359–60
    introduction, 333–34
    leased line/frame relay performance
        comparison, 357–59
    multidrop to DLSw migration, 361–65
    NetWare, 334, 353–60
    Novell SAPs performance
        impact, 353–56
    sizing bandwidth, 339–42
    SNA performance in presence of TCP/IP
        traffic, 365–69
    SNA-related, 334, 360–69
    SNMP polling, 334, 369–74
    TCP bulk data transfer, 337–39
    TCP/IP, 334–42
    troubleshooting performance
        problems, 349–53
    troubleshooting response times (customer
        client/server), 347–49
    troubleshooting response times (sales-aid
        application), 343–46
    types of, 333–34
    validating latency and
        throughput, 334–37
CheckPoint, 250
Citrix Winframe/Metaframe
        model, 283, 350
Class of service (COS) priorities, 326
Client build delay, 288
Client/server
    application reference
        connection, 281, 298
    application server, 282
    application WAN traffic
        characterization, 283–90
    bandwidth estimation
        guidelines, 292–300

data collection, 291–92
overview, 281–83
thin client technology and, 280, 300–308
three-tier architecture, 282
three-tier transaction example, 288–90
tiered architecture, 281
two-tier architecture, 282
two-tier traffic patterns, 284–88
WAN performance issues, 279–308
Client/server application case
    studies, 342–53
troubleshooting performance
    problems, 349–53
troubleshooting response times (custom
    client/server), 347–49
troubleshooting response times (sales aid
    application), 343–46
Committed information rate
    (CIR), 21, 23, 114
bursting above, 217
excess traffic, 156
packets exceeding, 26
sizing of, 156, 364
Component allocation, 134
Congestion shift
illustrated, 165
response to, 166–74
from routers into network, 165–66
Constant bit rate (CBR) connection, 33
Conversion, 17
Credit management schemes, 23–25
Cyclic redundancy check (CRC), 17, 20

Data collection, 127–32
for client/server applications, 290–91
LAN/WAN analyzers, 127–28
network management systems, 128–30
tools for performance engineering data
    networks, 130–32
Data link channel identifiers (DLCIs), 20
frame relay, 175
global vs. local, 174–76
number field length, 176
to remote IP address mapping, 175, 176
Data link switching (DLSw), 312–14, 318
defined, 312
illustrated, 313

packets, 355
popularity, 314
SSP, 313–14
See also System Network Architecture
    (SNA)
Data terminal equipment (DTE)
devices, 21
frames, 21
interfaces, 15
Dedicated Digital Data Service (DDS), 64
Definitions, 393–94
Delay(s)
bandwidth product, 97
burst, 358
client build, 288
components, 58
defined, 57, 79
dual insertion, 140–43
end-to-end, 150
input, 70
insertion, 145, 163
latency vs., 57
ping, 181, 184–89, 192–96
processing, 58, 64–66, 69
propagation, 57, 59–61, 69, 187
queuing, 58, 66–69, 378
round-trip, 84, 89, 160, 202
store-and-forward, 145, 149, 324
synopsis, 69
total, 77, 84
trade-offs in SNA migration, 324–26
transmission, 57, 61–64, 69
variation, 143–46
Delay versus bandwidth plot, 147
Delay worksheets
for isolated transaction delay, 162
reference connection, 158–59
Discard eligible (DE) bit, 21
defined, 126
marking support, 170–71
Distance vector routing, 40, 41
Dual insertion delay, 140–43

Echoplex delay
calculating, 201
frame relay, 243
private line, 243

Echoplex delay (continued)
    for satellite connection, 243–44
Egress pipelining, 142, 143
    defined, 142
    network delay variation impact on, 145
    timing diagram showing, 144
Encapsulation, 17
    direct, over frame relay, 315–16
    frame relay, on ATM, 51
    identifier, 48
    IP standard, 43
    MAC format, 42
    multiprotocol, 47–49
    packet, inside protocol for transport, 48
    scenario illustration, 47
    standard examples, 49
End-to-end delay, 150
End-to-end throughput, 84, 98
Enterprise networking, 1–4, 60
Enterprise resource management (ERP)
        applications, 279
Erlang loss model, 379

FEP-to-FEP links, 326
Forward error correction systems, 63
Forward explicit congestion notifications
        (FECNs), 21, 354, 355
    communication, 28
    DE bit marking support and, 170–71
Frame relay, 2, 3
    application impact of migration
        to, 147–48
    bandwidth entities, 215
    basis, 20
    BECNs, 21, 27–29
    best/worst case conditions, 216
    bulk data transfer analysis, 215–22
    bursting, 215–16
    closed-loop feedback scheme, 25–27
    congestion control, 20–21
    connection illustration, 20
    credit management schemes, 23–25
    delay variation, 143–46
    DLCIs, 175
    dual insertion delay, 140–43
    echoplex delay estimation, 243
    encapsulation on ATM, 51

enterprise networks and, 29–37
    explicit congestion notifications, 29
    FECNs, 21, 27–29
    with full bursting, 218
    global, connections, 148–50
    infinitely fast, network, 141
    insertion delay, 216
    large ping delay calculation, 194–95
    link management interface, 22–23
    link management protocol, 21
    NetWare formulas, 271–76
    network architecture, 19–29
    with no bursting, 220
    open-loop feedback scheme, 25–27
    packet format, 20
    performance issues, 137–78
    ping delay calculation, 186–87
    priority PVCs and, 169–70
    private line vs., 138–48
    public, service, 25
    reference connection, 22, 140, 223
    reference connection for formula
        development, 266
    replacement for SNA multidrop with
        routers, 363
    satellite access to, 221–22
    service pricing, 177
    sparse VC topology design, 156
    switch schematic, 28
    TCP file throughput for, 216–17
    TCP throughput for loaded
        links, 227–30
    TCP throughput formulas, 222–25
    for terminal server solution, 306–7
    traffic convergence, 170
    See also WAN technologies
Front-end processor (FEP), 326
FTP, 200
    mixing Novell traffic with, 276–77
    mixing telnet with, 244
    performance, 199
    telnet priority over, 247
    See also Transmission Control/Internet
        Protocol (TCP/IP)
Gilbert model, 383, 384
Global frame relay connections, 148–50
    illustrated, 148

ping use and, 190–91
See also Frame relay

Hybrid transactions, 288
  bandwidth estimation, 295–96
  examples, 289
  phases, 288
  response times for, 300
HyperText Markup Language (HTML)
    documents, 230, 232, 234
HyperText Transfer Protocol (HTTP), 199
  approaches, 231
  download, sniffer trace, 233
  HTTP/1.0 performance (with multiple
      TCP sessions) estimation, 238–39
  HTTP/1.0 performance
      estimation, 236–38
  HTTP/1.1 caution, 231–32
  HTTP/1.1 performance
      estimation, 239–41
  interactive nature, 200
  over WAN connection, 203
  performance, estimating, 234–42
  performance estimate comments, 241–42
  request size, 235
  transaction, sample trace, 232–34
  versions, 230
  WAN performance issues for, 230–42

Input delay, 70
Insertion delay, 145, 163
  dual, 140–43
  extra, 216
Integrated Services Digital Network
    (ISDN), 64
Interactive traffic performance, 161–64
Internal pipelining, 142
Internet Control Message Protocol
    (ICMP), 182, 183
  echo message format, 184
  header, 193
  messages, 183, 194
Internets
  addresses, 39
  defined, 38, 39
  QoS transport within, 42
  topological design, 45

Internetwork datagram protocol (IDP), 254
Internetworking protocol (IP), 1, 37–47
  architecture, 38–42
  development, 37
  header processing, 39
  links, 39
  model overview, 38
  multicasting, 42
  on ATM, 43–47
  on broadcast networks, 42–43
  packet format, 39
  packet forwarding, 40
  QoS capabilities, 41–42
  routers, 39
  routing, 40–41
  transport entity, 38
  See also IP on ATM; WAN technologies
Internetwork packet exchange (IPX), 254
  defined, 256
  header, 256, 257, 259
  packets, 259
  See also NetWare
Interpacket gap (IPG), 262
Interworking function (IWF), 50
Intranets
  defined, 39
  reference connection, 234
IP on ATM, 43–47
  PVC-based (WANs), 43
  SVC-based (LANs), 44
  SVC-based (WANs), 45–47
  See also Asynchronous transfer
      mode (ATM)

Keyboard data entry, 305

LAN emulation (LANE)
  802 MAC layer interface, 45
  defined, 44
LAN/WAN analyzers, 127–28
  defined, 127
  use of, 128
  See also Data collection
Latency
  defined, 57
  delay vs., 57

Latency (continued)
    impact on remote presentation
        approach, 305–6
    impact on TCP throughput, 213–15
    validation case study, 334–37
    verifying, with pings, 189–91
Latency sensitive applications, 101, 106–8
    defined, 107
    example trace, 108
    performance, 107
Latency-sensitive traffic
    separate PVC, 167–68
    separating, 166
Leaky bucket algorithm, 23, 90, 91
    components, 92
    data packet entry and, 157
Leased line connections, ping
        calculation, 184–85
Leased line/frame relay performance case
        study, 357–59
    issue, 357
    resolution, 358–59
    See also Case studies
Link management interface (LMI), 22–23
    capabilities defined through, 22
    defined, 22
    See also Frame relay
Link management protocol, 21
Link state routing, 40, 41
Little's law, 378
Loaded links
    TCP throughput (frame relay), 227–30
    TCP throughput (private line), 226–27
Load engineering, 102–6
    goal, 102
    port, 155–57
    successful, 106
    VC, 155–57
Local-area networks (LANs), 1
    FDDI, 35
    interfaces, 48
    IP on SVC-based ATM, 44–45
    protocols, 48
    servers, 259
Lossless throughput systems, 93–94

MAC
    address discovery, 42
    format encapsulation, 42
    LANE interface, 45
Maximum information rate (MIR), 26
M/D/1 model, 379
M/G/1 model, 103
M/M/1 model, 104, 378, 386, 387
Modems, 64
Mouse movements, 304–5
Multicasting, 42
Multidrop, 322, 325
    defined, 361
    frame relay replacement for, 363
    illustrated, 361
    migration to DLSw, 361–65
    See also System Network
        Architecture (SNA)
Multidrop to DLSw migration case
        study, 361–65
    defined, 361–62
    design process, 362–65
    response time calculation, 365
    See also Case studies
Multiplexed interfaces
    defined, 17
    as part of information collected, 128
Multiprotocol integration layer, 2

NetWare, 11
    bandwidth necessary to support, 253
    case studies, 353–60
    client/server architecture, 255
    deployment, 277
    frame relay formulas, 271–76
    implementations, 257
    interpacket gap (IPG), 82, 262
    link services protocol (NLSP), 260
    overhead/bandwidth
        considerations, 256–60
    overview, 254–56
    pre-release 3.11, 261
    protocols, 254
    protocol stack, 254, 255
    release 3.11, 262–64
    releases 3.12 & 4.0, 264–65
    releases, 254, 262–65

service transactions, 263–64
TCP/IP protocol stack support, 255
WAN performance
    considerations, 253–77
windowing system, 81
See also Novell
Network
    interworking, 50, 51
    redesign, 133–34
    servers, 128
Network computing (NC)
    approach, 307–8
    examples, 280
    WAN performance issues, 307–8
    See also Thin client
Network Computing Architecture
    (NCA), 280, 283, 307
Network congestion, 124
    explicit feedback, 125
    implicit feedback, 125
    packet loss and, 125
Network core protocol
    (NCP), 254, 256, 261
    functions, 256
    packet format (pre-release 3.11), 261
    packet format (release 3.11), 263
    packet overhead, 263
    transport capabilities, 256–57
    See also NetWare
Network delay, 56–69
    components, 58
    defined, 84
    latency vs., 57
    processing, 58, 64–66, 69
    propagation, 57, 59–61, 69
    queuing, 58, 66–69
    synopsis, 69
    transmission, 57, 61–64, 69
    See also Delay(s)
Network loadable modules (NLMs), 254
Network services access point (NSAP)
    addressing, 35
Network-to-network interfaces (NNIs), 148
    congestion/bandwidth management
        mismatch, 149–50
    delays, 178
    issues, 149–50

link management, 149
link status and failure recovery and, 149
store-and-forward delays and, 149
Next hop resolution protocol (NHRP), 46
Node Manager, 370
Nonbroadcast multiple access (NBMA)
    network design, 43
    subnetworks, 39
Novell
    burst-mode technology, 262
    file transfer issues, 357
    mixing TCP/IP and, 276–77
    NCOPY, 357, 359
    protocol suite, 253
    windowing schemes, 260–65
    See also NetWare
Novell networking case studies, 353–60
    increasing bandwidth results in worse
        performance, 359–60
    leased line/frame relay performance
        comparison, 357–59
    SAPs performance impact, 353–56
    See also Case studies
Novell SAPs performance impact case
    study, 353–56
    defined, 353
    rationale, 355–56
    resolution, 354–55
    See also Case studies

Operations, administration, maintenance,
    and provisioning (OAM&P), 150
Optimal window size
    best case analysis, 219
    latency impact and, 213
    satellite access to frame relay, 222
    satellite links, 215
    segment size impact and, 212
    TCP throughput formula, 224
    worst case analysis, 221
    See also Window size
Organization, this book, 5–7
Out-of-sequence receptions, 388
    causes, 388
    defined, 382
    See also Packet losses

Packet(s)
    arrival rate, 67
    burst, 262
    length measurement, 62
    pipelining, 77
    sequence numbers, 389
Packet assemblers/dissemblers
            (PADs), 12, 29
Packeteer, 250
Packet losses, 381–82
    buffer overflows and, 381–82, 385–87
    impact of, on throughputs, 388–92
    out-of-sequence receptions and, 382, 388
    throughput degradation in
            presence of, 391
    timing diagram, 390
    transmission errors and, 381, 382–85
    transmitter time-outs and, 382, 387–88
PacketShaper, 174
Performance analysis, 55–99
    pings for, 181–96
    of TCP/IP applications, 199–251
Performance engineering
    approach to, 4–5
    techniques, 135
Performance engineering tools, 130–32
    application analysis, 131
    capacity management, 131
    predictive modeling, 131–32
    sniffers, 130–31
Permanent virtual circuits (PVCs), 15, 16
    allocating, for telnet, 249
    backup, 320
    burst levels in, 196
    carrying latency-sensitive traffic, 166
    priority, 169–70
    sizing, 151–64
    SNA, 327
    splitting, 168
Ping delays
    average, 191
    calculating, 181, 184–89
    comparison, 188
    to estimate global connection delays, 192
    frame relay connection, 186–87
    for large pings, 192–96
    leased line connection, 184–85

as low-priority packets, 192
    maximum, 191, 195
    minimum, 185, 187, 191
    observations, 187–89
    propagation delay and, 187
    to represent application delays, 192
    See also Delay(s)
Ping-pong transactions, 284–87
    bandwidth estimation, 292–93
    example, 285
Ping program
    defined, 182
    echo_request packets, 182
    output, 182
    in TCP/IP host implementations, 183
    variables, 183
Pings, 181–96
    drawing conclusion from, 181
    large, delay calculation for, 192–96
    size, 192
    for throughput calculation, 195–96
    timing diagram, 193
    use of, 191–92
    for verifying network latency, 189–91
Pipelining, 74–79, 142
    defined, 56
    delay reduction and, 77
    egress, 142, 143
    internal, 142
    packet, 77
    process, 66
    timing diagram showing
            advantages of, 76
    tradeoff, 79
    transit delay improvement, 78
Point of presence (POP), 187–88
Poisson process, 377
Ports
    load engineering, 155–57
    sizing, 151–64
Predictive modeling tools, 131–32
Priority queuing, 115–21
    circuit ID, 120
    defined, 116
    high-level protocol identifier, 119
    implementations, 116–18
    incoming port/interface, 119–20

nonpreemptive, 117
packet length, 120–21
preemptive, 117
priority level ID, 118
protocol encapsulation type, 119
protocol ID, 118–19
schemes, 118, 121
source/destination address pair, 120
starvation, 121
three-level system, 117
See also Traffic discrimination methods
Private line
echoplex delay estimation, 243
frame relay vs., 138–48
reference connection, 138, 223
reference connection for formula
development, 266
TCP bulk data transfer analysis, 208–15
TCP throughput for loaded
links, 226–27
TCP throughput formulas, 222–25
Private network-to-network interface
(P-NNI)
architecture, 36
routing capability, 36
Processing delays, 64–66, 69
defined, 58, 64
improvement options, 95–96
summary, 66
See also Delay(s)
Processor sharing, 121–24
advantages, 123–24
defined, 121
maximal delay in, 123
packet-level system, 122
schemes, 123
server, 122
See also Traffic discrimination methods
Propagation delays, 59–61, 69
defined, 57, 59
field measurements, 59
handling, 61
in ping delay, 187
significance, 59
See also Delay(s)
Protocol overhead factor, 207–8, 212

Quality of Service (QoS)
IP, 41–42
parameters, 21
SNA issues, 322–28
Queuing
introduction to, 66–69
mathematical discussion of, 375–80
M/D/1 model, 379
M/G/1 model, 103
M/M/1 model, 104, 378, 386, 387
models, 103, 376, 378–79, 386–87
priority, 115–21
system characterization, 375
system components, 67
TCP rate control vs., 173–74
theory, 377
time, 102, 103
weighted fair, 122, 248
Queuing delays, 66–69, 102, 378
average, 69
defined, 58, 66
estimating, 68
expression from M/G/1 model, 103
as function of input process, 67
as function of utilization, 102
at low loads, 68
summary, 68
utilization and, 103
See also Delay(s)

Random bit error model, 382
Random early discard (RED), 126–27
Rate-based throughput systems, 90–93
defined, 90
overload and, 91
schematic, 90
summary, 94
variations, 91
See also Throughput
Rate throttling, 171
Real-time transport protocol (RTP), 42
Reference connection
background load and, 306
client/server application, 281, 298
delay worksheet for, 158–59
frame relay, 22, 140, 223, 266

Reference connection (continued)
    for intranet between client/Web
        server, 234
    network latency impact and, 305–6
    number of slowest links in, 97
    over frame relay WAN, 83
    private line, 138, 223, 266
    for protocol trace of sales-aid
        application, 285
    timing diagram, 70, 71
Remote presentation
    application characteristics and, 305
    approach, 300–307
    architecture for two-tier applications, 301
    blinking cursors and, 304
    compression and, 304
    examples, 280
    keyboard data entry and, 305
    LAN-like performance and, 301–3
    mouse movements and, 304–5
    number of colors used and, 304
    screen resolution and, 304
    WAN bandwidth issues, 303–5
    WAN performance issues, 305–7
Reservation service protocol (RSVP), 41
Response times
    approximate total, 298
    bulk data transfer transactions, 299–300
    computing, 298–99
    hybrid transactions, 300
    multidrop to DLSw migration
        calculation, 365
    SNA, 324–25, 366, 368
    troubleshooting, 343–49
RFC 1490, 315–16, 317
RMON probes, 129–30
    defined, 129
    RMON2, 129–30
Round-trip delay, 84, 89, 160
    bandwidth-sensitive
        applications and, 108
    computing, 160–61
    estimate of, 202
    See also Delay(s)
Route aggregation, 40
Routers, 128, 177
    prioritization for SNA, 327

    prioritize telnet over bulk data
        transfers at, 247–49
    SAP broadcasts and, 259
    SNA, 320
    traffic shaping at, 249–50
Routing
    distance vector, 40, 41
    interdomain, 40
    intradomain, 40
    IP, 40–41
    link state, 40, 41
    path vector, 41
    protocol classification, 40, 41
    TOS, 113–15
Routing information protocol (RIP), 254
    bandwidth requirement, 258
    defined, 257
    message format, 257
    packets, 258
    See also NetWare

Sales-aid application, 285, 343–46
SAP R3
    application example, 296–98
    implementations, 283, 284
    transactions, 290
    See also Service advertisement
        protocol (SAP)
Satellite links, 214–15
    access to frame relay, 221–22
    echoplex delay, 243–44
    minimum window size for, 214–15
    optimal window size for, 215
Screen resolution, 304
Selective discards, 126–27
Sequenced packet exchange (SPX)
        protocol, 254
    defined, 265
    SPX II, 265
    See also NetWare
Serial line speed, 146
Serial trunk interface, 128
Service advertisement protocol (SAP), 254
    bandwidth requirement, 260
    defined, 259
    information tables (SITs), 259
    message formats, 259

See also NetWare; SAP R3
Service interworking, 50, 51–52
Service transactions, 263–64
Simplex windowing systems, 82–88
    behavior, 84
    defined, 80
    end-to-end throughput, 86–87
    with multiple packets per window, 86
    multiple rotations, 84, 85
    summary, 93
    timing diagram, 83
    See also Windowing systems
Sizing bandwidth case study, 339–42
    analysis, 340–42
    methodology, 340
    See also Case studies
Sliding window algorithm, 203
Sliding windowing systems, 88–89
    defined, 80
    summary, 93–94
    timing diagram, 88
    window size and, 89
    See also Windowing systems
SNA performance in TCP/IP traffic case
            study, 365–69
    conclusions, 368–69
    defined, 365–66
    questions, 366
    resolution, 366–69
    See also Case studies
SNA-related case studies, 360–69
    multidrop to DLSw migration, 361–65
    SNA performance in presence of TCP/IP
            traffic, 365–69
    See also Case studies
Sniffers, 130–31
Sniffer trace, 233
    of client/server application, 348
    of load application, 347
    of login transaction, 344
    of user application, 345
SNMP polling
    assumptions, 369–70
    bandwidth impact analysis, 373–74
    polled objects/frequencies, 370
    traffic generated by, 370–72
    WAN bandwidth impact of, 369–74

See also Case studies
Store-and-forward delays, 145
    defined, 149
    in SNA migration, 324
Sustainable cell rate, 114
Switched virtual circuits (SVCs), 15, 16
    ATM capabilities, 35
    signaling capabilities, 115
Switch processors, 128
System Network Architecture
            (SNA), 1, 2, 11, 200
    application mixing concern, 311–12
    applications, 311
    ATM at central site and, 321–22
    bandwidth allocation and, 319–20
    bandwidth management, 328
    data center architecture issues, 319–22
    design and performance issues, 311–29
    disaster recovery, 320, 322
    FEP-to-FEP issues, 326
    gateways, 314–15, 318, 326
    migrating, 312, 318
    migration, delay trade-offs, 324–26
    multidrop, 322, 325, 361
    parallel routers and, 319
    performance in presence of TCP/IP
            traffic, 365–69
    polling overhead, 324–25
    protocol overhead, 324
    PVCs, 327
    QoS issues, 322–28
    redundancy issues, 320, 321
    response times, 324–25, 366, 368
    router, 320, 327
    router-based traffic shaping, 328
    traffic discrimination, 327–28
    translation, 316, 318
    transport methods, 312–19
    Web access, 316–19
    windowing system, 81

Tag switching, 46–47
    defined, 46
    proposal, 47
TCP bulk data transfers, 196, 200
    calculation assumptions, 207
    case study, 337–39

TCP bulk data transfers (continued)
  file transfer time calculations, 205–8
  frame relay analysis, 215–22
  impact on window size, 209–12
  performance, 200
  performance, variables affecting, 205
  prioritizing telnet over, 247–49
  private line analysis, 208–15
  throughput calculation formulas, 222–25
  time/throughput calculation, 204–25
  timing diagram, 209
  trace of, 204
TCP/IP case studies, 334–42
  sizing bandwidth, 339–42
  TCP bulk data transfer, 337–39
  validating network latency and
    throughput, 334–37
  See also Case studies
Telnet, 200
  allocating separate PVC for, 249
  echoplex characteristic, 242
  echoplex delays, calculating, 201
  mixing with FTP over frame relay, 244
  packets, 249
  performance, 199, 242–47
  prioritize over bulk data
    transfers, 247–49
  prioritize over FTP, 247
  remote PC as client, 243
  response times, 246
  separate, on its own PVC, 249
  use of TCP/IP services, 242
Terminal adaptation (TA) devices, 17, 29
  conversion, 17, 18, 29, 37
  encapsulation, 17, 29, 37
Thin client, 300–308
  approaches, 280, 300–308
  network computing approach, 307–8
  remote presentation approach, 300–307
  See also Client/server
Three-tier transactions, 288–90
  bulk data characteristics, 290
  characteristics, 288–89
  See also Client/server
Thresholds
  defining, 104, 106
  fixed, 105

Throughput, 56, 79–99
  calculation with large pings, 195–96
  classes, 114
  complex systems, 79
  defined, 79
  degradation in presence of packet
    losses, 391
  end-to-end, 84, 98
  estimate, 98, 158
  file application, 157–61
  limit, 98
  lossless, systems, 93–94
  in lossy environments, 381–92
  maximum, 160
  packet loss impact on, 388–92
  rate-based, system, 90–93
  simple systems, 79
  summary, 98–99
  TCP, 213–15
  TCP bulk data transfer, 204–25
  TCP calculation for loaded links, 225–30
  TCP formula calculation, 222–25
  validation case study, 334–37
  windowing systems, 82–93
  window limited, 88
Time-division multiplexing (TDM)
    networks, 13–15
  bandwidth sharing, 14–15
  as baseline network configuration, 15
  components, 14
  illustrated, 14
  static nature, 14
  switches, 14
  See also WAN technologies
Timing diagrams, 69–74
  defined, 56
  for file transfer traffic, 110, 111
  for frame relay reference connection, 159
  for frame relay with full bursting, 218
  for frame relay with no bursting, 220
  for HTTP/1.0, 237
  for HTTP/1.0 with multiple TCP
    sessions, 240
  for HTTP/1.1, 241
  illustrated, 73
  input delay, 70
  for isolated transaction, 163

packet flow and, 71–72
ping, 193
reference connection, 70, 71
round trip, for frame relay
    connection, 272, 273
showing advantages of data pipelining, 76
showing cross-application effects, 111
showing effect of egress pipelining, 144
showing effects of lost packets, 390
showing impact of network
    pipelining, 143
showing serial line speed effect, 146
showing telnet echoes mixing
    with FTP, 245
simplex windowing system, 83
sliding windowing system, 88
TCP file transfer, 209
utility of, 74
for window size, 211
TN3270, 316, 318
Token buffer, 92
  non-existence of, 92
  size effect, 93
  transaction waiting time, 93
Total delay, 77, 84
Traffic discrimination, 165–74
  congestion shift, 165–74
  SNA, 327–28
Traffic discrimination methods, 108–27
  adaptive controls, 124–25
  priority queuing, 115–21
  processor sharing, 121–24
  selective discards, 126–27
  service routing, 113–15
  for TCP/IP applications, 247–50
  window size tuning, 109–13
Traffic matrix, 133
  for four-location design, 153
  point-to-point, 133
Traffic shaping, 124, 171–74
  bandwidth management vs., 173
  defined, 171
  router-based, 172, 249–50, 328
  troubleshooting performance problems
    case study, 352–53
Transmission Control/Internet Protocol
    (TCP/IP), 2

application performance
  analysis, 199–251
bandwidth sizing, 339–42
encapsulation, 189
file transfer, 168–69
mixing Novell and, 276–77
See also TCP/IP case studies
Transmission Control Protocol (TCP)
  delayed ack, 203
  multiple sessions, HTTP/1.0
    performance and, 238–39
  operation, 201–4
  protocol overhead, 201
  rate control, 173–74
  segment sizes, 201
  session establishment, 201, 202
  sliding window algorithm, 203
  throughput calculation, 225–30
  windowing system, 81
  window size, 213
  See also TCP bulk data transfers
Transmission delays, 62–64, 69
  affecting, 63–64
  defined, 57, 58, 61
  packet length and, 62
  summary, 64
  See also Delay(s)
Transmission errors, 382–85
  defined, 381
  effects of, 384
  Gilbert model, 383, 384
  random bit, 382, 384
  statistics, 385
  summary, 385
  system, 383
  See also Packet losses
Transmitter time-outs, 387–88
  causes, 388
  defined, 382, 387
  See also Packet losses
Transmit timers, 389
Troubleshooting performance problems case
    study, 349–53
  analysis, 350–51
  context, 349
  local Internet access and, 351–52
  PVC for Winframe traffic, 352

Troubleshooting performance problems case
study (continued)
rearchitect network and, 352
recommendations, 351–53
traffic reshaping, 352–53
See also Case studies
Troubleshooting response times
(client/server application) case
study, 347–49
load application, 348
resolution, 349
See also Case studies
Troubleshooting response times (sales-aid
application) case study, 343–46
analysis, 343–46
conclusion, 344–45
context, 343
See also Case studies
Two-tier applications, 282
bulk data transfer transactions, 287–88
hybrid transactions, 288, 289
ping-pong transactions, 284–87
remote presentation architecture for, 301
traffic patterns, 284–88
Type of service (TOS) routing, 113–15
implementing, 114
low-delay, 115
multiple paths and, 115
performance and, 115
values, requested, 114
See also Traffic discrimination methods

User-to-network interface (UNI), 148

Validating latency and throughput case
study, 334–37
analysis, 335–37
background and context, 334–35
issue, 335
See also Case studies
Variable bit rate (VBR) connection, 33
Virtual circuit identifier (VCI), 17
Virtual circuits (VCs)
capacity, 155
CIR rate, 271
full mesh topology, 154
load engineering, 155–57

scaling issues, 176–77
size of, 153
sparse topology design, 156
Virtual loadable modules (VLMs), 254, 264
Virtual Private Networks (VPNs), 3

WAN technologies, 11–53
frame relay, 19–37
internetworking protocol (IP), 37–47
introduction, 11–13
link level interworking
agreements, 49–52
multiprotocol encapsulation, 47–49
for multiprotocol integration, 13
TDM, 13–15
X.25, 15–19
See also Wide-area networks (WANs)
Web access (SNA), 316–19
Weighted fair queuing (WFQ), 122, 248
Wide-Area Data Network Performance
Engineering
approach, 4–5
organization, 5–7
Wide-area networks (WANs), 1
analytical understanding of, 5
bandwidth issues, 303–5
IP on PVC-based ATM, 43
IP on SVC-based ATM, 45–47
performance issues for client/server
applications, 279–308
performance issues for HTTP, 230–42
performance issues for NetWare
networks, 253–77
performance issues for SNA
networks, 311–29
private line facility, 96, 112, 178
X.25 networks, 12
See also WAN technologies
Windowing systems, 82–93
AppleTalk Data Stream Protocol
(DSP), 81
calculated throughput for, 87
dynamic, 80
feedback to, 80
fixed, 80
IBM SNA, 81
Novell NetWare, 81

packet loss impact on, 388–92
rate-based throughput, 90–93
simplex, 80, 82–88
sliding, 80, 88–89
TCP, 81
Window size, 80, 89
   adjustments, 212–13
   file transfer startup, 95
   limiting, 168–69
   optimal, 94–98, 113, 212, 213, 215,
      219, 222, 224
   TCP, 213
   TCP bulk data transfer
      impact on, 209–12
   timing diagram, 211
   tuning, 109–13, 116

X.25, 3, 11
   configuration illustration, 18

connection illustration, 16
defined, 15
networking protocols, 17
network interconnection through X.75
   interface gateways, 19
networks, 12, 15–19
network service providers, 18
packet format, 17
PADs, 29
QoS parameter support, 16
signaling protocol, 16
standards, 11
switches, 15, 16
TA devices, 18
technology, 15–16
virtual circuit, 16
X.75 gateway links, 19
Xedia Systems, 250
Xerox Networking System (XNS), 254

Recent Titles in the Artech House Telecommunications Library

Vinton G. Cerf, Senior Series Editor

Access Networks: Technology and V5 Interfacing, Alex Gillespie

Achieving Global Information Networking, Eve L. Varma, Thierry Stephant, et al.

Advanced High-Frequency Radio Communications, Eric E. Johnson, Robert I. Desourdis, Jr., et al.

Advances in Telecommunications Networks, William S. Lee and Derrick C. Brown

Advances in Transport Network Technologies: Photonics Networks, ATM, and SDH, Ken-ichi Sato

Asynchronous Transfer Mode Networks: Performance Issues, Second Edition, Raif O. Onvural

ATM Switches, Edwin R. Coover

ATM Switching Systems, Thomas M. Chen and Stephen S. Liu

Broadband Network Analysis and Design, Daniel Minoli

Broadband Networking: ATM, SDH, and SONET, Mike Sexton and Andy Reid

Broadband Telecommunications Technology, Second Edition, Byeong Lee, Minho Kang, and Jonghee Lee

Client/Server Computing: Architecture, Applications, and Distributed Systems Management, Bruce Elbert and Bobby Martyna

Communication and Computing for Distributed Multimedia Systems, Guojun Lu

Communications Technology Guide for Business, Richard Downey, Seán Boland, and Phillip Walsh

Community Networks: Lessons from Blacksburg, Virginia, Second Edition, Andrew Cohill and Andrea Kavanaugh, editors

Computer Mediated Communications: Multimedia Applications,
Rob Walters

Computer Telephony Integration, Second Edition, Rob Walters

Convolutional Coding: Fundamentals and Applications,
Charles Lee

Desktop Encyclopedia of the Internet, Nathan J. Muller

*Distributed Multimedia Through Broadband Communications
Services,* Daniel Minoli and Robert Keinath

Electronic Mail, Jacob Palme

*Enterprise Networking: Fractional T1 to SONET, Frame Relay to
BISDN,* Daniel Minoli

FAX: Facsimile Technology and Systems, Third Edition, Kenneth R.
McConnell, Dennis Bodson, and Stephen Urban

Guide to ATM Systems and Technology, Mohammad A. Rahman

Guide to Telecommunications Transmission Systems,
Anton A. Huurdeman

A Guide to the TCP/IP Protocol Suite, Floyd Wilder

Information Superhighways Revisited: The Economics of Multimedia,
Bruce Egan

International Telecommunications Management, Bruce R. Elbert

Internet E-mail: Protocols, Standards, and Implementation,
Lawrence Hughes

Internetworking LANs: Operation, Design, and Management,
Robert Davidson and Nathan Muller

Introduction to Satellite Communication, Second Edition,
Bruce R. Elbert

Introduction to Telecommunications Network Engineering,
Tarmo Anttalainen

Introduction to Telephones and Telephone Systems, Third Edition,
A. Michael Noll

IP Convergence: The Next Revolution in Telecommunications,
Nathan J. Muller

LAN, ATM, and LAN Emulation Technologies, Daniel Minoli and
 Anthony Alles

The Law and Regulation of Telecommunications Carriers,
 Henk Brands and Evan T. Leo

*Marketing Telecommunications Services: New Approaches for a
 Changing Environment,* Karen G. Strouse

Mutlimedia Communications Networks: Technologies and Services,
 Mallikarjun Tatipamula and Bhumip Khashnabish, Editors

Networking Strategies for Information Technology, Bruce Elbert

Packet Switching Evolution from Narrowband to Broadband ISDN,
 M. Smouts

Packet Video: Modeling and Signal Processing, Naohisa Ohta

Performance Evaluation of Communication Networks,
 Gary N. Higginbottom

Practical Computer Network Security, Mike Hendry

Practical Multiservice LANs: ATM and RF Broadband,
 Ernest O. Tunmann

Principles of Secure Communication Systems, Second Edition,
 Don J. Torrieri

Principles of Signaling for Cell Relay and Frame Relay,
 Daniel Minoli and George Dobrowski

Pulse Code Modulation Systems Design, William N. Waggener

Service Level Management for Enterprise Networks, Lundy Lewis

Signaling in ATM Networks, Raif O. Onvural and Rao Cherukuri

Smart Cards, José Manuel Otón and José Luis Zoreda

Smart Card Security and Applications, Mike Hendry

SNMP-Based ATM Network Management, Heng Pan

Successful Business Strategies Using Telecommunications Services,
 Martin F. Bartholomew

Super-High-Definition Images: Beyond HDTV, Naohisa Ohta

Telecommunications Department Management, Robert A. Gable

Telecommunications Deregulation, James Shaw

Telemetry Systems Design, Frank Carden

Teletraffic Technologies in ATM Networks, Hiroshi Saito

Understanding Modern Telecommunications and the Information Superhighway, John G. Nellist and Elliott M. Gilbert

Understanding Networking Technology: Concepts, Terms, and Trends, Second Edition, Mark Norris

Understanding Token Ring: Protocols and Standards, James T. Carlo, Robert D. Love, Michael S. Siegel, and Kenneth T. Wilson

Videoconferencing and Videotelephony: Technology and Standards, Second Edition, Richard Schaphorst

Visual Telephony, Edward A. Daly and Kathleen J. Hansell

Wide-Area Data Network Performance Engineering, Robert G. Cole and Ravi Ramaswamy

Winning Telco Customers Using Marketing Databases, Rob Mattison

World-Class Telecommunications Service Development, Ellen P. Ward

For further information on these and other Artech House titles, including previously considered out-of-print books now available through our In-Print-Forever® (IPF®) program, contact:

Artech House	Artech House
685 Canton Street	46 Gillingham Street
Norwood, MA 02062	London SW1V 1AH UK
Phone: 781-769-9750	Phone: +44 (0)20 7596-8750
Fax: 781-769-6334	Fax: +44 (0)20 7630-0166
e-mail: artech@artechhouse.com	e-mail: artech-uk@artechhouse.com

Find us on the World Wide Web at:
www.artechhouse.com